NanoScience and Technology

NanoScience and Technology

The series NanoScience and Technology is focused on the fascinating nano-world, mesoscopic physics, analysis with atomic resolution, nano and quantum-effect devices, nanomechanics and atomic-scale processes. All the basic aspects and technology-oriented developments in this emerging discipline are covered by comprehensive and timely books. The series constitutes a survey of the relevant special topics, which are presented by leading experts in the field. These books will appeal to researchers, engineers, and advanced students.

Sliding Friction
Physical Principles and Applications
By B.N.J. Persson
2nd Edition

Scanning Probe Microscopy
Analytical Methods
Editor: R. Wiesendanger

Mesoscopic Physics and Electronics
Editors: T. Ando, Y. Arakawa, K. Furuya, S. Komiyama, and H. Nakashima

Biological Micro- and Nanotribology
Nature's Solutions
By M. Scherge and S.N. Gorb

Semiconductor Spintronics and Quantum Computation
Editors: D.D. Awschalom, N. Samarth, D. Loss

Semiconductor Quantum Dots
Physics, Spectroscopy and Applications
Editors: Y. Masumoto and T. Takagahara

Nano-Optoelectonics
Concepts, Physics and Devices
Editor: M. Grundmann

Noncontact Atomic Force Microscopy
Editors: S. Morita, R. Wiesendanger, E. Meyer

Nanoelectrodynamics
Electrons and Electromagnetic Fields in Nanometer-Scale Structures
Editor: H. Nejo

Single Organic Nanoparticles
Editors: H. Masuhara, H. Nakanishi, K. Sasaki

Epitaxy of Nanostructures
By V.A. Shchukin, N.N. Ledentsov and D. Bimberg

Nanostructures
Theory and Modeling
By C. Delerue and M. Lannoo

Nanoscale Characterisation of Ferroelectric Materials
Scanning Probe Microscopy Approach
Editors: M. Alexe and A. Gruverman

Magnetic Microscopy of Nanostructures
Editors: H. Hopster and H.P. Oepen

Applied Physics of Carbon Nanotubes
Fundamentals of Theory, Optics and Transport Devices
Editors: S.V. Rotkin and S. Subramoney

S.V. Rotkin · S. Subramoney (Eds.)

Applied Physics of Carbon Nanotubes

Fundamentals of Theory, Optics and Transport Devices

With 200 Figures

Springer

Dr. Slava V. Rotkin
Lehigh University
Department of Physics
16 Memorial Drive East
Bethlehem, PA 18015, USA
E-mail: rotkin@lehigh.edu

Dr. Shekhar Subramoney
DuPont Company
Experimental Station
Wilmington, DE 19880-0228, USA
E-mail: shekhar.subramoney@usa.dupont.com

ISSN 1434-4904
ISBN-10 3-540-23110-2 Springer Berlin Heidelberg New York
ISBN-13 978-3-540-23110-3 Springer Berlin Heidelberg New York

Library of Congress Control Number: 2004117730

Springer is a part of Springer Science+Business Media.
springeronline.com

Printed in Germany

Cover background image: "Single-wall nanotube METFET inverter: organic molecules wrap around a metallic tube channel and modify it to a semiconductor while the second tube is used for gating". Concept: S.V. Rotkin, Physics Department, Lehigh University, image: B. Grosser, Imaging Technology Group, Beckman Institute, UIUC.

Typesetting and prodcution: PTP-Berlin, Protago-TEX-Production GmbH, Berlin
Cover design: *design& production*, Heidelberg

Printed on acid-free paper SPIN: 11008958 57/3141/YU - 5 4 3 2 1 0

To my wife, Lolita, and my son, Lev, for their loving support and patience, and for my mother and grandfather.

Slava V. Rotkin

Foreword

Back in 1991 Sumio Iijima first saw images of multi-walled carbon nanotubes in the TEM. Two years later, he and Donald Bethune synthesized the first single-walled nanotubes (SWNTs). Since then, we have seen tremendous advances in both the methods for nanotube synthesis and in the understanding of their properties. Currently, centimeter-long SWNTs can be readily grown at selected positions on a solid substrate, and large quantities of nanotubes can be produced for industrial applications. Significant progress has been made in producing nearly homogeneous samples of nanotubes of only a few diameters/chiralities. It is expected that the development of techniques for the synthesis of a single type of nanotube is not far away. At the same time, physical and chemical procedures for the separation of nanotube mixtures are being demonstrated. In addition to pure nanotubes, derivatized nanotubes with attached chemical or biochemical groups are being prepared. Nanotubes acting as containers for atoms, molecules (such as the "peapods") and chemical reactions are attracting significant attention.

In parallel with the synthetic effort there has been a race to decipher the properties of these materials. It is now clear that nanotubes possess unique mechanical, electrical, thermal and optical properties. Scientists and engineers around the world are exploring a wide range of technological applications that make use of these properties. For example, the outstanding mechanical properties of NTs are used in the fabrication of new, strong composites; their field-emission properties are employed to fabricate flat panel displays; the ballistic character of electronic transport in SWNT has been utilized to demonstrate SWNT transistors that outperform corresponding state-of-the-art silicon devices; while the sensitivity of their electrical characteristics on interactions with their environment is being used to produce chemical and biological sensors. Some of these technologies already have matured enough to enter the market place; others will require much more time. New uses of carbon nanotubes are continually being proposed, and it would not be an exaggeration to say that NTs are destined to become the key material of the 21st century.

This book, written by recognized experts in their areas, provides an up-to-date review of the science and technology of NTs. In the "theory and modeling" section, S. Rotkin discusses the classical and quantum mechanical behavior of different single-walled nanotube (SWNT) devices. He analyzes the current-voltage characteristics of long channel SWNT field-effect transistors operating in the quasi-diffusive regime, and he derives analytical expressions for both the geometrical and quantum capacitance of SWNTs. He investigates the changes in electronic structure resulting from the interaction (charge-transfer) between the SWNT with its substrate and the resulting breaking of the axial symmetry of the SWNT. He also discusses the possibility of "band-gap engineering" by external electric fields. He finds that an electric field can open a band-gap in a metallic nanotube, and conversely close the gap of a semiconducting tube. Ideas for new electronic devices are also presented. Damnjanovic et al. provide a detailed symmetry-based analysis of the electronic structure of both single-walled and double-walled nanotubes (DWNT). The results of simple tight-binding theory and density functional theory are compared. Using symmetry arguments again, they discuss the optical absorption spectra of nanotubes, SWNT phonons and their Raman and IR spectroscopies. Finally, the interactions between the walls of DWNTs are discussed. Analytical continuum models of the acoustic and optical phonon modes of finite length NTs are provided in the chapter by Stroscio et al., where both dispersion relations and mode amplitudes are given.

In the "synthesis and characterization" section of the book, Huang and Liu discuss the latest developments in the controlled synthesis of SWNTs. While the heterogeneous NT mixtures produced by various synthetic routes can be used in a number of applications, high technology applications, such as those in electronics, require control over the diameter, orientation and length of the SWNTs. The authors demonstrate the strong relation between catalyst particle size and the diameter of the resulting SWNTs in CVD growth. They go on to show that oriented growth of NTs can be induced simply by the laminar flow of the reaction gases. The CVD methods of Hunag and Liu are based on fast flow of the reaction gases coupled with a fast heating of the reacting mixture of gases and catalysts. This version of CVD leads not only to directional growth that allows SWNT structures such as crossbars to be generated, but also produces extraordinarily long nanotubes in the range of centimeters. Then, Okazi and Shinohara discuss the synthesis and properties of peapods, i.e. the compounds formed by the occlusion of fullerenes by SWNTs. Occlusion of both simple fullerenes and endohedral metallo-fullerenes is considered, and structural data and electrical properties of peapods provided by techniques such as STM, EELS and electron diffraction are presented. The use of SWNTs as containers for confined chemical reactions is also discussed. Strano et al. discuss how to use spectroscopic measurements, absorption, fluorescence and Raman, to study covalent and charge transfer interactions between small molecules and SWNTs. Examples discussed include the selective reaction of metallic SWNTs with diazonium molecules to form aryl C-C bonds and functionalize the side walls of SWNTs, and the selective protonation of NTs in the presence of oxygen.

In the section on "optical spectroscopy", Simon et al. focus on double-walled CNTs (DWNTs). They discuss the structure of these CNTs using Raman spectroscopy. They analyze the mechanism of the process by which SWCNTs incorporating C60, i.e. peapods, are converted into DWNTs through high energy electron beam irradiation. From Raman spectra they conclude that the walls of the inner tubes of DWNTs are structurally perfect and use the splitting of the radial breathing modes to decipher the interaction between the two carbon shells. A detailed account of the emission spectra of NTs is provided by B. Weisman. Fluorescence originating from the lowest excited state (E11) of SWNTs is readily observed upon resonant or higher state (e.g. E22) excitation. Weisman explains how, by combining, fluorescence, fluorescence excitation and resonant Raman spectra of SWNTs dispersed using surfactants, one can deduce the (n,m) indices that describe their structure. He points out that the energies of the individual transitions and the ratios between them show important deviations from expectations based on single electron tight-binding theory. The environment of SWNTs is found to affect both the widths and position of the excitation peaks. Such effects are predicted by theories that account for the many electron effects and exciton formation. (See also chapter by Avouris et al.)

In the section on "transport, electronic, electro-optical and electromechanical device applications", Avouris, Radosavljevic and Wind discuss the electronic structure, electrical properties and device applications of SWNTs. Special emphasis is placed on SWNT field-effect transistors (SWNT-FETs). The fabrication, switching mechanism, scaling properties and performance of p-, n- and ambipolar SWCNT-FETs are analyzed and compared with conventional silicon metal-oxide-semiconductor field-effect transistors (MOSFETs). The key role of Schottky barriers and the critical effects of the environment on the performance of SWNT-FETs are stressed. Then the nature of the excited states of nanotubes and their optoelectronic properties are discussed, and single SWNT light emitting and light detecting devices, both based on the SWNT-FET structure, are demonstrated. The authors suggest the possibility that a future integrated electronic and optoelectronic technology based on SWNTs may be possible. Jagota et al. discuss the interaction of SWNTs with biological systems. Such interactions are of interest because they (a) allow the manipulation and sorting of SWNTs, and (b) they can be used as the basis of sensors for biomolecules. The solubilization of SWNTs by complexing with DNA is discussed. This interaction is used in the separation of metallic from the semiconducting SWNTs. The authors also provide evidence that the separation of SWNTs according to their diameter may be possible by the same technique. A different separation method based on selective protonation of SWNTs is also discussed. The use of SWNTs as bio-sensors is demonstrated using the detection of cytochrome C as an example. Finally, Cumings and Zettl discuss a variety of mechanical and electrical experiments on multiwall NTs (MWNTs) and boron nitride nanotubes (BNNT). These include the peeling, sharpening and telescoping of MWNTs inside the TEM. They use MWNT telescoping as a means to study nanofrictional forces and determine the static and dynamic frictional components involved. They also use the same process to determine the length dependence of the conductance of an NT shell. Under their conditions, they find an exponential dependence of the resistance on length which they attribute to localization phenomena.

The mechanical properties of NTs are the subject of the chapter by Fisher et al. The authors discuss in some detail the construction of nanomanipulator systems which allow the manipulation of NTs and other nanostructures in 3D. They also describe measurements of mechanical properties, such as the tensile loading of single-wall and multi-wall NTs, the mechanics of carbon nanocoils, and the results of pull-out tests of single NTs from NT-polymer matrices.

Yorktown Hights, December 2004 *Phaedon Avouris*

Preface

Since the discovery of carbon nanotubes about a decade and a half ago by Sumio Iijima, the scientific community involved in various aspects of research related to carbon nanotubes and related technologies has observed a steady progress of the science, as is typical for any new and novel material. Right from day one, it was apparent to the scientists working on carbon nanotubes that the chirality of individual nanotubes would dictate their electronic properties, besides the well-established knowledge that individual sheets of sp2-bonded carbon had extremely attractive physical and mechanical properties. So, the field of carbon nanotubes took a giant leap in 1993 when research groups at NEC and IBM almost simultaneously discovered the single-walled variant of carbon nanotubes. Since then, we have observed the progress of science and technology as it relates to carbon nanotubes changing from the discovery of various methods to synthesize them to their structure-property relationships to how one might synthesize them in bulk quantities.

A number of edited books have been published in the last five to eight years outlining a variety of topics of current interest related to carbon nanotube research. The chapters in these books deal with topics ranging from synthesis methods to large-volume production concepts to the studies of the unique physical and mechanical properties of carbon nanotubes. Some of the chapters in these books deal with what might be unique about carbon nanotubes and where one might apply them to real-world commercial applications of the future. In fact, it is becoming very evident that carbon nanotubes (specifically single-walled nanotubes with unique electronic properties) will eventually replace silicon in electronic devices that dominate our present information/data driven world. Having stated that, the challenges to selectively obtain and manipulate carbon nanotubes into desired positions in these devices are enormous, and conventional silicon-based technologies will essentially be useless to achieve these goals. On a cumulative basis, the chapters in this book deal with a number of these very new challenges related to carbon nanotubes – how one might go about synthesizing nanotubes of specific chiralities and/or electronic properties and possible experimental routes to separate out the desirable nanotubes, and unique and novel measurement tools to characterize the chiralities of nanotubes. Other chapters deal with the measurement of the electronic properties of carbon nanotubes and how these may be used in real devices. Clearly, the authors who have contributed to this book have done an outstanding job in their respective arenas of interest. From a set-theory point of view, it is our sincerest hope that the reader will benefit immensely from the wealth of information from the individual sets (chapters) as well as from the intersection of the various sets.

Urbana–Champaign, Wilmington
January 2005

Slava V. Rotkin
Shekhar Subramoney

Contents

Part I Theory and Modelling

Part II Synthesis and Characterization

Part IV Transport and Electromechanical Applications

List of Contributors

Phaedon Avouris
IBM T.J. Watson Research Center
Yorktown Heights, NY 10598, USA
avouris@us.ibm.com.*

S. Baik
118 Roger Adams Laboratory, Box C-3, 600 South Mathews Avenue, Urbana, IL 61801, USA

P.W. Barone
118 Roger Adams Laboratory, Box C-3, 600 South Mathews Avenue, Urbana, IL 61801, USA

Salah Boussaad
DuPont Central Research and Development, Experimental Station, Wilmington, DE 19880, USA

Xinqi Chen
Northwestern University, Department of Mechanical Engineering, 2145 Sheridan Road, Evanston, IL 60208, USA

John Cumings
Department of Physics, Stanford University, Stanford, CA 94305, USA
cumings@stanford.edu.*

M. Damnjanović
Faculty of Physics, POB 368, Belgrade 11001, Serbia and Montenegro

Dmitriy A. Dikin
Northwestern University, Department of Mechanical Engineering, 2145 Sheridan Road, Evanston, IL 60208, USA

Bruce A. Diner
DuPont Central Research and Development, Experimental Station, Wilmington, DE 19880, USA

E. Dobardžić
Faculty of Physics, POB 368, Belgrade 11001, Serbia and Montenegro

Mitra Dutta
Department of Electrical & Computer Engineering, Department of Physics, University of Illinois at Chicago, Chicago, IL 60607, USA

Frank T. Fisher
Department of Mechanical Engineering, Stevens Institute of Technology, Hoboken, NJ 07030, USA

D.A. Heller
118 Roger Adams Laboratory, Box C-3, 600 South Mathews Avenue, Urbana, IL 61801, USA

M. Holzweber
Institute of Materials Physics, University of Vienna, Strudlhofgasse 4, 1090 Vienna, Austria

Shaoming Huang
Department of Chemistry, Duke University, Durham, NC 27708, USA

Anand Jagota
Department of Chemical Engineering, Iacocca Hall, Lehigh University, Bethlehem PA 18015, USA
anj6@lehigh.edu

C. Kramberger
Institute of Materials Physics, University of Vienna, Strudlhofgasse 4, 1090 Vienna, Austria

H. Kuzmany
Institute of Materials Physics, University of Vienna, Strudlhofgasse 4, 1090 Vienna, Austria

Jie Liu
Department of Chemistry, Duke University, Durham, NC 27708, USA
j.liu@duke.edu

I. Milošević
Faculty of Physics, POB 368, Belgrade 11001, Serbia and Montenegro

B. Nikolić
Faculty of Physics, POB 368, Belgrade 11001, Serbia and Montenegro

Marko Radosavljević
Novel Device Group, Intel Corporation Hillsboro, OR 97124
marko.radosavljevic@intel.com.*

Toshiya Okazaki
National Institute of Advanced Industrial Science and Technology (AIST), Tsukuba, 305-8565, Japan
toshi.okazaki@aist.go.jp.*

R. Pfeiffer
Institute of Materials Physics, University of Vienna, A-1090 Vienna, Strudlhofgasse 4., Austria

Amit Raichura
Department of Electrical & Computer Engineering, University of Illinois at Chicago, Chicago, IL 60607, USA
araich2@uic.edu

Slava V. Rotkin
Physics Department, Lehigh University, 16 Memorial Drive East, Bethlehem PA 18015, USA
rotkin@lehigh.edu.*

Rodney S. Ruoff
Northwestern University, Department of Mechanical Engineering, 2145 Sheridan Road, Evanston, IL 60208, USA
r-ruoff@northwestern.edu

Hisanori Shinohara
Department of Chemistry and Institute for Advanced Research, Nagoya University, Nagoya 464-8602, Japan
and
CREST, Japan Science Technology Corporation, c/o Department of Chemistry, Nagoya University, Nagoya 464-8602, Japan
noris@cc.nagoya-u.ac.jp.*

F. Simon
Institute of Materials Physics, University of Vienna, A-1090 Vienna, Strudlhofgasse 4., Austria
fsimon@ap.univie.ac.at

M.S. Strano
118 Roger Adams Laboratory, Box C-3, 600 South Mathews Avenue, Urbana, IL 61801, USA

Michael A. Stroscio
Department of Electrical & Computer Engineering, Department of Physics, Department of Bioengineering, University of Illinois at Chicago, Chicago, IL 60607, USA
stroscio@uic.edu

M.L. Usrey
118 Roger Adams Laboratory, Box C-3, 600 South Mathews Avenue, Urbana, IL 61801, USA

T. Vuković
Faculty of Physics, POB 368, Belgrade 11001, Serbia and Montenegro

R. Bruce Weisman
Department of Chemistry, Center for Nanoscale Science and Technology, and Center for Biological and Environmental Nanotechnology, Rice University, 6100 Main Street, Houston, TX 77005, USA
weisman@rice.edu.*

Shalom J. Wind
Department of Applied Physics and Applied Mathematics Center for Electron Transport in Molecular Nanostructures Columbia University New York, NY 10027, USA
sw2128@columbia.edu.*

Alex Zettl
Department of Physics, University of California at Berkeley and Materials Sciences Division, Lawrence Berkeley National Laboratory, Berkeley, CA, 94720
azettl@physics.berkeley.edu.*

Ming Zheng
DuPont Central Research and Development, Experimental Station, Wilmington, DE 19880, USA
ming.zheng@usa.dupont.com

Part I

Theory and Modelling

1 From Quantum Models to Novel Effects to New Applications: Theory of Nanotube Devices

S.V. Rotkin

Classical and quantum effects in the physics of nanotube devices are presented. In particular, weak screening in one–dimensional systems is shown to essentially modify textbook theory of field–effect devices and lead to an interesting dependence of the device characteristics on geometrical factors. The capacitance of a nanoscale device has two main components: a classical geometric capacitance and a quantum term. The latter is related to a finite density of states of the nanosystem. Derivation of this density of states in the presence of external perturbations is a difficult task. We present some examples of the modification of the nanotube bandstructure by external perturbations. Electric fields can be used for band gap engineering in nanotubes, which may be translated into the device function. The concept of the Metallic Field–Effect Transistor is proposed. This device shows, at least theoretically, metallic conductance in the ON state and insulating behavior in the OFF state, which may be important for applications.

1.1 Introduction: Classical vs. Quantum Modelling

One of the expectations of nanotechnology, an area foreseen by Dr. Richard P. Feynman in 1959 [1], is that we may be able to access quantum properties of materials, which may ultimately lead to new applications and new device operations, which are not possible at the macroscale. To enable this new technology a theory that can make both qualitative and quantitative predictions is needed, whether it be a classical or quantum theory or a combination of both. In this chapter we give a few examples and present the quantum vs. the classical approach using recent results from our modelling of nanotube based devices.

Carbon nanotubes (NTs), discovered in 1991 [2], nowadays represent a new class of electronic materials. The electronic properties of NTs depend on their symmetry [3]. This is not unusual but for a single–wall nanotube (SWNT) there are just a couple of geometrical parameters: a curvature radius, R, and a helicity angle (the measure of chirality of the SWNT lattice), which solely define transport [4], optical [5] and even, to some extent, chemical [6,7] properties of a SWNT.

Knowledge of these two parameters will allow us to divide all possible NTs into several distinct classes. Two thirds of SWNTs have a forbidden band gap, which makes them semiconductors. The band gaps of the semiconductor SWNTs are in the optical region (near–IR/visible), depending on the value of R. The experimental fact that, over a wide range of R, the energy gap is proportional to the curvature, $1/R$ [5], is a clear manifestation of a simple quantum effect of a space quantization of an electron. When winding around the tube circumference, the electron acquires a phase. After making a full turn the phase has to be 2π, which results in a so–called quantization condition. The quantization energy sets the separation between the conduction and valence bands, and therefore the optical gap. In a similar way, as the atomic size (atomic number) of an element in the Periodic Table solely defines the properties of the substance, the curvature radius and the helicity angle of a SWNT define its electronic material properties.

One third of SWNTs are either metals or narrow gap semiconductors (often called "quasi–metals" in NT literature). The difference between the gap size in the last two SWNT classes appears in the second order of the curvature, $1/R^2$. The gap in the quasi–metallic SWNT scales as γ/R^2, where $\gamma \simeq 2.7\,\text{eV}$ is the hopping integral which gives the NT energy scale. In the same second order of the curvature, $1/R^2$, the first SWNT class (semiconductors) splits into two sub–classes by their chirality. All this constitutes a specific "Periodic Table" of nanotubes.

The other important property of nanotubes relates to their third dimension: we have already considered the radius and chirality of the tube but not the length. Recent success in NT synthesis (see also [8]) allowed experimental study of NTs with $R \sim 1\,\text{nm}$ and lengths of about several hundreds to thousands of microns, which implies an aspect ratio of 1:100,000 and greater. Certainly, this object must show physics similar to the physics of a one–dimensional (1D) wire, for example, a weak screening. Below we demonstrate that the weak 1D screening properties of NTs have important consequences for electronic devices.

The depth and wide scope of the physics of low–dimensional structures is due to strong correlations. This is because of the weak screening of the Coulomb interaction in low–dimensional systems. Indeed, the lowering of the system dimension from 3D to 2D is typically defined by allowing the Debye screening length to be larger than one of the system dimensions (the width of the 2D layer in this case). This results in the underscreening of an external potential [9], which is further enhanced in the 1D case [10], when the Debye length is larger than any transverse size of the 1D wire. This effect, though classical, will be shown to manifest itself and result in non–classical device behavior.

In the next section we consider two types of 1D objects: semiconductor nanowires (NWs) and carbon NTs on a common basis. Because the screening properties of these systems are very similar, a universal device theory can be developed under certain conditions.

The quantum approach is most appropriate in the modelling of nanoscale systems. Though much simpler, classical models may be also very useful to understand both material properties and device behavior if these models are corroborated by a microscopic theory [11]. To give an example, the modelling of the nanotube/nanowire Nano–Electromechanical Systems (NEMS) [12] requires a knowledge of the elasticity of the material. Clearly, classical elastic theory may break down at this scale. The use of an atomistic molecular dynamics to simulate the mechanical response allowed us to parameterize the stiffness of the tube [13]. Later, this stiffness may be used for classical continuum level NEMS modelling. Conjugating the atomistic/quantum level calculations with the lower level continuum/classical modelling constitutes a *multiscale approach*, which we will exemplify below.

1.2 Classical Terms: Weak Screening in 1D Systems

Many nanotube or nanowire electronic devices have a common geometry of a 1D Field–Effect Transistor (FET): the 1D channel connects two metal electrodes and is separated by an insulating layer from a backgate (conducting electrode). An important characteristic of such devices is the ratio between the transverse size of the nanowire/nanotube (diameter of the channel) and the Debye screening length in the material. It is typical for a NT or NW that the Debye screening length is larger than the diameter. Thus, the electrostatics of the system is the electrostatics of a 1D wire. A textbook bulk device model may fail to obtain carrier densities and electric potential distributions in the 1D channel properly. In this section we put forward a self–consistent model to simulate the charge density in 1D FET, stressing the role of the *weak screening* in 1D.

In order to isolate and clarify the effect of the weak screening we do not address ballistic transport models in this section. Some of the *short–channel* SWNT FETs were found to be accurately described by the ballistic model (cf. transport in a Schottky barrier FET in [4]). The ballistic picture may hold unless very high bias and/or gate voltages are applied and the scattering length becomes shorter than the channel length. In contrast, results of this section may be applicable for NT FET, providing *the channel is sufficiently long* [14], and for the majority of NW FETs [15–18] where the scattering length is always short.

It is known that, because of the weak 1D screening, the geometry of the leads/gates is important for device characteristics. A specific model geometry for the 1D FET, which we consider here includes the source ($x < -L/2$) and drain ($x > L/2$) electrodes connected by a 1D channel of length L and a gate electrode separated by a thin dielectric layer of thickness d. We assume the channel (semiconductor nanowire) to be uniformly doped with a 1D specific density of impurities, N. The gate voltage V_g changes the charge density in the channel controlling the FET transport. We employ the drift–diffusion model assuming that the scattering rate in the channel is sufficiently high to support a local charge equilibrium. In the opposite (ballistic) limit (see [4] and references [12,14,34–36] therein), the channel conductance is rather unimportant for the current as compared to the Schottky barriers at the leads.

1.2.1 Drift–Diffusion Equation and Quasi–equilibrium Charge Density

We measure all potentials from the wire midpoint ($x = 0$), so that the source and drain potentials are $\mp V_d/2$. In this case, the potentials and charge distributions along the wire caused by V_g together with the contact potentials (even) and by V_d (odd) are, respectively, even and odd functions of x and are denoted by the subscripts s and a: $\phi_{s,a}(x)$ and $n_{s,a}(x)$.

The potentials $\phi_{s,a}(x)$ can be divided into two parts: $\phi^0_{s,a}(x)$ created by the electrodes and found from the Laplace equation containing no electron charge, and $\phi^1_{s,a}(x)$ caused by the electron charge in the wire $-en_{s,a}(x)$. We assume that the characteristic length of charge variation along the wire, $l = \min\{L, 2d\}$, noticeably exceeds the wire radius R. In this case, the relationship between $\phi^1(x)$ and $n(x)$ is approximately linear [20–23] and the current j containing both drift and diffusion components can be written for the semiconducting NW with non–degenerate carriers in the form [21]:

$$\frac{j}{e\mu} = n(x)\frac{d\phi^0}{dx} - \left[\frac{2e}{\varepsilon}\ln\left(\frac{l}{R}\right)n(x) + \frac{kT}{e}\right]\frac{dn}{dx}, \tag{1.1}$$

where $n = n_s + n_a$, $\phi^0 = \phi^0_s + \phi^0_a$, μ is the carrier mobility and ε is the ambient dielectric permittivity. For a nanotube with $N = 0$ and degenerate carriers, kT should be replaced by the concentration–dependent Fermi energy. As a result, the terms inside the square brackets in (1.1) are replaced by $en(x)C_t^{-1}$ with C_t being the nanotube total capacitance derived in [22]. To obtain this total capacitance one has to go beyond the classical approximation. In this section we take C_t as a parameter of the model to be obtained at a different level of the multiscale approach as detailed in Sect. 1.3 below.

We solve the differential equation (1.1) with the boundary conditions $n(\pm L/2) = n_c$ assuming a constant charge density is maintained at the contacts, independently of the applied voltage. Two possible boundary conditions allow us to determine the value of current j, so far considered as some unknown constant. The case $n_c = N$ corresponds to ohmic contacts not disturbing electric properties of the wire. $n_c > N$ describes the situation where the carriers are supplied by electrodes, which is often the case for nanotubes, and $n_c < N$ corresponds to Schottky contacts. In the latter case, j is determined by the contact regions with the lowest charge density n_c and mostly independent of V_g. Thus for the structures adequately described by the classical drift–diffusion theory (obeying (1.1)), the transconductance will be very small. The only situation of interest is that when the Schottky barrier has a noticeable tunnel transparency strongly dependent on V_g[1]. This situation has been recently considered in [19] and will not be discussed below.

1.2.2 Linear Conductivity and Transconductance

In the first order in V_d, (1.1) can be linearized in n_a and easily integrated. The knowledge of $n_a(x)$ gives us the expression for j, which for the NW FET becomes especially simple for $A \equiv (2e^2N/\varepsilon kT)\ln(l/R) \gg 1$ resulting in the ordinary Kirchhoff law:

$$j = \frac{V_d}{R}, \tag{1.2}$$

where the resistance is

$$R = \frac{2}{e\mu}\int_0^{L/2}\frac{dx}{n_s(x)}, \tag{1.3}$$

which is not surprising since the condition $A \gg 1$ is equivalent to neglecting the diffusion component of current. The same equation is valid for the NT FET, where

$$n_s = -C_t\varphi_s. \tag{1.4}$$

Thus, a common description can be given for both NT and NW FET in this case. At arbitrary V_d, the described linear approach fails and simulation of the current–voltage characteristic (IVC) $j(V_d)$ requires solution of the non–linear equation (1.1), which can only be performed numerically and is discussed in the next section.

The linear device characteristic depends solely on the equilibrium charge density profile or, in other words, on the potential $\phi^0(x)$. The most intriguing result of this study is that this profile and hence all device properties depend on the geometry of the structure, particularly of the source and drain contacts. In the model reported in this chapter we consider bulk contacts with all three dimensions noticeably exceeding the characteristic lengths R, d, and L (see [24] for further details on the contacts of different geometry). Calculation of $\Phi(x, y)$, the potential created by this system of electrodes, is rather cumbersome even for this model geometry, and to obtain relatively simple analytical results we assume additionally that the relation $d \ll L$, often realized in 1D FETs, is fulfilled. Then, we solve the Laplace equation $\Delta\Phi = 0$ in the semi–infinite strip $-L/2 < x < L/2$; $y > 0$ with the boundary conditions: $\Phi(y = 0) = V_g$; $\Phi(x = \pm L/2) = \pm V_d/2$. Assuming $y = d$, we obtain $\phi^0_{s,a}(x)$. For $n_c \neq N$ the potential contains an additional term $\phi_c(x)$ that is

[1] In the case of the NT FET the Schottky barrier width (and the barrier transparency, therefore) is sensitive to the charge density in the channel (and to V_g).

proportional to $(N-n_c)$ and describes the potential of the uniformly charged wire [25]. This potential has been derived in [21]. With all the terms included explicitly the expressions fcr the even and odd components of the potential are given by:

$$\phi_s^0(x) = \frac{8e(N-n_c)L}{\pi^2 \varepsilon R} \sum_{n=0}^{\infty} \frac{(-1)^n \left\{ \mathrm{K}_0\left[\frac{\pi R}{L}(2n+1)\right] - \mathrm{K}_0\left[\frac{2\pi d}{L}(2n+1)\right] \right\}}{(2n+1)^2\, \mathrm{K}_1\left[\frac{\pi R}{L}(2n+1)\right]}$$
$$\times \cos\left[\frac{\pi x(2n+1)}{L}\right] + \frac{4V_g}{\pi} \sum_{n=0}^{\infty} \frac{(-1)^n}{(2n+1)} \cos\left[\frac{\pi x(2n+1)}{L}\right] \exp\left[-\frac{\pi d(2n+1)}{L}\right]; \tag{1.5}$$

$$\phi_a^0(x) = V_d \left[\frac{x}{L} + \sum_{n=1}^{\infty} \frac{(-1)^n}{\pi n} \sin\left(\frac{2\pi x n}{L}\right) \exp\left(-\frac{2\pi d n}{L}\right) \right], \tag{1.6}$$

where K_0 and K_1 are Bessel functions of an imaginary argument [26].

For the linear case in the limit $A \gg 1$, (1.2) gives the explicit expression for the dimensionless channel conductance $\sigma = jL/(n_c e\mu V_d)$:

$$\sigma = \left[2 \int_0^{1/2} \frac{dt}{1 + g\Psi(t)} \right]^{-1}, \tag{1.7}$$

where

$$g = \frac{2\varepsilon V_g}{[\pi e n_c \ln(l/R)]} \tag{1.8}$$

is the dimensionless gate voltage and for $n_c = N$

$$\Psi(t) = \sum_0^{\infty} \frac{(-1)^n}{(2n+1)} \cos[\pi t(2n+1)] \exp[-\pi d(2n+1)/L]. \tag{1.9}$$

The $\sigma(g)$ dependence has a cut–off voltage $g_0 = -\Psi^{-1}(0)$ characterized by vanishing σ. The exact behavior of σ near the cut–off can be calculated analytically. It is determined by the point $x = 0$ (where the charge density is at a minimum) and hence by the properties of $\Psi(t)$ at small t, which can be expanded as:

$$\Psi(t) \simeq \frac{\pi}{2} - \arctan[\exp(-\pi d/L)] - \frac{\pi^2 t^2}{2} \frac{\sinh(\pi d/L)}{\cosh^2(\pi d/L)}, \tag{1.10}$$

allowing us to perform integration in (1.7) and to obtain

$$i = \frac{\sqrt{(g-g_0)\sinh(\pi d/L)}}{\sqrt{2}\cosh(\pi d/L)}. \tag{1.11}$$

Here the dimensionless cut-off gate voltage is

$$g_0 = -\left\{ \frac{\pi}{2} - \arctan[\exp(\pi d/L)] \right\}^{-1}. \tag{1.12}$$

Thus the transconductance di/dg at $T = 0$ diverges at the cut–off: $\sim (g - g_0)^{-1/2}$. This theoretical prediction has been recently confirmed in the experiments on the long–channel NT FETs [14].

We stress that the non–classical result in (1.11) is not in contradiction to the textbook physics of Field–Effect devices, but extends the classical theory to the case of low–dimensional channels. It is a specific property of the 1D FET, namely, the weak screening of the Coulomb potential, which results in the divergence of the 1D FET transconductance at the cut–off voltage.

The IVCs of the 300 μm long NT devices have recently been measured [14] and are in qualitative agreement with the prediction of our model. The square root dependence of the subthreshold current on the gate voltage (1.11) was seen experimentally as was the divergence of the transconductance at the threshold, smeared at finite temperature by thermal excitation events. Further analysis will be required to explain why this subthreshold behavior can be seen not only close to the cut–off gate voltage but in the range of voltages up to 10 V.

If $n_c \neq N$, the function $\Psi(t)$ contains an additional contribution from $\phi_c(x)$. This function, studied in more detail in [21], is not analytical at $x \to \pm L/2$ but, similarly to $\phi_g(x)$, has an extremum at $x = 0$ and can be expanded in this point. This modifies the value of g_0 and the coefficient in i but retains unchanged the square–root character of $i(g)$.

The simplified expressions (1.2), (1.7) neglect the diffusion effects, which is equivalent to the limit $T = 0$, then $n_s = 0$ for all points where $\phi_s^0(x) < -C_t^{-1} n_c$. The potential ϕ_s^0 and the charge density acquire their minimal values at $x = 0$ and, hence, in the linear approximation, the cut–off voltage g_0 corresponds to the condition $\phi_s^0(0) = -C_t^{-1} n_c$ and at lower g the current is exactly zero. It is evident that at $T \neq 0$ an activation current will flow at $g < g_0$:

$$j \sim \exp(-\Delta/kT),$$

where

$$\Delta = e\left(-C_t^{-1} n_c - \phi_s^0(0)\right) \propto V_g. \tag{1.13}$$

Since $\phi_s^0(0)$ depends linearly on V_g (see (1.5)), the activation energy Δ is directly proportional to $g_0 - g$. This means that the above–mentioned singularity of di/dg is fictitious and real $i(g)$ has some maximum to the right of g_0 with a sharp, temperature–dependent decrease at lower g.

1.2.3 Numerical Results and Discussion

By measuring charge density in units of n_c, length in units of L, potential in units of en_c/ε, and current in units of $e^2 n_c^2 \mu/(L\varepsilon)$, the basic equation (1.1) acquires the dimensionless form

$$j = n(x)\frac{d\phi}{dx} - \left[2\ln\left(\frac{l}{R}\right) n(x) + \tau\right]\frac{dn}{dx}, \tag{1.14}$$

where $\tau = \varepsilon kT/(e^2 n_c)$ is the dimensionless temperature. The potential consists of three parts: $\phi(x) = \phi_c(x) + \phi_g(x) + \phi_a(x)$ describing the influence of the contact work function, gate voltage and source–drain voltage, each component being proportional, respectively, to $N - n_c, V_g$ and V_d. The particular form of each component depends on the geometry of the contacts and for bulk contacts is given by (1.5) and (1.6). The dimensionless version of (1.1) for nanotubes can be easily derived from (1.14) by assuming $\tau = 0$ and replacing $2\ln\left(\frac{l}{R}\right) \to \varepsilon C_t^{-1}$. Equation (1.14) should be solved with the boundary conditions: $n(\pm 1/2) = 1$.

We perform numerical calculations for two situations: ohmic contacts with $n_c = N$, and an undoped nanowire (nanotube) with injecting contacts: $N = 0$. For $n_c = N$ and for the chosen set of parameters: $d/L = 0.3$, the dimensionless threshold gate voltage $V_{g0} \simeq -12.8$. Figure 1.1a shows IVCs at two gate voltages (in units of en_c/ε): $V_g = -13.2$ (below the threshold) and $V_g = -12$ (above the threshold), which are superlinear because the high driving voltage V_d tends to distribute carriers uniformly along the channel. In our conditions, when powerful contact reservoirs fix the charge density at the points where it is maximal, such a redistribution will increase the minimal value of n at $x = 0$ and hence increase the conductivity. Such superlinear behavior is experimentally observed in NW FETs [15, 17, 27, 28] and differs noticeably from a sublinear IVC typical for bulk FETs and ballistic short–channel nanotube [29–31] structures.

Above the threshold, the channel conductivity is almost temperature–independent. The IVC curves for $V_g = -12$ (Fig. 1.1a) at different temperatures do not deviate from the dashed line corresponding to $\tau = 0.2$ by more than 10%. For V_g below the threshold, Fig. 1.1a demonstrates a strong temperature dependence of the current shown in more detail for $V_d = 0.1$ in Fig. 1.1b. While the two upper curves, corresponding to above–threshold V_g, have no noticeable temperature dependence, the two lower curves demonstrate such a dependence with the activation energy growing with $|V_g|$, in accordance with the analytical predictions of Sect. 1.13. At high V_d, where contact injection tends to create uniform charge density equal to n_c, different IVC curves merge and temperature dependence collapses.

The case of $N = 0$ formally differs from that of ohmic contacts only by the presence of $d\phi_c(x)/dx$ in (1.14). As can be seen from (1.5), this derivative has singularities at the contacts, which embarrasses numerical calculations. To get rid of these singularities, we use the following trick. In the closest vicinity of contacts the first term in the right side of (1.14) tends to infinity so that we can neglect the coordinate–independent left side. The remaining terms correspond to the quasi–equilibrium carrier distribution with ϕ_c playing the role of ϕ_s^0. This formula gives us the charge density profile in the vicinity of contacts to be matched with the solution of (1.14) far from the contacts. Since in this case $\phi_c(x) < 0$ (or, in other words, the electron density is lower due to the absence of doping), we obtain a lower absolute value of the cut–

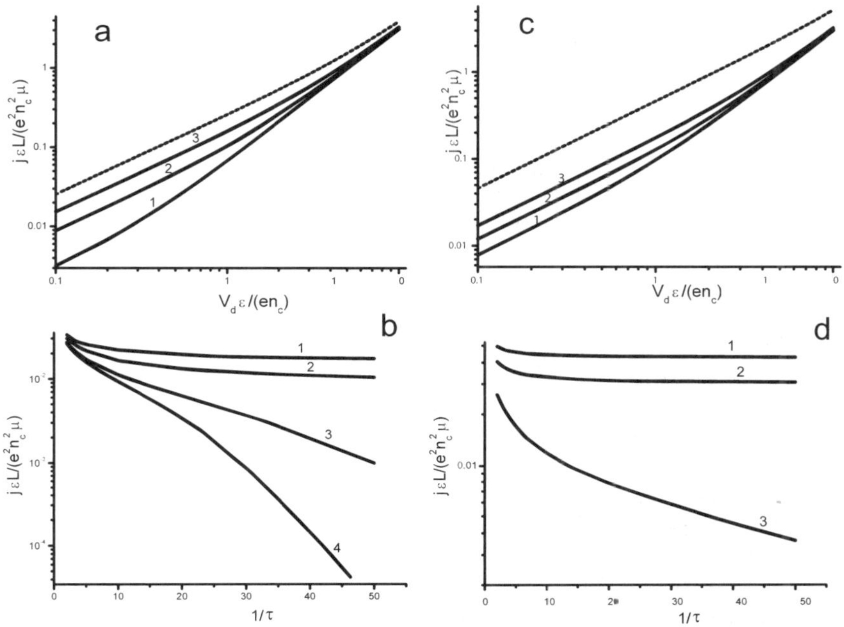

Fig. 1.1. Calculated characteristics of a NW FET with $d/L = 0.3$ and Ohmic contacts (**a, b**) and injecting contacts (**c, d**). (**a, c**) IVC for the temperatures $\tau = 0.05$ (1); 0.1 (2), and 0.2 (3): $V_g = -12$ (*dashed line*) and $V_g = -13.2$ (*solid lines*) in (**a**); $V_g = -6$ (*dashed line*) and $V_g = -10$ (*solid lines*) in (**c**). (**b, d**) Temperature dependence of linear conductance (at $V_d = 0.1$) for the same NW at $V_g = -12$ (1); -12.5 (2); -13 (3); -13.2 (4) in (**b**); -6 (1); -7.6 (2); -10 (3) in (**d**)

off voltage V_{g0} and lower transconductance as compared to $n_c = N$. For the same parameters as above, (1.12) gives $V_{g0} = -7.64$. Figures 1.1cd present the results of numerical calculations for this case. Qualitatively they are similar to Fig. 1.1ab but the dependencies on V_g and temperature are weaker. The above–threshold curve in Fig. 1.1c ($V_g = -6$) is practically temperature–independent, as in Fig. 1.1a, with the difference in currents between $\tau = 0.05$ and $\tau = 0.2$ being less than 5%.

1.3 Quantum Terms. I. Quantum Capacitance

In this section we introduce the concept of the quantum capacitance of a SWNT, the quantity which depends intimately on the Density of States (DoS) of a nanosystem. The DoS in the case of a bulk material is often assumed to be infinite. We note that the importance of the finite DoS for calculating the electrostatic response of low–dimensional systems (two–dimensional in this case) has been known for decades [9, 20]. One of first papers on practical applications of this effect was [32]. In this section we perform simple quantum mechanical calculations to obtain DoS of a SWNT for two different situations: the nanotube in vacuum (air) and the nanotube on a surface, interacting with the polarizable substrate. These calculations are required for modelling various electronic devices with nanotubes, but not only for electronic devices.

The calculation of the equilibrium charge density for the SWNT with a moderate mechanical deformation has been required to support our recent modelling of nanotube electromechanical systems [13, 33–35]. Knowledge of the induced charge allows one to calculate the electrostatic energy of the nanotube cantilever, which can be rewritten in terms of a distributed capacitance, and ultimately the electrostatic forces. In addition, the induced charge density and corresponding induced potential define the band profile and modify the conductance of the nanotube electronic devices as shown in Sect. 1.2.1.

The atomistic capacitance of the SWNT has two contributions: a purely geometrical term and another one, specific for the tube. It is very natural to call the second term "a quantum capacitance" as a similar definition was proposed for the two–dimensional (2D) electron gas system in [32] and used for one–dimensional electrons in SWNTs in [22].

In order to calculate the quantum capacitance one needs to know a charge on the nanotube surface as a function of an applied voltage. It was found that a statistical description (similar to what was used in [36]) is valid and gives a fairly good estimate for the charge density as compared to the full quantum mechanics (see also Fig. 1.5 and the related text below). The nanotube radius is the smallest important length scale of the model, including the characteristic range of the variation of the electrostatic potential: $l_g \sim \varphi/\nabla\varphi$. This allows one to consider φ as a *long–range potential* and to average the action of φ over the motion of the (quantum mechanical) electron. Thus, a perturbation theory can be developed to include effects of the potential on the electronic structure of the nanotube. For example, the self–consistent energy level shift due to the flat electrostatic potential is the same for every subband of the SWNT, to a first approximation [37]. One can then treat the influence of the external potential via the shift of an electrochemical potential (for all valence electrons) and calculate the charge distribution within the statistical approximation.

The applicability of the statistics and the macroscopic electrostatics to an *equilibrium* charge distribution has been discussed in [22, 23]. The same arguments may hold for a system which is slightly away from an equilibrium, e.g., for the description of a current carrying device. The reason for this "classical" behavior of such a small quantum object is two–fold. Firstly, the statistical approach gives a correct result as far as that the correct DoS is substituted in the Boltzmann equation. This DoS must be computed quantum

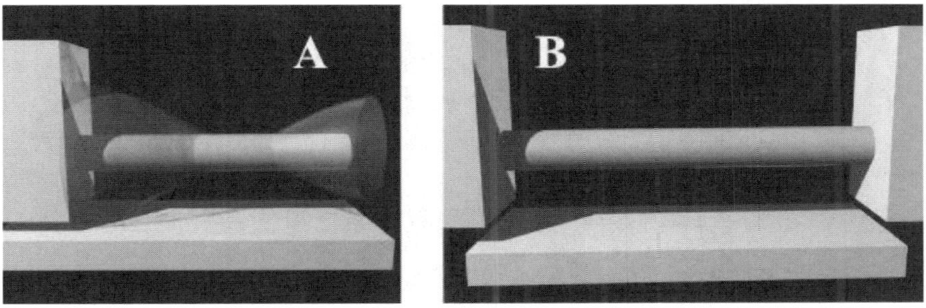

Fig. 1.2. Schematic geometry of the two single–wall nanotube devices studied in the section: (**a**) cantilever NEMS, and (**b**) string NEMS. The nanotube is shown sufficiently wider (not to scale)

mechanically. Secondly, the electron structure of the SWNT, in contrast to bulk materials, cannot substantially disturb/change the external electric field of the leads. Thus, the channel of the nanotube device is much closer to a definition of an ideal electric probe than any other system. This simplifies obtaining a *self–consistent* solution for the device transport characteristics.

1.3.1 Statistical Approach to Calculating Self-Consistent Charge Density in SWNT in Vacuum

Let us consider an example of a specific nanoelectromechanical device, a SWNT switch, and derive its quantum capacitance. We define a cantilever geometry of the nanotube electromechanical switch as follows: a straight SWNT connected to a side electrode and suspended over a ground plane electrode (Fig. 1.2A), and a "string" geometry of a SWNT device, as follows: the straight nanotube fixed (suspended without a slack) between two metal side electrodes over a backgate electrode (Fig. 1.2B).[2] The side electrodes are kept at the same potential with respect to the backgate. This design is standard for electromechanical systems, and the first experimental realizations of a SWNT NEMS appeared recently [38, 39].

As we discussed in the Introduction, depending on the symmetry of the tube, three different situations can be realized: (i) the armchair SWNT has two subbands crossing at the Fermi-point (Fig. 1.3 right). In this case the SWNT is metallic. (ii) The zigzag/chiral nanotube of a certain diameter and chirality has subbands that are separated by a non–zero band gap (Fig. 1.3 left, semiconductor tubes). (iii) One-third of zigzag and chiral nanotubes have a very small gap (quasi–metallic tubes). All three cases are almost equivalent for NEMS applications in the case of degenerate injection/doping, if the

[2] If the nanotube geometry is different from the ideal one because of small NT deformations one can still use the same theory, as will be explained in Sect. 1.3.3.

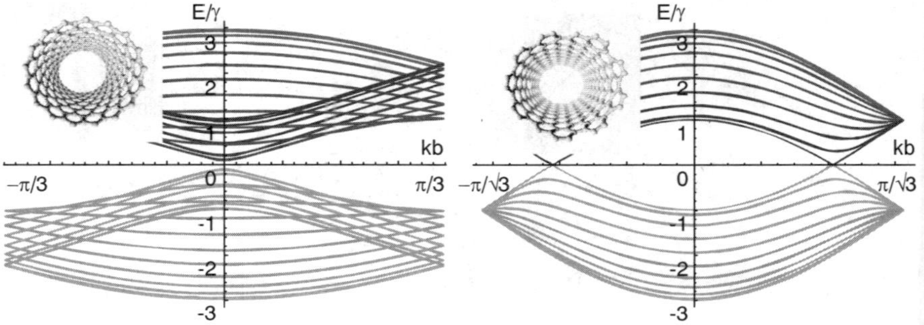

Fig. 1.3. Bandstructure of an armchair metallic SWNT [10,10] (right) and a zigzag semiconducting SWNT [17,0] (left) within the first Brillouin zone. Energy is in units of $\gamma \simeq 2.7\,\mathrm{eV}$, hopping integral. Insets: geometric structure of the same tubes

doping level is not too high (see also Sect. 1.4.2 for arbitrary doping levels). For the sake of simplicity we consider a metallic tube in this section.

In order to calculate the charge distribution of the straight metallic SWNT as a function of the total acting potential the latter is represented as a sum of the external and induced potentials:

$$\varphi^{\text{act}} = \varphi^{\text{xt}} + \varphi^{\text{ind}}. \tag{1.15}$$

Within the statistical model the induced charge is an integral over the nanotube DoS from a local charge neutrality level ($E = 0$) to a local electrochemical potential (see Fig. 1.4). The local electrochemical potential (which becomes a Fermi level at zero temperature) follows the local acting potential within the statistical model approximation. Great simplification is achieved in the case of the metallic nanotube, if the device is operating at a low voltage, then the Fermi level shifts within the lowest subband. Since the electron dispersion is linear, the density of states is constant (see Fig. 1.4) and equals

$$\nu_M = \frac{8}{3\pi b\gamma}. \tag{1.16}$$

Here $b \simeq 1.4\,\text{Å}$ is the interatomic distance and $\gamma \simeq 2.7\,\mathrm{eV}$ is the hopping integral. Within this approximation of the linear energy dispersion, the induced charge density reads as:

$$\rho(z) = -e^2 \nu_M \varphi^{\text{act}}(z). \tag{1.17}$$

Although, (1.17) gives the exact charge density in the case of the Fermi level crossing only the first (lowest) subband of the metallic nanotube, we note that the other singularities of the one–dimensional DoS are integrable. This means that for practical applications the approximate linear dependence of the charge density on the acting potential holds across the whole voltage

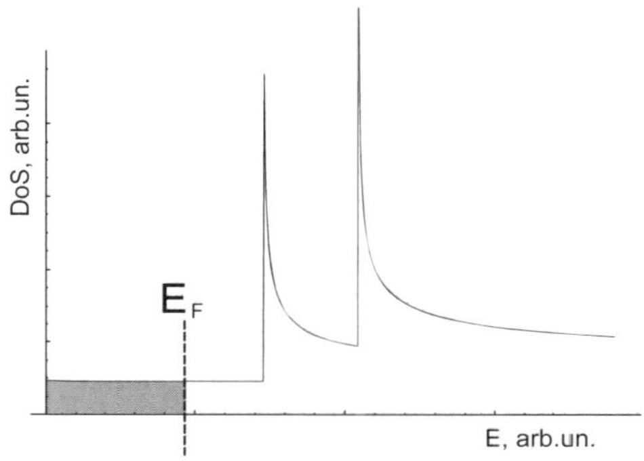

Fig. 1.4. Schematic DoS of a metallic SWNT. First (massless) subband contributes to a constant DoS at the $E = 0$. An injected/induced charge is proportional to the shaded area and is a linear function of the Fermi level, E_F, when E_F is lower than the second (massive) subband edge

(Fermi energy) range, with the possible exception in a small region at the subband edge.

We conclude this subsection by rewriting (1.17) as $\rho = -C_Q\varphi^{act}$ and identifying the combination $e^2\nu_M$ with the distributed quantum capacitance. In the 1D case the distributed capacitance is dimensionless. For a SWNT with a single massless (4–fold degenerate) subband it reads as:

$$C_Q = \frac{8e^2}{3\pi b\gamma}. \tag{1.18}$$

Let us note that in 1D (1.17) indicates a linear dependence of the charge density on the potential; while in 2D the charge density is proportional to the electric field (first derivative of the potential) by the Gauss–Ostrogradskii theorem, and in 3D the charge density is proportional to the Laplacian of the potential via the Poisson equation:

$$\begin{aligned} \rho_{2D} &\propto \nabla\varphi, \\ \rho_{3D} &\propto \nabla^2\varphi. \end{aligned} \tag{1.19}$$

1.3.2 Green's Function Approach for Geometric Capacitance

In order to obtain the self–consistent solution for the charge density the induced potential is needed, which can be calculated with the use of a Coulomb operator Green's function, $G(\mathbf{r}, \mathbf{r}')$:

$$\varphi^{\rm ind}(\mathbf{r}) = 4\pi \int G(\mathbf{r}, \mathbf{r}')\rho(\mathbf{r}')d\mathbf{r}'. \tag{1.20}$$

Green's function of a 1D system is known to have a logarithmic singularity at large distance unless some external screening is considered. In the case of a nanotube device this screening is provided by the closest gates/contacts. An equation implicitly giving the nanotube charge density follows from (1.15, 1.17, 1.20) and reads as:

$$-\frac{\rho(\mathbf{r})}{e^2\nu_M} - 4\pi \int G(\mathbf{r},\mathbf{r}')\rho(\mathbf{r}')d\mathbf{r}' = \varphi^{\rm xt}(\mathbf{r}). \tag{1.21}$$

This equation can be inverted analytically for simple cases. In general, it allows only numerical solutions or may be expressed as a series.

The numerical study of (1.21) has been performed in [23]. An interesting result is that the nanotube may be divided into three parts: two contact regions and a "central" region. The side parts are the regions near the side contact (or near the NT end if no external contact screens the electrostatic potential) of a length about several h long (several R long if no contact is present), where h is the distance to the screening gate. The aspect ratio of the state–of–the–art NEMS and electronic devices is very large, which means that the length of the nanotube, L, is much longer than h. The central region of the nanotube then covers most of the device length.

The electrostatics of the central region is elementary and allows an analytical solution for (1.21). Because of the screening of the Coulomb interaction by the backgate and the depolarization of the valence electrons of the nanotube, the corresponding Green's function is short–ranged. Therefore (at a distance about $2-3h$ from the contact), the self–consistent charge density is given by a simple expression:

$$\rho \simeq \rho_\infty = -\frac{\varphi^{\rm xt}}{C_g^{-1}+C_Q^{-1}} \simeq -\varphi^{\rm xt}\, C_g \left(1-\frac{C_g}{C_Q}\right) = \rho_{cl}\left(1-\frac{C_g}{C_Q}\right). \tag{1.22}$$

where ρ_∞ is an equilibrium charge density of the SWNT, calculated at the distance from the side electrode much larger than the screening length (distance to the backgate, h). $\rho_{cl} = -C_g\varphi^{\rm xt}$ is the classical charge density in a metallic cylinder with the same geometry. For the straight SWNT in vacuum: $C_g^{-1} = 2\log\left(\frac{2h}{R}\right) > 2\log 2 \sim 1.4$ and the quantum term $C_Q^{-1} = 1/(e^2\nu_M) \sim 0.3$ is much smaller, which allows us to expand the total capacitance in series in $C_g/C_Q \ll 1$.

Equation (1.22) clearly shows that: (1) the quantum mechanical correction to the classical charge density is small for a SWNT in vacuum/air. For a typical NT NEMS geometry it constitutes 10 % or less, which corroborates post factum our perturbative approach. Also we note that (2) the quantum correction is inversely proportional to the DoS and thus disappears in the classical limit of an infinite DoS of a 3D bulk material. (3) The ratio of the classical capacitance to the quantum capacitance is *not geometry (device) independent*. This means that the classical term $2\varepsilon\ \ln\ (2h/R)$ has an explicit dependence on the screening length (device geometry) and also on the

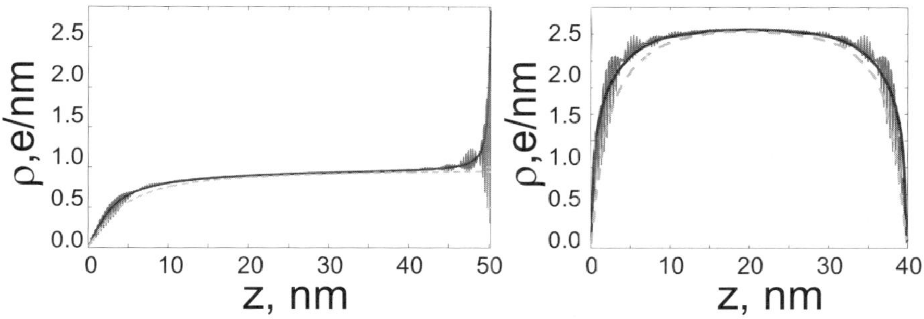

Fig. 1.5. Specific charge density for two devices: (right) string and (left) cantilever NEMS. The *thin solid oscillating curve* is a result of the quantum mechanical calculation. The *thick solid line* is a solution of joint Poisson and Boltzmann equations. The *dashed line* is the result of the analytical approximation (see Color Plates, p. 339)

screening properties of the surrounding matrix and/or the substrate (via an effective dielectric function, ε).

1.3.3 Results and Discussion

Green's functions have been derived for several realistic device geometries and the self–consistent charge densities have been calculated, using the statistical approach, detailed in [23, 34]. These charge densities were compared with the results of the quantum mechanical computation. In order to obtain the latter, joint Schroedinger and Poisson equations have been solved for the valence π electrons of a metallic armchair [10,10] SWNT in one subband approximation, *i.e.* in full neglecting the intersubband or sigma–pi mixing which has been estimated to be of minor importance for this problem. Apart from such purely quantum effects as quantum beatings at the ends of the finite length nanotube (Friedel oscillations), the statistical, semi–classical and quantum mechanical charge distributions are almost identical (a cross check has been done with the use of periodic boundary conditions to exclude the finite length effects). Figure 1.5 left shows the typical charge density distributions calculated with the use of a tight–binding (TB) theory and the Boltzmann equation for the cantilever SWNT of 50 nm long. One must conclude that a simple statistical description works fairly well for the case of a straight ideal single wall nanotube. A similar result has been obtained for the string SWNT with two side contacts (Fig. 1.5 right).

Edge effects were also studied to reveal the dependence of the total capacitance on the geometry of the finite length NT. One effect is in the nonuniform distribution of the classical charge density along the cylinder of the finite length, also known as fringe field effects. This was taken into account by solving numerically the electrostatic part of the problem by Green's function technique. Although, for general cases, there is no analytical solution for this problem, the numerical computation is straightforward [23].

The other contribution to the total capacitance of the device is due to the change of the DoS in a finite system. This problem is complicated and has recently been partly addressed in [40].

The equation for the equilibrium charge density is valid for a distorted nanotube as well as for an ideal straight nanotube. For the case of a slightly bent SWNT, one has to use in (1.22) the capacitance of the bent metallic cylinder, $C_g^{-1}(h(z))$, instead of the logarithmic capacitance, which is valid only for a straight tube. We assume that the mechanical perturbation is weak so it results in no appreciable change in the DoS. Thus, the quantum capacitance of the distorted SWNT remains the same and the total capacitance depends on the tube shape via the geometrical term only:

$$C(z) \simeq \frac{1}{C_g^{-1}(h(z)) + C_Q^{-1}} \simeq C_g(z)\left(1 - \frac{C_g(z)}{C_Q}\right). \tag{1.23}$$

This analytical form of solution for the device electrostatics is very useful for calculating electrostatic forces in various NEMS devices.

1.4 Quantum Terms. II. Spontaneous Symmetry Breaking

Now, we present a microscopic quantum mechanical theory for a charge transfer between a SWNT and a conductive substrate (and/or metallic leads). The finite value of the DoS of the SWNT at the Fermi level was shown to result in a quantum term in the self–consistent calculation of the induced charge density, for example, in the quantum capacitance, C_Q, as detailed in the last section. Thus far, C_Q has not been explicitly calculated but, within the statistical approach, has been derived from the given expression for the nanotube DoS. Most importantly, the DoS was assumed to be independent of the external potential (and other external perturbation). We noted already that the total capacitance of the NT device may depend on the geometry via the classical term. This section deals with the dependence of the quantum capacitance term on the polarization of the environment.

1.4.1 Splitting of SWNT Subband Due to Interaction with the Substrate

In general, the charge transfer (charge injection) may essentially change the DoS and can therefore modify the bare electronic properties of the nanotube material. This charge injection results from a natural work function difference between the nanotube and the substrate and/or from an external potential

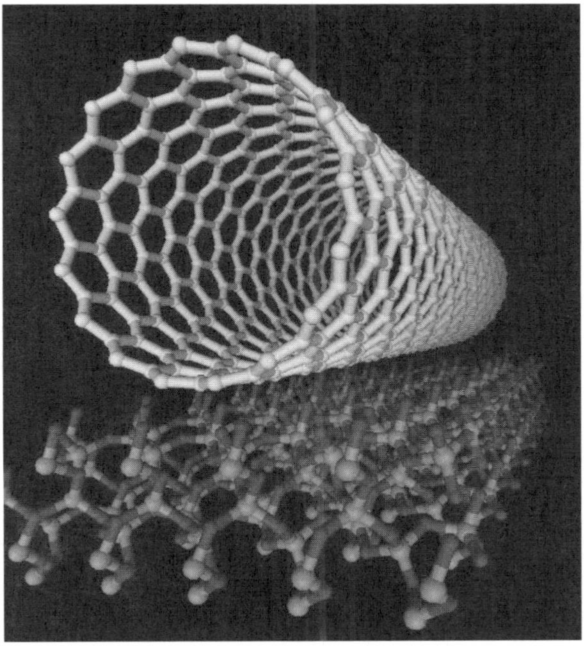

Fig. 1.6. Zigzag [17,0] nanotube on a surface of SiO_2 substrate. Geometry has been relaxed with Molecular Mechanics

applied between the two. The surface charge density on the SWNT is calculated below self–consistently within an envelope function formalism of TB approximation.

What is the influence of this charge transfer on the electronic structure of the SWNT? The most important consequence is a spontaneous breaking of the axial symmetry of the system. Because of this symmetry breaking the DoS changes qualitatively: degenerate subbands $\pm m >$ (where $m \neq 0, n$) split. It has a simple physical interpretation — the electrons with x and y polarizations are no longer equivalent as their attraction to the substrate is different. This effect can be related to a degenerate level Stark effect with an appropriate choice of external field of the image charge. The gap between the new x and y subbands is constant in k–space (for the external field which is uniform along the tube), so it shows up dramatically at the subband edge. The Van Hove singularity splits and we present analytical expressions for the splitting.

The depolarization of the charge density in the SWNT and *intrasubband* splitting will be studied for the setup shown in Fig. 1.6: a single SWNT lying on the insulator substrate. This is the typical geometry of the Field–Effect device. We also consider here a conductive substrate or a thin insulating layer separating the nanotube from the conductor, which resembles the situation of the contact and/or the top gate. We assume that the nanotube is connected to electron reservoirs, which may be the other leads or the conductor substrate itself. A transverse external electric field and/or a work function difference between the SWNT and the substrate/contact induce a non–zero electron/hole charge density in the nanotube. This extra charge density polarizes the substrate, which breaks the axial symmetry of the nanotube. We will demonstrate that the direct effect of the *uniform* external electric field is of minor importance as compared to the *nonuniform* field of surface charges on the substrate.

To calculate the splitting and shift of the electron energy levels one needs to know the matrix elements of the perturbation potential between corresponding wave–functions. In our case, the perturbation is a (self–consistent) Coulomb potential that describes the interaction between the probe electron and: (1) the extra charge density on the SWNT, and (2) the polarization charge density on the substrate surface[3]:

$$V = e\int_{-L/2}^{L/2} dZ \int_{0}^{2\pi} R d\beta \left(\frac{\sigma}{\sqrt{(z-Z)^2 + (R\cos\alpha - R\cos\beta)^2 + (R\sin\alpha - R\sin\beta)^2}} \right. \tag{1.24}$$

$$\left. + \frac{\sigma^*}{\sqrt{(z-Z)^2 + (R\cos\alpha - R\cos\beta)^2 + (R\sin\alpha + 2h - R\sin\beta)^2}} \right) .$$

Both the probe electron and the nanotube surface charge are taken on a cylinder of radius R. Then z and α are the electron coordinates in the cylindrical coordinate system. σ is the 2D surface charge density which does not depend on the coordinate Z along the nanotube because we assume translational invariance (although the theory can be easily extended for the case of slow variation of σ along the axis). We will show later that one can neglect the dependence of σ on the angle β in a *linear response* theory (in higher orders of perturbation theory a direct transverse polarization must be taken into account [41]). σ^* is an image charge density which is equal to $-\sigma$ for the metallic substrate.

The first term of (1.24) is the Hartree term for the SWNT in vacuum (without charge injection). The second term in (1.24) is the energy of interaction of the electron with the image charge. The separation between the SWNT axis and the surface of the conductor is h. In the case of the metallic substrate it is about the nanotube radius, R, plus the Van der Waals distance for graphite: $h \sim R + 0.34\,\text{nm}$.

The matrix element of the Coulomb operator (1.24) is calculated with the wave–functions of a TB Hamiltonian. We use envelope wave–functions, obtained similarly to [42], in the one–band scheme (π electrons only):

[3] In the original paper [37] this equation had the wrong sign in the term "$+2h$", corrected below.

$$\langle z, R, \alpha | \psi_{m,k,\zeta} \rangle = \frac{1}{\sqrt{2\pi L}} \frac{\langle \boldsymbol{r} - \boldsymbol{R}_I | \varphi \rangle + \zeta c_{mk} \langle \boldsymbol{r} - \boldsymbol{R}_{II} | \varphi \rangle}{\sqrt{2}} e^{ikz} e^{im\alpha}, \quad (1.25)$$

here the complex coefficients c_{mk} are to be found as eigenvectors of the TB Hamiltonian; the index m denotes the angular momentum of the electron and labels orbital subbands; k is the longitudinal momentum; $\zeta = \pm 1$ is the pseudospin. The components of the pseudospinor vector are atomic–like wave–functions, defined on two atoms of the unit cell, $\varphi_{I/II}$. We consider that the electron is confined to the surface of a cylinder. Here L is the length of the tube.

We assume that our potential is smooth at the scale of the single unit cell (0.25 nm). Then one may neglect transitions with the pseudospin flip (transitions between sublattices). With use of the normalization and orthogonality relation between the spinor components, it yields:

$$\langle m | V | n \rangle = -\frac{8\pi e R \sigma}{|m-n|} i^{m-n} \left(\frac{R}{2h} \right)^{|m-n|}, \qquad m \neq n \quad (1.26)$$

$$\langle m | V | m \rangle = 4\pi e R \sigma \log \left(\frac{2h}{R} \right), \qquad m = n \quad (1.27)$$

where σ, the surface charge density, has to be defined later in a self–consistent way. If σ has no dependence along the NT equator it becomes the 1D specific charge density we used before: $\rho = 2\pi R \sigma_o$.

Equations (1.26) and (1.27) are obtained by a direct Fourier transformation of (1.24) and describe the energy level shift when $m = n$ and the mixing of different subbands at $m \neq n$. The most interesting term with $n = -m$ is the mixing between the degenerate electron states within the same subband. By solving a secular equation for the intrasubband mixing of the electron doublet we obtain the splitting of the Van Hove singularity at the subband edge (Fig. 1.7). The new subband energy separation reads as:

$$\delta E_m = \frac{8\pi e R \sigma}{m} \left(\frac{R}{2h} \right)^{2m}. \quad (1.28)$$

Let us now calculate the injected/induced charge density σ which will allow us a numerical estimation for the δE_m splitting.

1.4.2 Charge Injection due to the Fermi Level Shift

Equations (1.26)–(1.28) are written for the given charge density σ, which will be derived in this section. When the SWNT is not considered in vacuum, one must include the work function difference between the nanotube and the contact or the conducting substrate. An external electric potential may be applied to the substrate as well. The total potential shifts the Fermi level in the SWNT [22]. As a result, the positive/negative charge is injected into the nanotube:

$$\sigma = \frac{e}{2\pi R} \int \left[f(E - \mu(\sigma)) - f(E) \right] \nu(E) dE \simeq \frac{e}{2\pi R} \int_0^{\mu(\sigma)} \nu(E) dE, \qquad (1.29)$$

here $\nu(E)$ is a bare one–dimensional DoS (independent of σ in linear response theory); $\mu = \Delta W - e\varphi^{xt} - e\varphi^{ind}(\sigma)$ is the total shift of the electrochemical potential of the SWNT (with respect to a charge neutrality level $E = 0$), which depends on: the work function difference ΔW, the external potential φ^{xt}, applied between the nanotube and the reservoir, and the potential φ^{ind}, induced by the charge density of the nanotube σ. To obtain the r.h.s. of (1.29) we approximate the Fermi distribution, $f(E)$, by the unit step function (which is exact if $T \to 0$).

We follow [22, 43] in the derivation of σ: the induced potential is obtained by direct integration of the charge density along the SWNT. This self–consistent equation for σ, (1.29), is readily solved analytically if the electrochemical potential is below the second subband edge. Then the charge is a product of the constant DoS and the electrochemical potential, μ. The solution of (1.29) is as follows:

$$\sigma_A = \frac{\Delta W - e\varphi^{xt}}{2\pi R\, e \left(2\log(2h/R) + C_Q^{-1} \right)}, \qquad (1.30)$$

and for a semiconductor SWNT, which has the DoS $\propto$

$$C_Q E\, \theta(E - \Delta)/\sqrt{E^2 - \Delta^2},$$

the charge is as follows:

$$\sigma_Z = \sigma_A\, \theta(\Delta W - e\varphi^{xt} - \Delta)$$
$$\times \frac{\sqrt{\left(\frac{\Delta}{\Delta W - e\varphi^{xt}} \right)^2 \left(4\log^2(2h/R) - C_Q^{-2} \right) + C_Q^{-2}} - 2\log(2h/R)}{2\log(2h/R) - C_Q^{-1}}. \qquad (1.31)$$

Here $\theta(x)$ is the Heaviside unit step function, Δ is 1/2 of the energy gap, and C_Q is the quantum capacitance of the SWNT given by (1.18).

From (1.28) and (1.30) we obtain the splitting of the degenerate subbands $|\pm m\rangle$ of the metallic SWNT (when the Fermi level is within the first subband):

$$\delta E_m = \frac{4\,(\Delta W - e\varphi^{xt})}{m \left(2\log(2h/R) + C_Q^{-1} \right)} \left(\frac{R}{2h} \right)^{2m}. \qquad (1.32)$$

The splitting decreases with m exponentially, hence, the effect is likely to be observable for the lowest degenerate subband. Then, for the following parameters: the SWNT radius $R \simeq 6.7\,\text{Å}$, the distance to the metal substrate $h = 10.1\,\text{Å}$, and the quantum capacitance $C_Q^{-1} \simeq 0.31$ [22], we obtain a numerical estimate for the subband splitting $\delta E_\pm \simeq 0.15\,(\Delta W - e\varphi^{xt})$. Experimental data for the work function of SWNTs ranges from 4.9 to 5.05 eV [44, 45]. For the SWNT on the gold substrate we use $\Delta W \sim 0.3\,\text{eV}$ as an estimate. In absence of the external potential, this work function difference results in an $\sim 46\,\text{meV}$ gap between two split peaks of the density of states (Fig. 1.7), which is larger than kT at room temperature. We also calculated the contribution of all other subbands, which shifts the doublet as a whole, but it is negligible in the splitting. As a result, the two new peaks in Fig. 1.7 appear asymmetrically with respect to the original DoS singularity.

The splitting of the $\pm m$ doublet is an analogue of a degenerate level Stark effect for the nanotube in a multipole potential of the image charge. The lower subband has x symmetry and the upper subband has y symmetry (with corresponding wave–functions $|x\rangle = 1/\sqrt{2}(|+m\rangle + |-m\rangle$ and $|y\rangle = 1/\sqrt{2}(|+m\rangle - |-m\rangle)$ because the energy of attraction of the electron to its image charge is lower for the second combination.

We predict a similar effect for the semiconductor nanotube, although the total external potential causing the charge density injection must be larger than one half of the gap in this case. In this section we study only the effect that is linear in the external potential, therefore all high order terms in (1.31) have to be discarded.

1.4.3 Dipole Polarization Correction

The charge injection in the nanotube may be readily achieved by applying an external electric field. Although the external field itself can break the band-structure symmetry and result in the direct splitting of the SWNT orbital doublet $\pm m$, for *uniform* electric fields this is forbidden by symmetry. The matrix element for an intrasubband splitting in the uniform external field $\mathcal{E}_{xt}$ equals zero by parity: $\langle m|e\mathcal{E}_{xt}y|-m\rangle = 0$.

In Sect. 1.4.1 we assumed that the charge density σ has no dependence on the angular coordinate β along the nanotube circumference. This is an accurate approximation since a dipole (and higher multipole) component of σ is small in comparison to what is given by (1.30) and (1.31). Let us prove this assumption for the dipole polarization of the SWNT.

The non–uniform external potential causes a deviation of the surface density from the uniform equilibrium value, σ, which is given by following expression:

$$\delta\sigma(\beta) = e\sum_{i\neq j}\frac{(f_i - f_j)\langle i|V|j\rangle}{E_i - E_j}\langle j|\beta\rangle\langle\beta|i\rangle, \tag{1.33}$$

where f_i are the occupation numbers, and $\langle i|V|j\rangle$ is the matrix element, given by (1.26) and (1.27) for the potential $V = \mathcal{E}_{xt}h$, where h is the distance

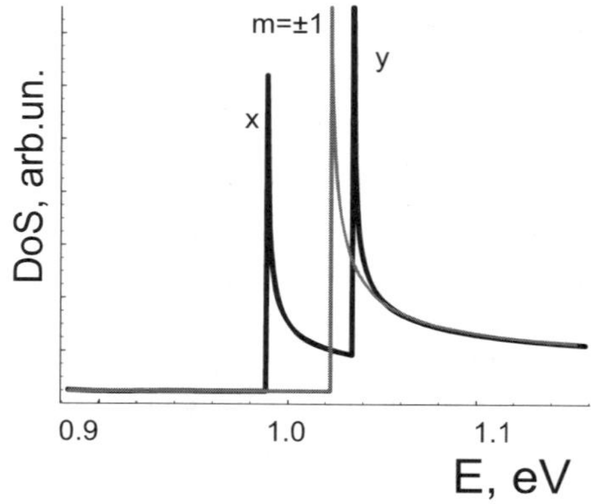

Fig. 1.7. DoS of the [10,10] armchair nanotube in vicinity of first Van Hove singularity (black color). Charge injection in the NT due to work function difference (see the text) results in a splitting of a doublet, which is clearly seen as compared to bare DoS of neutral NT (light gray color)

between the axis of the tube and the metal surface. E_i are the energies of subbands and $\langle\beta|i\rangle$ are the corresponding wave–functions.

We define the nonuniform dipole part of the charge density of a SWNT as $\delta\sigma_1 \equiv \int_0^{2\pi} \sin\beta \; \sigma(\beta) d\beta$. Then, the dipole component of the surface charge is as follows:

$$\delta\sigma_1 = \frac{ie}{8\pi R^2} \sum_i \frac{(f_i - f_{i\pm1})\langle i|V|i\pm1\rangle}{E_i - E_{i\pm1}}. \qquad (1.34)$$

Let us recall that according to (1.26) $\langle i|V|i\pm1\rangle = -i8\pi R^2 e\sigma/(2h)$.

In the case where the electrochemical potential equals zero (no charge in the nanotube), the transverse polarization includes transitions from the valence to the conduction band only, $\langle v|V|c\rangle$ (the details of the calculation are presented elsewhere [46]). Here we study an extra component of the polarization which is due to the induced charge density. Thus, we need to consider only transitions from the levels above the charge neutrality level, $E = 0$, and below the Fermi level, $E = E_F$ (the shaded area in Fig. 1.4). Hence, the dipole polarization is proportional to the net charge density σ, and the dipole charge density of the armchair SWNT is given by the following expression:

$$\delta\sigma_1 = \frac{\sqrt{3}C_Q^2}{32\pi} \frac{(2\pi R\sigma_A)^2}{e} \frac{R}{h} \log\frac{2h}{R} \propto \mathcal{E}_{xt}^2. \qquad (1.35)$$

We single out the term $2\pi R\sigma_A$, which is the specific one–dimensional charge density of the SWNT, ρ, proportional to the external potential and thus to the external field, $\mathcal{E}_{xt}$.

Equation (1.35) shows that the effect of the transverse polarization on the bandstructure is quadratic in the external field, in good agreement with a plain dielectric response theory [47,48], while the effect of the image charge is linear in $\mathcal{E}_{xt}$. Thus, the degenerate level splitting due to the dipole component of the polarization will be less important than the splitting due to a uniform component: $\sigma_0 \equiv \int_0^{2\pi} \sigma(\beta) d\beta$, at least, in the weak field regime discussed in this section. This proves post factum that our assumption of σ to be independent of β is valid.

We assumed in this section that the perturbation theory in a linear approximation in μ (or equivalently in σ) is applicable. One restriction that may follow from this assumption is that the external potential has to be small. We neglect here the dipole and higher multipole terms in the induced charge density. This is equivalent to a weak intersubband mixing, an assumption which may not hold for wide nanotubes or strong external fields. The effect of the strong field on the bandstructure is discussed in the next section.

So far we developed a quantum theory for the modification of the SWNT DoS by the potential induced by the *equilibrium charge density*, based on the assumption that (1.29) is valid. One may consider transport devices in the same way, as long as the NT charge distribution is still given by the quasi–equilibrium density. For a non–zero current flowing through the nanotube one must use an expression for the charge density which differs from (1.29). For example, for a stationary quasi–equilibrium $n(x)$ one can use (1.5) and (1.6) as described in the first section.

1.5 Quantum Terms. III. Band Structure Engineering

In semiconductor nanostructures, which are the main building blocks of modern nanotechnology, some electronic properties may depend on the lateral size of the structure, *e.g.*, the radius of the SWNT. For certain applications, *e.g.*, switches, it is important to be able to change the electronic properties by applying an external signal, which may be an electric signal, light, etc.

Interaction with the substrate, as discussed in the last section, may lead to modification of the nanotube DoS due to the spontaneous breaking of symmetry of the quantum mechanical Hamiltonian. When an external voltage, which may also break the symmetry, is applied across the system, the changes are even more interesting, leading to new effects and possible applications.

We propose a way to control the properties of SWNT electronic materials by engineering their band structure. For standard bulk semiconductor materials one obtains this by doping, making compound materials and/or mechanically stressing the material. In this section we demonstrate the use of external electric fields as an approach to band structure engineering in SWNTs.

1.5.1 Band Gap Opening and Closing in Uniform Fields

We consider first the influence of a *uniform* electric field on the band structure of nanotubes. We emphasize that this electric field is assumed to be uniform at the size scale of the nanotube diameter, about 1 nm. Almost any external electrode size is much larger than this length. Therefore, if no special effort to produce a strong field gradient is performed, in the typical Field–Effect device geometry, the fields of the gate and contacts will be uniform at the nanometer size scale. We note that the charged tube itself is a source of a nonuniform field if placed close to the polarizable surface, but this effect has been considered in Sect. 1.4 above.

The band structure changes can be classified according to the nanotube symmetry. In the strong uniform electric field the armchair tubes remain metallic, while metal–insulator transition is induced in other quasi–metallic nanotubes. This means that with an increasing external field the constant DoS of the quasi–metallic SWNT near the Fermi level changes. At some field the DoS decreases to zero at the Fermi level and the forbidden gap opens. The gap grows with increasing the field strength to some critical field, and then decreases again. The critical field strength is inversely proportional to the square of the nanotube radius as we will discuss in more detail below.

The method used here to model the effect of the transverse electric field on the band structure is a direct diagonalization of a single π orbital TB Hamiltonian, together with a self-consistent electrostatics calculation to account for the screening effect. Within the limit of relatively weak fields, the electrical potential variation along the SWNT circumference is much smaller than the typical energy distance between the closest subbands, $V = e\mathcal{E}R \ll \Delta E_m \sim \hbar v_F/R$, where m is the subband index; $v_F = 3b\gamma/2$ is the Fermi velocity, and γ and b are the hopping integral and carbon–carbon bond length as above. Then a perturbation theory may be used for an analytical estimate for the band gap dependence on the applied field. When the field is strong enough to essentially mix the subbands but not too strong to distort the π orbitals itself, *i.e.* $e\mathcal{E}R \geq \Delta E_m$ and $e\mathcal{E}b < \gamma$, the band structure changes qualitatively.

Figure 1.8 shows the energy bands of a [10, 10] armchair nanotube in electric fields of different strength ranging from 0 to 0.3 V/Å. The two lowest subbands are always crossing, even at very large fields, when the band structure has already been noticeably modified [46]. At $\mathcal{E} \simeq 0.1$ V/Å (see Fig. 1.8b), the Fermi points shift toward the Γ point ($k = 0$) and the two lowest subbands are flattened near the Fermi points. At the same time, all states that were degenerate with respect to the angular momentum, $\pm m$, split. The splitting becomes more obvious closer to the lowest subbands, similarly to the spontaneous splitting described in the last section. As the field strength increases (Fig. 1.8(c,d)), the two lowest subbands show multiple nodes, while the first node moves closer to $k = 0$.

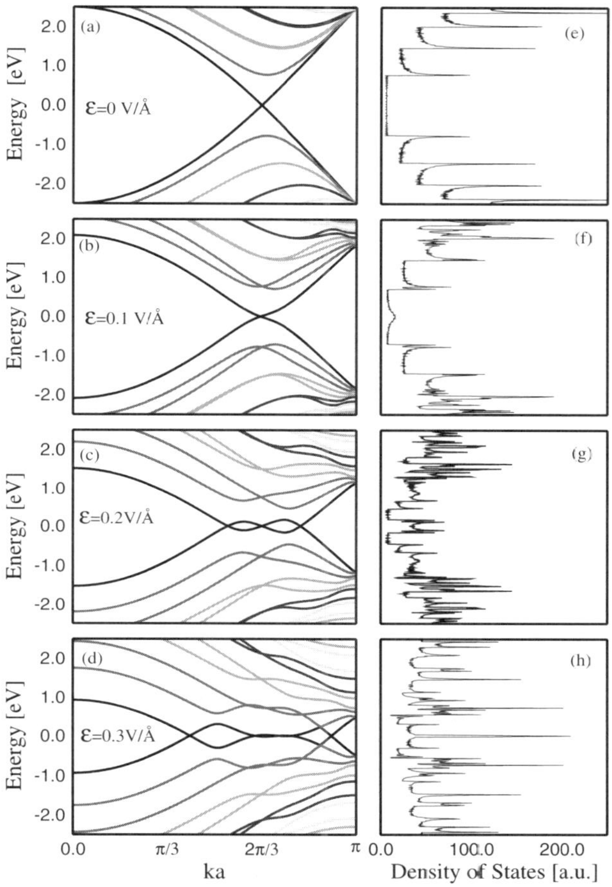

Fig. 1.8. Band structure (**a,b,c,d**) and DoS (**e,f,g,h**) of a [10,10] armchair tube at various perpendicular electric fields: $\mathcal{E} = 0, 0.1, 0.2$, and $0.3\,\mathrm{V/\AA}$

This band structure modification is clearly seen in the density of states of the nanotube, as shown in Fig. 1.8 (d–h). A bump appears in the low energy plateau and increases with the field strength. The enhanced DoS near $E = 0$ is due to the flattening and bending of the two lowest subbands. On the other hand, the lifting of $\pm m$ degeneracy of all doublets shifts and splits the single Van Hove singularity peaks into multiple ones.

For quasi-metallic zigzag tubes, the physics is quite different. At weak fields, a band gap opens at the Fermi point $k = 0$, as shown in the inset of Fig. 1.9 for a [18,0] nanotube. The gap increases quadratically with the field strength until it reaches a critical field $\mathcal{E}_c$. The value of the critical field depends on the radius and an analytical estimate is:

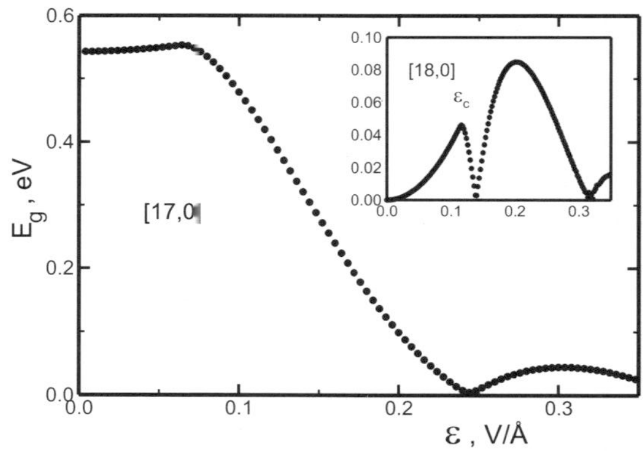

Fig. 1.9. Band gap variation of a [17,0] zigzag nanotube with increasing field strength. Inset: same plot for a [18,0] nanotube

$$e\mathcal{E}_c R \sim \Delta E_m \sim \hbar v_F / R, \tag{1.36}$$

when the total potential drop becomes comparable with the energy distance between neighboring subbands. As the field increases beyond $\mathcal{E}_c$, the energy band minimum shifts away from the original Fermi point $k = 0$, and splits into multiple valleys. As a result, the gap does not exceed some critical value beyond which a further increase of the field closes the gap (see Fig. 1.9). The maximum band gap opening at $\mathcal{E}_c$ is obtained within the perturbation theory as follows:

$$E_g \sim \frac{(e\mathcal{E}_c R)^2}{6\gamma} \sim \frac{v_F^2}{6\gamma R^2}. \tag{1.37}$$

The maximum gap decreases with the SWNT radius and becomes almost negligible for nanotubes with $R > 8$ Å. On the other hand, we notice that for very narrow SWNTs, the $\sigma - \pi$ mixing may result in the opening of secondary gaps, which may modify the above TB results.

Semiconductor zigzag nanotubes also exhibit a drastic change in the presence of a perpendicular electric field. Figure 1.9 shows the band gap variation of a [17,0] nanotube with increasing field strength. The band gap first remains almost constant at weak fields. When the perturbation is comparable to the original band gap, *i.e.* $e\mathcal{E}_c R \sim E_g \approx 0.53$ eV, both the conduction band bottom and the valence band top shift away from the original Fermi point $k = 0$ and the gap closes. At fields larger than the critical field the gap opens again. The critical field strength $\mathcal{E}_c$ is inversely proportional to R^2, as in the case of a quasi–metallic nanotube. We note that for two different types of semiconductor SWNTs (mentioned in the Introduction) the weak field regime is different. Figure 1.9 shows that the gap of a [17,0] SWNT has a maximum at the weak field ~ 0.7 V/Å. In contrast, the gap of a [19,0] SWNT will monotonically decrease with the field [47].

We stress that many experimental techniques, ranging from Raman scattering to scanning tunnelling spectroscopy use high electric fields to probe the electronic properties of a nanotube, which may perturb the electronic structure. Our theoretical results may help to understand the disagreements between experimental measurements and predictions for band gap, local density of states (LDoS), effective masses and the locations of Van Hove singularity peaks. On the other hand, the novel behaviors of SWNTs under large fields provide an effective way to electronically tune the SWNT properties and modulate the device conductance as we will discuss in the last section.

1.6 Novel Device Concepts: Metallic Field–Effect Transistor (METFET)

In the last section we proved, at least theoretically, that one can modulate electronic properties of SWNTs by applying an external field, thereby opening the band gap, transforming the nanotube from a metallic state at zero field to a semiconductor (insulator at low temperature). This new quantum effect may be used for designing an electronic device [49, 50].

The Field–Effect Transistor is one of the basic semiconductor devices of modern electronics. The scaling trend to nanometer dimensions calls for ever higher doping and channel ON state conductances for FETs [51]. Ultimately it leads to a conductance close to that of a metal, which requires higher levels of doping. This might be technologically challenging because as the device sizes shrink, the conductivity degrades due to quantum effects, non–uniformities of the material, etc.

Although a metallic channel would be appealing, metallic conductance prevents the penetration of the electric field except for extremely short distances (skin–layer) that are too short to achieve the FET function. In other words, the screening length is small in the bulk metals, thus one can neither create an electrostatic barrier for electrons nor impede electron flow by strong scattering.

On the other hand, the FET conductance in the OFF state has to be as low as possible to minimize OFF current. This requires material properties close to an insulator, which is in visible contradiction with the need for a metallic ON state, unless we consider changing the material properties during the device operation. It was proposed that the use of exotic materials (which can undergo a phase transition from a metallic to insulating state [52]) may present a solution to the problem. However, in earlier studies bulk and spatial uniform materials were always used and, therefore, uniform gates and channels were considered. In contrast, we recently proposed an approach to control the carrier transport in the *one–dimensional* metallic FET (1D METFET) using the nonuniform electric field created by a low–dimensional *local gate*. It is clear that the field can penetrate into the low–dimensional structure, unlike bulk materials. The Debye screening length is larger than the transverse size of the 1D channel (e.g., nanotube diameter), therefore one may argue that a low-dimensional system represents only "surface" to external electric fields and hope that a Field–Effect might be observable for low–dimensional systems.

It is known that electron transport may reach the ballistic limit in metal/semiconductor wires and particularly in metallic carbon nanotubes, because of restricted phase space for electron backscattering. Furthermore, in the armchair SWNT the special symmetry of the lattice results in a strong suppression of the backscattering at the Fermi level at zero doping, which enhances the conductance.

We propose to use both (a) 1D weak backscattering, and (b) 1D weak screening to design a metallic 1D channel that may be electrostatically gated. Though several low–dimensional structures/materials can be proposed as possible candidates for making a 1D METFET, in this chapter we consider only the example of a carbon SWNT.

The use of metallic single–wall nanotubes (M–SWNTs) to realize the possibility of 1D Metal Field–Effect is appealing because of the extremely small size of the tube (typically, the SWNT radius is $\sim 0.7\,\mathrm{nm}$). In addition, following the view of IBM researchers on combining metallic and semiconducting nanotubes (S–SWNT's) in circuits, where the M–SWNT's would serve as interconnects and the S–SWNT's as active devices [19, 53–55], there exists a the possibility of using the *metallic tubes* as transistors. The METFET will add an ambipolar device to the predominantly p–type S–SWNT's without using extra doping and/or complicated control on a contact work function. The METFETs could be combined to circuits in analogy to CMOS.

An ideal SWNT theoretically has very high conductance, ultimately approaching the ballistic limit. As we have shown in the last section, certain SWNTs can undergo the metal–semiconductor transition under the action of a uniform electric field. This metal–semiconductor (metal–insulator) transition allows us to obtain a transistor action.

Next we introduce the concept of *local gating*. Electric fields with large gradients may lead to a stimulated symmetry breaking (as compared to the spontaneous symmetry breaking considered in Sect. 1.4). As a result, the band gap may open, even in the true metallic armchair SWNT. Also, the use of a highly localized gate allows one to obtain a strong enhancement of the electric field in a narrow region. Thus, any depletion of charge is easier to achieve since it covers extremely small volumes. In addition, the spatial non-uniformity of the electric field represents a source of scattering which can modulate the conductivity in the gated region while a homogeneous gate influences the whole channel and contributes little to scattering.

1.6.1 Symmetry and Selection Rules in Armchair Nanotubes

The use of M–SWNT's for the METFET is advantageous due to a low phonon scattering-rate [56] and an equally low rate of scattering by impurities with a

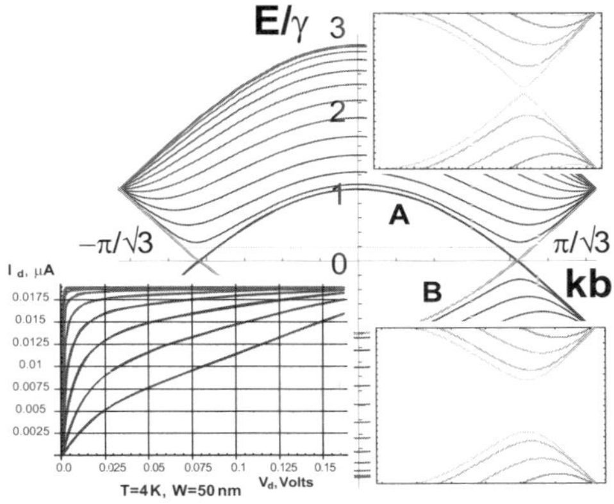

Fig. 1.10. Bandstructure of an armchair [10,10] M–SWNT. Subbands of orbital quantization $m = 0, 1 \ldots 9$ are shown from top to bottom in an conduction band (from bottom to top in the valence band). The two closest massless subbands, A and B, have the same $m = 10$. The upper right inset shows a zoom view of the Fermi point with the opening of a gap in the M–SWNT due to perturbation as described in the text. The lower left inset shows how the gap grows linearly with applied potential. The upper left inset shows IVC for METFET with gate width 15 nm

long–range potential [57,58]. This follows from a special symmetry (selection rules, cf. [59] and references therein) applicable to the transitions between the closest subbands of the metallic armchair SWNT's. It results in the ballistic conductance of armchair M–SWNT's for the low doping levels. These selection rules are also important for our study and will be considered next.

The wave–function of the electron in the conduction/valence band can be written as the product of the envelope function and the pseudospinor amplitude in the unit cell, (1.25). We recall that the pseudospin has the opposite sign, ± 1, for states of the valence and conduction band, which are otherwise symmetrical[4] with respect to the Fermi level at zero doping/injection:

$$E^{(C)}_{m,k,+1} = \zeta E_{mk\zeta} = -E^{(V)}_{m,k,-1} . \tag{1.38}$$

For each pair of symmetric electron and hole subbands of the armchair nanotube with $m = 0, 1 \ldots (n-1)$ (see Fig. 1.10), the angular and axial momentum quantum numbers are the same but the pseudospin has the opposite sign. If the unit cell orbits are chosen to be orthogonal: $\langle \varphi_I | \varphi_{II} \rangle$, the overlap integrals of the symmetric states of the conduction and valence bands are exactly

[4] The extra (electron–hole) symmetry is model dependent. For example, by including the effect of curvature or higher order terms in the TB approach in the Hamiltonian, one lifts this symmetry. This may result in some modification of the expressions given in the chapter (to be discussed elsewhere).

zero: $\langle\psi_{m,k,-1}|\psi_{m,k,+1}\rangle = 0$. The same holds for the overlap matrix elements between the two (A and B) massless subbands (crossing at the Fermi level).

A common approximation of solid state theory is that the spatial variation of a (long–range) external potential is small at the scale of a unit cell (effective mass theorem). The matrix element of the potential is therefore a product of the unit cell overlap integral and the Fourier component of the potential:

$$\langle\psi|U|\psi\rangle = \langle\varphi|\varphi\rangle U_q. \tag{1.39}$$

It is evident that this approximation leads to zero scattering (and also mixing, see below) between two closest subbands of the armchair tube as $\langle\varphi|\varphi\rangle \equiv 0$. This is suggestive of ballistic transport in the armchair SWNT channel[5]. Thus, the ideal armchair SWNT is a great candidate for a 1D METFET as its conductance in the ON state has to be about $2G_o = 2e^2/h$, 4 times the conductance quantum (for 2 spin and 2 space channels), the maximum conductance for a 1D circuit with macroscopic leads [6].

In order to generate a metal–semiconductor transition and open a forbidden gap in the nanotube band structure one has to achieve a depletion of the DoS at the Fermi level. There are two crossing subbands that contribute to the DoS at this energy, A and B, as shown in Fig. 1.10. From simple quantum mechanics we know that two energy levels may cross if there is at least one quantum number different for these levels. In the case of the armchair nanotube this is the pseudospin quantum number. Thus, one has to break this symmetry and mix crossing subbands. Any Field–Effect would be proportional to the gap opening due to this mixing.

1.6.2 Gap Opening and Switching OFF: Armchair SWNT

The gap, E_g, that may be generated by breaking the symmetry of the armchair SWNT with respect to the pseudospin quantum number, is proportional to the the matrix element of the external potential between subbands A and B. This matrix element is the product of the unit cell overlap integral (which is non–zero only for the potential of odd symmetry with respect to the mirror reflection plane between neighbor atoms I and II) and the lattice Fourier transform of the potential, $U_{q,\delta m}$. The $U_{q,\delta m}$ is non–zero when the potential is a full scalar with respect to the rotations of the symmetry group of the nanotube [59], *i.e.* for $\delta m = 2Qn$, where $Q = 0, 1, \ldots$ and n is the index of [n,n] armchair SWNT. That is why a *direct mixing* of the subbands A and B happens only if the high–multipole components are presented in the Fourier

[5] Although there is a possibility of scattering to the other valley of the same subband, this scattering amplitude is small due to a large momentum transfer $q \simeq 2\pi/\sqrt{3}b$ (see thin line between left and right Fermi points in Fig. 1.10).

[6] The lower resistance can be expected in the case of entirely nanotube circuit [53]. The quantum contact resistance will not limit anymore the ON current in this case. That device can fully exploit all advantages of the METFET.

transform of the external potential and, thus, has not been observed before. The gap is linear in the applied gate potential (see Inset in Fig. 1.10). There is no upper limit for the magnitude of the opened gap in this case, except for a natural condition that the external field must not cause an electric breakdown. In addition, when the field becomes comparable with the atomic one, the simple TB model is no longer applicable.

The direct mixing of the crossing subbands is difficult because one needs to break the mirror reflection symmetry *and* keep the full angular symmetry of the applied potential such that the transition between states with the same angular momentum $m = n$ is allowed. This requires a very high multipole moment of the potential, at least $\delta m = 2n$ [46]. Although such potential may be created by applying the gate voltage to a chemically modified surface of the SWNT (to be discussed elsewhere [60]), this is a challenging problem for future technology.

However, if only the first condition is fulfilled (mirror symmetry breaking), the gap may open by *indirect* mixing of the subbands in a higher order of applied potential [46]. In particular, by applying a uniform field (dipole component of the potential) across the armchair SWNT *together* with a higher order (quadrupole) component we obtained a gap which scales as a third power of the gate potential. Perturbation theory for nanotubes (similar to what was used in [41]) predicts a maximum gap that depends only on the size of the tube and scales as R^{-2} or n^{-2} [46].

1.6.3 Switching OFF Quasi–metallic Zigzag Nanotube

While opening the energy gap in the metallic armchair SWNT requires a special symmetry of the gate potential, quasi–metallic tubes are the best candidates for band gap engineering. As we already noticed in Sect. 1.5, the uniform external electric field applied across a zigzag quasi–metallic nanotube opens the gap, as a result of the Stark effect. Therefore, the gap is proportional to the square of the potential (cf. the third power for the non–linear Stark effect in the armchair tubes). Though similar to the case of the Stark effect in the armchair tube, this gap will eventually close with increasing external potential. Thus, there is a maximum gap opening, which is defined solely by the tube radius $E_g^{(c)} \sim R^{-2}$, given by (1.37). This analytical prediction was confirmed by numerical calculations in [41,46]. We note here that even though the band gap opening is easier to obtain for a quasi–metallic nanotube, because there are no special selection rules for a transition (mixing) between the closest subbands, the same argument results in a stronger scattering of carriers in this tube, similar to the case of a semiconductor tube [58]. When engineering an effective electronic switch one has to trade between these two factors. The scattering rate depends on a specific mechanism of scattering. It is important therefore to have only long–range scattering potential and a short–range electrostatic potential of the local gate to modulate effectively the FET channel.

1.6.4 Modulation of Ballistic Conductance

After these discussions on the possible mechanisms of the metal–insulator transition, we consider the application of a very inhomogeneous electric field localized in a relatively small volume of the NT METFET channel. This inhomogeneous field is capable of opening a semiconductor gap in the DoS of the armchair and/or quasi–metallic nanotube. The electron flow (at the Fermi level, which has to be in the middle of the forbidden gap in the gated region) is then classically suppressed (at $T = 0$). A quantum mechanical tunnelling mechanism has to be invoked in the current calculation because the effective mass of the electron in the SWNT is very light. Thus, the rate of tunnelling through the classically forbidden region is high. We describe it by transmission and back–scattering coefficients calculated within the Wentzel–Kramers–Brillouin approximation.

Any opening of the gate induced gap along the METFET channel will result in the potential barrier, which impedes conduction. This may be added by a Coulomb blockade at the interface, as will be discussed later. The transmission coefficient is an exponential function of the height of the barrier, *i.e.* of the gap, $E_g/2$, and of the barrier length. Thus the OFF/ON ratio will also be controlled by this parameter E_g and the total length of the gated region.

The inset of Fig. 1.10 shows typical[7] IV curves (IVC) for a METFET made with an armchair metallic nanotube of radius 1.4 nm ([10,10] SWNT) at $T = 4$ K. We assume that a local gate with a width of 15 nm can be used to apply a non–uniform potential to the 1D channel and open the gap. The upper IVC corresponds to a zero gate voltage (no gap). The channel is fully open and the current is determined by an injection from the Ohmic contact and, thus, by the conductance quantum $2G_o$ (see footnote 6). With increasing gate voltage one observes a substantial decrease of the current due to the opening of the gap and the depletion of electrons in this region, which is nothing but a field-effect. We present IVCs at several gate voltages corresponding to the opened gap up to 1 eV for the lowest curve.

One mechanism of the suppression of conductivity that we study in this chapter is the ballistic scattering of the electron by the potential barrier created by the field of the local gate, or the reduction of the total transmission coefficient (dimensionless conductance). We plot the dimensionless conductance of the METFET channel (G divided by the maximum conductance $2G_o$ (see footnote 6) as a function of the drain voltage in lower left inset of Fig. 1.11. Several families of curves are given for different temperatures. In the upper family of curves (room temperature) the suppression of transmission through the gated region is clearly seen as a function of the increased

[7] We intentionally did not include here any scattering (resistance) due to the contacts and/or the parts of the 1D channel other than the gated region. This allowed us to present clearly fundamentals of the METFET device and study its maximum performance. All these contact and material aspects can be included in a straightforward way.

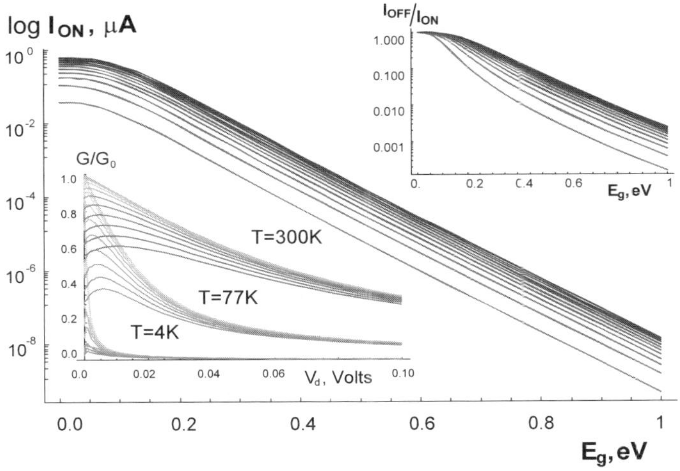

Fig. 1.11. Logarithm of the room temperature OFF/ON current ratio versus the opened gap. The width of a local gate is 50 nm. The upper right inset shows the logarithm of the OFF/ON current ratio at $T = 4$ K and the gate width 15 nm. Each curve from bottom to top corresponds to increasing drain voltage from 1 to 20 meV. In the lower left inset the drain voltage dependence of the conductance at given temperature ($T = 4$ K, 77 K and 295 K from bottom to top) and gate voltage (from 0 to 0.2 eV from top to bottom) is presented (see Color Plates, p. 339)

gap: from 0 to 0.2 eV (from top to bottom). For other families at lower temperatures T =77 K and 4 K this effect is even sharper.

1.6.5 Results and Discussion

Next we discuss the OFF/ON ratio of our proposed METFET and show that the theoretical limit for a specific design of SWNT METFET's can be as small as 10^{-6}. Figure 1.11 shows the room temperature (300 K) ballistic current through a METFET as a function of the gap, E_g (which is proportional to the gate voltage). This was calculated for a wide gate device (50 nm). For a half eV band gap, the OFF/ON ratio at room temperature is $\sim 10^{-3}$ for $V_d \sim 20$ meV and may reach $\sim 10^{-5}$ for $V_d \sim 1$ meV. A high OFF/ON ratio was also observed for S–SWNT in standard and Schottky barrier FET's [19,61]. Note however, the large difference in the zero bias (zero gate voltage) conductance that can be obtained for the different types of nanotubes [54,62]. The inset of Fig. 1.11 shows the current of a device with a short gate (15 nm) at 4 K normalized to the ON current at zero applied gate voltage (thus, at zero gap).

The channel of the 1D METFET has very high density of states (high conductance) similar to a normal metal in the absence of the gate voltage. Thus, one can expect good scaling of the METFET.

The local gate of the 1D METFET may be fabricated in a different way. A scanning tip of a microscope may represent a local gate, as well as a very narrow lead fabricated closely to the nanotube, or special (electro–)chemical function group(s) at the tube sidewalls and/or inside the tube, etc. In this study we modelled IV curves for the METFET's with local gates with widths of the order of 5–50 nm. The lower limit is the typical size of a scanning tip, which may touch the M–SWNT channel, while the upper limit is about the size of a metal nanointerconnect [63] that can be lithographically fabricated close to the SWNT channel.

We note that at the points of the metal–semiconductor junction (at the edge of the gated region) a charge accumulation may happen. In a sense, this metal–semiconductor junction boundary becomes a side of a 1D capacitor with a capacitance $\sim W/2\ln(W/R)$. Thus, if we neglect the logarithmic term, a classical charging energy of this capacitor is $\sim e^2/2W$, which may amount to a large energy for a narrow gate. However, in this work we did not consider a Coulomb blockade because it is known that the effective gap is exponentially renormalized at a high conductance of the tunnel junction [64]. The specifics of the nanotube metal–semiconductor junction is that for SWNT band parameters (very light effective mass) the tunnelling length is large (from sub–nm to tens of nm) and the tunnelling rate is high. For high tunnelling coupling between sides of the metal–semiconductor junction, quantum fluctuations of the charge wash out the Coulomb correlation and destroy the Coulomb blockade. On the other hand, by applying the gate voltage one can decrease the conductance through the junction and therefore increase Coulomb correlation. This effect may further decrease the METFET OFF current.

This leads us to also propose granular nanowires [65] as another possible material for 1D METFET. These wires were shown to exhibit the Coulomb blockade at low temperature as well as a quantum localization due to Coulomb correlation between grains in wide range of temperatures [63]. These wires are expected to change their conductivity (undergo metal–disordered–semiconductor transition) under the application of an external potential. Using the local gate versus standard uniform backgate may lead to the local suppression of charge fluctuations and the opening of the Coulomb gap in the IVC.

We note that use of dual gates (both local and backgate) may be beneficial for 1D METFET because the "global" backgate controls the Fermi level (charge density) while the local one controls the conductivity and switches OFF/ON the channel.

In contrast to standard Si electronics, the nanotube METFET will not be sensitive to radiation. The reason is that collision cross sections are small [66] and thermal conductivity is extremely high. As the NT device performance does not rely on the doping level of the material, creation of dislocations [67] is also not a critical problem. We note that the nanotube device shows fairly good performance at low (helium) temperatures and we thus envision that NT based METFET's may be appealing for space applications.

Acknowledgements

The author is indebted to Professor K. Hess and Professor N.R. Aluru both for introducing the subject of MEMS and for fruitful discussions of obtained results, and to Dr. A.G. Petrov, Dr. I. Zharov, Ms. Y. Li and Mr. K.A. Bulashevich for collaboration on the electronic structure computations and fruitful discussions. The author acknowledges support through DoE grant DE-FG02-01ER45932, NSF grant ECS-0210495, NSF grant ECS–04-03489, the Feigl Scholarship from Lehigh University and the Beckman Fellowship from the Arnold and Mabel Beckman Foundation.

References

1. R.P. Feynman: There's plenty of room at the bottom: An invitation to enter a new field of physics. In: *Engineering and Science*, vol XXIII, No. 5, (Caltech 1960) pp 22–36
2. S. Iijima: Nature **354**, 56 (1991)
3. M. Damnjanović, I. Milošević, T. Vuković, B. Nikolić and E. Dobardžić: "Symmetry Based Fundamentals on Carbon Nanotubes", Chapter 2, in this volume.
4. Phaedon Avouris, Marko Radosavljević and Shalom J. Wind: "Carbon Nanotube Electronics and Optoelectronics", Chapter 9, in this volume.
5. R. Bruce Weisman: "Fluorescence Spectroscopy of Single-Walled Carbon Nanotubes", Chapter 8, in this volume.
6. Anand Jagota, Bruce A. Diner, Salah Boussaad, and Ming Zheng: "Carbon Nanotube – Biomolecule Interactions: Applications in Carbon Nanotube Separation and Biosensing", Chapter 10, in this volume.
7. M.S. Strano, M.L. Usrey, P.W. Barone, D.A. Heller and S. Baik: "The Selective Chemistry of Single Walled Carbon Nanotubes", Chapter 6, in this volume.
8. S. Huang and J. Liu: "Direct Growth of Single Walled Carbon Nanotubes on Flat Substrates for Nanoscale Electronic Applications", Chapter 4, in this volume.
9. T. Ando, A.B. Fowler, F. Stern: Rev. Mod. Phys. **54** (2), 437(1982)
10. J. Voit: Rep. Prog. Phys. **57**, 977 (1995)
11. N.R. Aluru, J-P. Leburton, W. McMahon, U. Ravaioli, S.V. Rotkin, M. Staedele, T. van der Straaten, B.R. Tuttle and K. Hess: "Modeling Electronics on the Nanoscale", in *"Handbook of Nanoscience, Engineering and Technology"*, Eds.: W. Goddard, D. Brenner, S. Lyshevski, G.J. Iafrate; (CRC Press 2002)
12. Dmitriy A. Dikin, Xinqi Chen, Frank T. Fisher and Rodney S. Ruoff: "Nanomanipulator Measurements of the Mechanics of Nanostructures and Nanocomposites", Chapter 12, in this volume.
13. M. Dequesnes, S.V. Rotkin, N.R. Aluru: Journal of Computational Electronics **1** (3), 313(2002)

14. T. Durkop, S.A. Getty, E. Cobas, and M.S. Fuhrer: Nano Letters **4**, 35 (2004)
15. Y. Cui, X. Duan, J. Hu, and C.M. Lieber: J. Phys. Chem. B **104**, 5213 (2000)
16. H. Hasegawa and S. Kasai: Physica E **11**, 149 (2001)
17. J.-R. Kim et al.: Appl. Phys. Lett. **80**, 3548 (2002)
18. Y. Zhang, A. Kolmakov, S. Chretien, H. Meitu and M. Moskovits: Nano Letters **4** (3), 403(2004)
19. S. Heinze, J. Tersoff, R. Martel, V. Derycke, J. Appenzeller, Ph. Avouris: Phys. Rev. Lett. **89**, 106801 (2002)
20. N.S. Averkiev and A.Y. Shik: Semiconductors **30**, 112 (1996)
21. H. Ruda and A. Shik: J. Appl. Phys. **84**, 5867 (1998)
22. K.A. Bulashevich and S.V. Rotkin: JETP Lett. **75**, 205 (2002)
23. S.V. Rotkin, V. Srivastava, K.A. Bulashevich, and N.R. Aluru: International Journal of Nanoscience **1**, 337 (2002)
24. S.V. Rotkin, H. Ruda, A. Shik: International Journal of Nanoscience **3** (1/2), 161 (2004)
25. S.V. Rotkin, H.E. Ruda, and A. Shik: Appl. Phys. Lett. **83**, 1623 (2003)
26. M. Abramovitz and I.A. Stegun: Handbook of Mathematical Functions (Dover, New-York, 1964)
27. T. Maemoto, H. Yamamoto, M. Konami, A. Kajiuchi, T. Ikeda, S. Sasa, and M. Inoue: Phys. Stat. Sol. (b) **204**, 255 (1997)
28. G.L. Harris, P. Zhou, M. He, and J.B. Halpern: *Lasers and Electro–Optics, 2001, CLEO'01.* Technical Digest, p.239
29. S.J. Wind, J. Appenzeller, R. Martel, V. Derycke, and P. Avouris: Appl. Phys. Lett. **80**, 3817 (2002)
30. X. Liu, C. Lee, and C. Zhou: Appl. Phys. Lett. **79**, 3329 (2001)
31. F. Leonard and J. Tersoff: Phys. Rev. Lett. **88**, 258302 (2002)
32. S. Luryi: Appl. Phys. Lett. **52** (6), 501 (1988)
33. M. Dequesnes, S.V. Rotkin, N.R. Aluru: Nanotechnology **13**, 120(2002)
34. S.V. Rotkin: in *Microfabricated Systems and MEMS - VI*, vol. PV 2002-6, Symposium - the Electrochemical Society Proceedings, P.J. Hesketh, S.S. Ang, J.L. Davidson, H.G. Hughes, and D. Misra, Eds. (ECS Inc., Pennington, NJ, USA 2002) pp. 90–97.
35. S.V. Rotkin, "Theory of Nanotube Nanodevices", in *Nanostructured Materials and Coatings for Biomedical and Sensor Applications*, Editors: Y.G. Gogotsi and Irina V. Uvarova (Kluwer Academic Publishers: Dordrecht-Boston-London 2003) Vol. **102**, pp. 257–277
36. A. A. Odintsov, Y. Tokura: Journal of Low Temperature Physics, **118**, 509 (2000)
37. A.G. Petrov, S.V. Rotkin: Nano Letters **3** (6), 701(2003)
38. Y. Yaish, J.-Y. Park, S. Rosenblatt, V. Sazonova, M. Brink, and P. L. McEuen: Phys. Rev. Lett., **92**, 046 401–1(2004)
39. S. Akita, Y. Nakayama, S. Mizooka, et al.: Appl. Phys. Lett. **79** (11), 1691(2001)
40. D. Lu, Y. Li, S.V. Rotkin, U. Ravaioli, and K. Schulten, Nano Letters **4** (12), 2383 (2004)
41. Y. Li, S.V. Rotkin, and U. Ravaioli: Nano Letters, **3**, no.2, 183(2003)
42. J. Gonzalez, F. Guinea, MAH. Vozmediano: Nuclear Physics B. **406** (3), 771 (1993) D.P. DiVincenzo, E.J. Mele: Phys.Rev. **B 29**(4), 1685 (1984)

43. S.V. Rotkin, K.A. Bulashevich, N.R. Aluru: in Procs.–ECS **PV 2002–12**, P.V. Kamat, D.M. Guldi, and K.M. Kadish, Eds. (ECS Inc., Pennington, NJ, USA 2002) pp. 512–519
44. S. Suzukia, Ch. Bower, Y. Watanabe, O. Zhou: Appl. Phys. Lett. **76** (26), 4007 (2000)
45. M. Shiraishi, M. Ata: Carbon **39**, 1913 (2001)
46. Y. Li, S.V. Rotkin, and U. Ravaioli: Appl. Phys. Lett. **85** (18), 4178 (2004)
47. S.V. Rotkin: Proceedings of SPIE, Vol. **5509**, 145 (2004)
48. F. Leonard, J. Tersoff: Appl. Phys. Lett. 81, 4835 (2002)
49. S.V. Rotkin, and K. Hess, Appl. Phys. Lett. **84** (16), 3139(2004)
50. S.V. Rotkin, and K. Hess, in *Technical Proceedings of the 2003 Nanotechnology Conference and Trade Show*, Volume 1-3, (March 7-11, 2004), Boston, Massachusetts, USA
51. K. Hess: *Advanced Theory of Semiconductor Devices* (New York: IEEE Press 2000)
52. F. Chudnovskiy, S. Luryi, and B. Spivak: in *Future Trends in Microelectronics: The Nano Millennium*, ed. by S. Luryi, J.M. Xu, and A. Zaslavsky (Wiley Interscience, New York 2002) pp. 148–155
53. J. Appenzeller: Device Research Conference, Utah, June 26, 2003
54. J. Appenzeller, R. Martel, V. Derycke, M. Radosavljevic, S. Wind, D. Neumayer, and P. Avouris: Microelectronic Engineering **64**, 391 (2002)
55. Ph.G. Collins, Ph. Avouris: Sci.Am. **12**, 62(2002) Ph. Avouris: Chemical Physics **281**, 429(2002)
56. H. Suzuura and T. Ando: Phys. Rev. **B 65**, 235412 (2002)
57. T. Ando and T. Nakanishi: J. Phys. Soc. Jpn. **67**, 1704 (1998) T. Ando, T. Nakanishi and R. Saito: J. Phys. Soc. Japan **67**, no. 8, 2857 (1998)
58. A.G. Petrov, S.V. Rotkin: Phys. Rev. **B 70** (3), 035408 (2004)
59. T. Vukovic, I. Milosevic, M. Damnjanovic: Phys. Rev. **B 65**(04), 5418 (2002)
60. S.V. Rotkin, G. Wilson, J.S. Moore, and K. Hess: unpublished
61. J. Appenzeller, J. Knoch, R. Martel, V. Derycke, S.J. Wind, P. Avouris: IEEE Transactions on Nanotechnology, **1** (4), 184 (2002)
62. P.L. McEuen, M. Bockrath, D.H. Cobden, Y-G. Yoon, and S.G. Louie: Phys. Rev. Lett. **83**, 5098 (1999)
63. L. Rotkina, J.-F. Lin, J.P. Bird, Appl. Phys. Lett. **83** (21), 4426 (2003)
64. K.A. Matveev: Phys. Rev. **B 51**, 1743 (1995)
65. J.-F. Lin, J.P. Bird, L. Rotkina, P.A. Bennett: Appl. Phys. Lett. **82** (5), 802 (2003); J.-F. Lin, J.P. Bird, L. Rotkina, A. Sergeev and V. Mitin, Appl. Phys. Lett. **84** (19), 3828 (2004)
66. A.V. Krasheninnikov, K. Nordlund, and J. Keinonen: Phys. Rev. **B 65**, 165423 (2002)
A.V. Krasheninnikov, K. Nordlund, M. Sirvio, E. Salonen, and J. Keinonen: Phys. Rev. **B 63**, 245405 (2001)
67. F.Z. Cui, Z.J. Chen, J. Ma, G.R. Xia and Y. Zhai: Physics Letters A **295** (1), 55 (2002)

2 Symmetry Based Fundamentals of Carbon Nanotubes

M. Damnjanović, I. Milošević, E. Dobardžić, T. Vuković, and B. Nikolić

Due to pioneering work of Eugen Wigner, Symmetry is well recognized as a clue to understanding physical processes. However, except in Particle Physics, role of Symmetry is mostly conceptual, to give *a posteriori* profound interpretation of the obtained results. For instance, in Solid State Physics, more than a century known crystal symmetries are applied to describe phase transitions, selection rules, tensor shapes. Probably the only exception is the Bloch theorem, which gives *a priori* general forms of the electronic quantum states and ionic displacements. However, this theorem treats only the translational symmetry, reducing the calculations to the elementary cell, which is, in many cases, far bellow the maximal possible reduction.

Over the past decade nanotubes have become one of the most attractive subjects in Condensed Matter Physics. Elementary cell of nanotubes often contains large number of atoms, which makes the Bloch theorem based calculations very robust, and frequently too demanding. Nevertheless, besides the translational symmetry, quasi-1D-crystals have helical and additional symmetries (like horizontal two-fold axes and/or mirror planes), enabling to generate the whole compound from the so called symmetry cell (symcell), being only a small part of the conventional unit cell. It is intuitively obvious that the physical properties of the system are determined by its symmetry group and by the symcell. Modified group projector technique offers an algorithm to maximally generalize the Bloch theorem to the full symmetry group. It also emphasizes that symmetry itself is the most essential property of nanotubes which proves to be sufficient to predict their unique physical properties.

2.1 Introduction

In this chapter, within a profound and efficient symmetry based approach, an overview of the electro-optical and vibrational properties of carbon nanotubes (NT) is presented.

In Sect. 2.2 the basics of single-wall tubes (SWNT) are introduced: configuration, chirality, symmetry and conserved quantum numbers. The stable configurations of narrow tubes are obtained by the symmetry preserving density functional relaxation. Afterwards the configuration of double-wall tubes is described.

Section 2.3 starts with an outline of the concept necessary to exploit full symmetry in the band structure calculations and proceeds by giving a unified description of electron and phonon dispersions: band degeneracy, symmetry assignation, systematic density of state singularities and selection rules for various processes. The corresponding generalized Bloch eigenfunctions are also listed. Finally, dynamical models are reviewed and discussed: p- and sp^3-orbitals are tight-binding and density functionals for electrons, while phonons are studied by intra-layer force constants and (for double-wall tubes) inter-layer Lenard-Jones interactions.

Electro-optical properties of SWNTs are studied in Sect. 2.4. Polarized optical absorption of conventional (with the diameters $8 - 16\,\text{Å}$) and template grown ($4\,\text{Å}$-diameter) SWNTs are calculated within first order time-dependent perturbation theory, using exactly evaluated optical transition matrix elements out of the full-symmetry adapted Bloch eigenfunctions. The results obtained are discussed and compared to the measured spectra.

Vibrational properties of NTs are discussed in Sect. 2.5. After description of the phonon dispersions of SWNTs (emphasizing some particularly interesting modes), double-wall tube branches are considered. The so called rigid layer modes are interpreted with the help of the simple perturbative model. Also, some related properties, such as heat capacity, sound velocities, overbending, Raman and infrared activity are discussed.

Finally, Sect. 2.6 is devoted to the interaction between the walls in double-wall NTs, which is calculated by the use of the expansion over symmetry based harmonics. Besides the high efficiency, the method profoundly correlates interaction with the breaking of the symmetry of the layers, explaining experimentally verified weak inter-layer interaction, and yielding, in some ideal cases, super slippery sliding.

2.2 Configuration and Symmetry

Symmetry of SWNTs is considered from their very discovery [1] due to its relevance for understanding physical properties of SWNTs. At first, only the rotational axis of the related fullerene molecule is observed [3]; later, in the electronic band calculations the translational [4], as well as the helical [5, 6] symmetry was used. Finally, the full symmetry groups, including parities in addition to the roto-helical symmetries, were found [7]. These considerations are based on infinite, defect-free model of SWNTs corresponding to simply rolled up graphene layer.

2.2.1 Single-Wall Nanotubes

Symmetry

The graphene layer is a highly symmetric structure, with translational periodicity in two directions, a six-fold principle rotational axis and six mirror

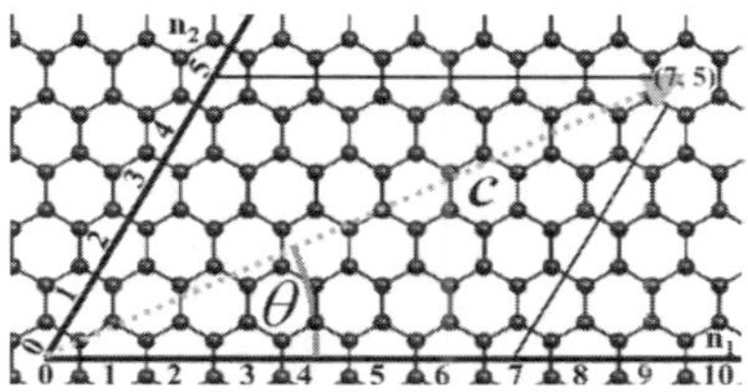

Fig. 2.1. Graphene honeycomb with depicted chiral vector $\boldsymbol{c}$ and chiral angle θ

planes perpendicular to the layer (the horizontal plane and U-axes are not relevant in this context). To obtain the SWNT (n_1, n_2), one takes the chiral vector $\boldsymbol{c} = (n_1, n_2)$ on the graphene (Fig. 2.1), and rolls up the layer so that this vector becomes the circumference of the cylinder. The special cases $(n, 0)$ and (n, n) are known as the achiral zig-zag $(\mathcal{Z})$ and armchair $(\mathcal{A})$ tubes, while the other ones are called chiral $(\mathcal{C})$ tubes. The folding changes the symmetry of the structure: the translations perpendicular to the chiral vector remain translations of the tube, those along $\boldsymbol{c}$ become rotational symmetries of the tube, while the combined ones give nanotube helical symmetries. After folding, the graphene principle axis is no longer symmetry, except that the rotations for π become the U axis of the tube. Finally, only in the achiral cases, two perpendicular mirror planes are preserved, becoming vertical (σ_v) and horizontal (σ_h) mirror planes of the zig-zag and armchair tubes. Combining all the symmetry transformations of the obtained tube results in its full symmetry group.

Since the translations perpendicular to the chiral vector (n_1, n_2) on graphene remain the translational symmetries of the tube (n_1, n_2), SWNTs are quasi one-dimensional crystals, and therefore their symmetries are gathered in the line groups describing the symmetries of such crystals [8, 9]. A straightforward calculation [7] shows that the obtained line group is:

$$\boldsymbol{L}_{\mathcal{C}} = \boldsymbol{T}_q^r(a)\boldsymbol{D}_n = \boldsymbol{L}q_p22, \quad \boldsymbol{L}_{\mathcal{ZA}} = \boldsymbol{T}_{2n}^1(a)\boldsymbol{D}_{nh} = \boldsymbol{L}2n_n/mcm, \tag{2.1}$$

where the parameters n (order of the principle axis), a (translational period), q (one half of the number of atoms per period) and r (helicity) are:

$$n = \mathrm{GCD}(n_1, n_2), \qquad a = \frac{\sqrt{3(n_1^2 + n_2^2 + n_1 n_2)}}{n\mathcal{R}} a_0 \tag{2.2}$$

$$q = 2\frac{n_1^2 + n_1 n_2 + n_2^2}{n\mathcal{R}}, \quad r = \frac{n_1 + 2n_2 - (\frac{n_2}{n})^{\phi(\frac{n_1}{n})-1} q\mathcal{R}}{n_1\mathcal{R}} \pmod{\frac{q}{n}} \tag{2.3}$$

($\mathcal{R} = \mathrm{GCD}(2n_1 + n_2, n_1 + 2n_2)/n$, $a_0 = 2.46\,\text{Å}$ is graphene period, $\phi(x)$ is the Euler function, i.e. the number of coprimes not greater than x). In international notation helicity is characterized by

$$p = n\left(r^{\phi(\frac{q}{n})-1} \pmod{\frac{q}{n}}\right). \tag{2.4}$$

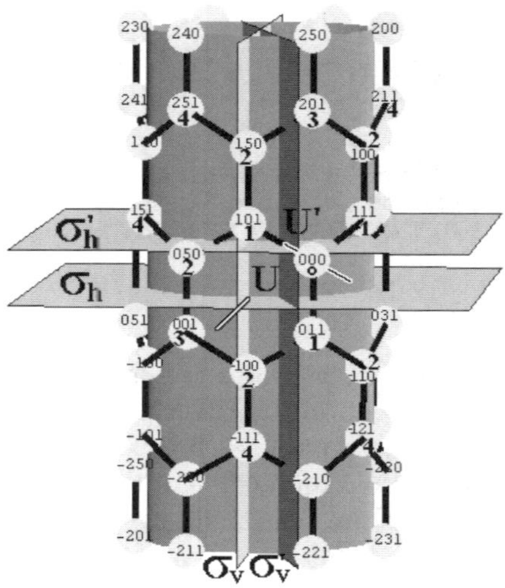

Fig. 2.2. Nanotube (6,0). Symmetries: vertical and horizontal mirror planes σ_{v} and σ_{h} intersect along the horizontal rotational axis U; glide plane σ'_{v} and roto-reflectional plane σ'_{h} intersect along U'. Atoms: labelled by tsu indices from (2.9). Neighbors of C_{000}: first to fourth level denoted by 1-4. Coordinate frame: x-axis coincides with U (z-axis is the tube axis)

Note that for achiral tubes $q = 2n$, $r = 1$, and $a_{\mathcal{Z}} = \sqrt{3}a_0$, $a_{\mathcal{A}} = a_0$. The helical group $\boldsymbol{T}_q^r(a)$ is generated by $(C_q^r|\frac{n}{q}a)$, i.e. the rotation for $2\pi r/q$ around the tube axis (z-axis) followed by the translation for na/q along it. The generators of the point group $\boldsymbol{D}_n$ are the rotation $\boldsymbol{C}_n$ for $2\pi/n$ around the tube, and the rotation U for π around the x-axis (through the center of the carbon hexagon, Fig. 2.2). The achiral tubes point group $\boldsymbol{D}_{nh}$ has the additional generator σ_{v}. Therefore, the symmetry transformations of SWNT are:

$$\ell_{tsu} = \ell_{tsu}^{\mathcal{C}} = (C_q^r|\frac{n}{q}a)^t C_n^s U^u, \quad \ell_{tsuv}^{\mathcal{ZA}} = \ell_{tsu}\sigma_{\mathrm{v}}^v, \tag{2.5}$$

with $t = 0, \pm 1, \dots$, $s = 0, \dots, n-1$ and $u, v = 0, 1$. All rotational, translational and helical symmetries (appearing due to the graphene translations) form the roto-translational subgroup $\boldsymbol{L}^{\mathrm{R}} = \boldsymbol{T}_q^r(a)\boldsymbol{C}_n$ including the elements ℓ_{ts0}. Among other transformations there are also radial U axes bisecting the C-C bonds, as well as various glide and roto-reflection planes (of achiral tubes). Applying only elements ℓ_{0su} onto the initial atom, one gets a monomer, i.e. the part of SWNT of the length na/q, from which the helical transformations alone generate the whole tube.

SWNT symmetry groups are non-symmorphic, and their isogonal point groups, being important for the description of various physical tensors, are:

$$\boldsymbol{P}_{\mathcal{C}}^{\mathrm{I}} = \boldsymbol{D}_q, \quad \boldsymbol{P}_{\mathcal{ZA}}^{\mathrm{I}} = \boldsymbol{D}_{2nh}. \tag{2.6}$$

Configuration

The number of symmetries in the chiral nanotubes is equal to the number of atoms, while in the achiral cases it is twice as big. This is extremely large symmetry. In fact, SWNTs are maximally symmetric structures in the sense that they are generated by symmetry operations from an arbitrary initial atom. This property can be used to parameterize the configuration in terms of symmetry. At first, the diameter and the linear density are:

$$D = \frac{an\mathcal{R}}{\pi\sqrt{3}}, \qquad \lambda = \frac{2q}{a} = cD \tag{2.7}$$

($c = 14.375\,\text{amu/Å}$). Taking for the initial atom C_{000} with the cylindrical coordinates (Fig. 2.2)

$$\boldsymbol{r}_{000} = (\frac{D}{2}, \varphi_{000} = 2\pi\frac{n_1 + n_2}{nq\mathcal{R}}, z_{000} = \frac{n_1 - n_2}{\sqrt{6}nq\mathcal{R}}a_0), \tag{2.8}$$

any other atom C_{tsu} is obtained by the action of ℓ_{tsu} on C_{000}:

$$\boldsymbol{r}_{tsu} = \ell_{tsu}\boldsymbol{r}_{000} = (\frac{D}{2}, (-1)^u\varphi_{000} + 2\pi(\frac{rt}{q} + \frac{s}{n}), (-1)^u z_{000} + t\frac{n}{q}a). \tag{2.9}$$

The described graphene folding structural model tacitly assumes that the SWNTs chemical bonds are based on the graphene sp^2 hybridization. However, due to curvature, sp^3 hybridization occurs (see Sect. 2.3.4). Therefore, the rolled up configuration is not stable, and should be relaxed (slightly changed) to achieve the local minimum of the total energy. Relaxation cannot change the symmetry: according to the famous topological theorem [10], the extremes of the energy functional correspond to the most symmetric configurations. Consequently, equations (2.1) and (2.9) refer to the relaxed structures as well, only the values of D, φ_{000}, z_{000} and a differ. Any change of these parameters preserves the symmetry of chiral tubes, except that $\varphi_{000} = \pi/n$ for zig-zag and $z_{000} = 0$ for armchair tubes must be fixed. Obviously, the curvature induced rehybridization is significant only for narrow tubes. This is confirmed by precise density functional (DF) methods: only for tubes with a diameter of less than 7 Å does the relaxed configuration differ from the rolled up one by a few percent (see Table 2.1), while the properties of thicker tubes are well described within the purely rolled up model.

2.2.2 Double-Wall Nanotubes

Double-wall tube W@W′ is assumed to be a pair of coaxially arranged single-wall tubes: W$= (n_1, n_2)$ is the inner wall and W′ $= (n_1', n_2')$ is the outer wall. The inter-wall separation $\Delta = (D' - D)/2$ is close to the graphite inter-layer distance 3.44 Å within some tolerance δ (we use $\delta = 0.25$ Å). Fixing

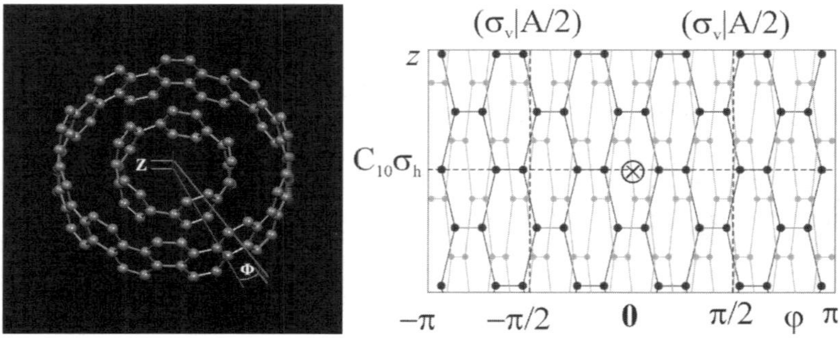

Fig. 2.3. Left: DWNT configuration. Right: unfolded tube (5,5)@(10,10) with symmetry elements

Table 2.1. Ultra narrow tubes considered in Sect. 2.4.2. Folded configuration parameters (raw F) given by (2.8) and (2.2) are compared to the relaxed ones obtained by density functional tight-binding (raw DFTB, used in this chapter) and local density (raw DFLD taken from [11]) methods.

SWNT	D[Å]	a[Å]	φ_{000}[°]	z_{000}[Å]
(5,0) F	3.92	4.26	36	0.71
DFTB	4.11	4.23	36	0.70
DFLD	4.12	4.30	36	0.73
(3,3) F	4.07	2.46	40	0
DFTB	4.19	2.48	40	0
DFLD	4.24	2.50	40	0
(4,2) F	4.15	11.28	39	0.27
DFTB	4.29	11.27	39	0.26
DFLD	4.34	11.34	39	0.28
(5,1) F	4.36	23.73	35	0.51
DFTB	4.51	23.67	35	0.51
DFLD	4.56	24.18	35	0.52

the interior wall, the configuration is completely determined by the twisting angle Φ and the axial shift Z of the outer wall: the x-axis of the interior wall (defined in Fig. 2.2) should be rotated for Φ and shifted upward by Z to get the x-axis of the outer wall (Fig. 2.3). Therefore, the atomic coordinates are (2.9) for W, while for W′:

$$\boldsymbol{r}_{t's'u'} = (\frac{D'}{2}, (-1)^{u'}\varphi'_{000} - 2\pi(\frac{r't'}{q'} + \frac{s'}{n'}) + \Phi, (-1)^{u'}z'_{000} + t'\frac{n'}{q'}a' + Z). \tag{2.10}$$

Here, q', r', n' and a' are the parameters (2.2) of the symmetry group (2.1) of the outer wall.

The stable configuration of W@W′ in such a model corresponds to the values of Φ and Z minimizing the inter-layer potential:

$$V(\Phi, Z) = \sum_{t's'u'} \sum_{tsu} v(\boldsymbol{r}_{tsu}, \boldsymbol{r}'_{t's'u'}). \tag{2.11}$$

where $v(\boldsymbol{r}_{tsu}, \boldsymbol{r}'_{t's'u'})$ is the pairwise interaction of the interior- and outer-wall carbon atoms. Irrespectively of v, according to the topological argument [10], the optimal relative position is among the maximally symmetric ones. In fact, the symmetry group of W@W′ contains only those transformations which simultaneously leave both walls invariant. Note that the common rotations, translations and helical symmetries are not affected by the relative positions of the walls, to differ from U, σ_{v}, σ_{h}, glide and roto-reflection symmetries appearing only when the corresponding axes and planes of the two walls coincide. Consequently, the candidates for the stable position are only those pairs (Φ, Z) enabling as much as possible coinciding parities. This singles out rather few *a priori* known possible stable configurations, reducing greatly the sensitivity of the choice of the potential model. This emphasizes the importance of symmetry of DWNTs.

At first, we consider Φ and Z independent roto-translational symmetry, described by the intersection of the walls' roto-translational subgroups [7,12, 13]:

$$\boldsymbol{L}^{\mathrm{R}}_{\mathrm{WW'}} = \boldsymbol{L}^{\mathrm{R}}_{\mathrm{W}} \cap \boldsymbol{L}^{\mathrm{R}}_{\mathrm{W'}} = \boldsymbol{T}^{R}_{Q}(A)\boldsymbol{C}_{N}. \tag{2.12}$$

The parameters of this group are [13]:

$$N = \mathrm{GCD}(n, n'),\ A = \hat{a}'a = \hat{a}a',\ Q = \frac{N\sqrt{\tilde{q}\tilde{q}'}}{\tau},\ R = \frac{(r\hat{a}\tau + s\tilde{q})Q}{q}, \tag{2.13}$$

where $\tilde{q} = q/n$, $\tilde{q}' = q'/n'$, $\hat{a} = \sqrt{\tilde{q}/\mathrm{GCD}(\tilde{q}, \tilde{q}')}$, $\hat{a}' = \sqrt{\tilde{q}'/\mathrm{GCD}(\tilde{q}, \tilde{q}')}$, while $\tau = \sqrt{\tilde{q}\tilde{q}'}/\mathrm{GCD}(\frac{r'\hat{a}n - r\hat{a}'n'}{N}, \sqrt{\tilde{q}\tilde{q}'})$. Further, the equation for the helicity R involves $s = \tau(r'\hat{a}'q - r\hat{a}q')((n'/N)^{\phi(n/N)} - 1)/n\tilde{q}\tilde{q}'$.

Particularly, if the periods of the two walls are incommensurate (i.e. a/a' is irrational), W@W′ possesses no translational periodicity ($A = \infty$), reducing the symmetry to the point group $\boldsymbol{L}^{\mathrm{R}}_{\mathrm{WW'}} = \boldsymbol{C}_N$. In fact, the commensurate tubes (CDWNTs) are very rare, about 0.5% of all DWNTs allowed by the inter-layer distance. Precisely, with 1280 SWNTs having diameters 2.8-50 Å altogether 42236 pairs satisfy requirements for Δ and δ; among all these DWNTs only 240 are commensurate. Note here that if both walls are chiral, the tubes $(n_1, n_2)@(n'_1, n'_2)$ and $(n_1, n_2)@(n'_2, n'_1)$ must be independently considered, since the spatial inversion does not relate them (as in the single-wall case, which enables us to consider only $n'_1 \geq n'_2$ SWNTs). This increases the number of different CDWNTs found to 318. Among them, with achiral walls are zig-zag and armchair series $\mathcal{ZZ}_n = (n, 0)@(n + 9)$ and $\mathcal{AA}_n = (n, n)@(n + 5, n + 5)$.

Finally, to find numerically the stable configurations and full symmetry groups, the choice of the potential must be made. It is well known that the pairwise interaction v is of the Van der Waals type, and several concrete forms are proposed in the literature [14, 15]. Here we use the simplest Lenard-Jones one [14]:

$$v(\boldsymbol{r}) = -\frac{18.5426}{|\boldsymbol{r}|^6} + \frac{29000.4}{|\boldsymbol{r}|^{12}}. \tag{2.14}$$

The results are in accordance with the topological prediction (some details of calculations are clarified in Sect. 2.6): in all the cases there are coinciding U axes of the walls, and when both walls are achiral there is also a symmetry plane (mirror, glide or roto-reflectional) of one wall coinciding with a symmetry plane of another wall. Particularly, for incommensurate DWNTs $\Phi = 0$, while Z is arbitrary (thus the super-slippery sliding of the walls is enabled, as discussed in Sect. 2.6). For CDWNTs with at least one chiral wall $\Phi = Z = 0$; for $\mathcal{ZZ}$ and $\mathcal{AA}$ tubes $\Phi = 0$ and $Z = A/4$, except that $\Phi = \pi/4N$ and $Z = A/4$ for $\mathcal{ZZ}_9$ and $\mathcal{AA}_5$. Accordingly, the symmetry groups of commensurate tubes are line groups:

$$\boldsymbol{L}^{\mathcal{C}}_{\mathrm{WW}'} = \boldsymbol{T}^R_Q(A)\boldsymbol{D}_N, \quad \boldsymbol{L}^{\mathcal{ZZ}_9,\mathcal{AA}_5}_{\mathrm{WW}'} = \boldsymbol{T}_c(A)\boldsymbol{S}_{2N}, \quad \boldsymbol{L}^{\mathcal{ZZ},\mathcal{AA}}_{\mathrm{WW}'} = \boldsymbol{T}(A)\boldsymbol{D}_{Nd}. \tag{2.15}$$

The specific results for more than 200 CDWNTs, given in Table 2.2, show that DWNTs have drastically reduced roto-translational symmetry in comparison to their walls.

Table 2.2. CDWNTs with collinear chiral vectors. The outer diameter of the series shown in column 1 is within 9.6-50 Å) only for n given in column 4 (without braced values). Columns 2 and 3: line and isogonal groups; translational periods are in the units of $a_0 = 2.46$ Å.

CDWNTs	Line Group	Isogonal	n
$(n,0)@(n+9,0)$	$\boldsymbol{T}(\sqrt{3})\boldsymbol{D}_{1d}$	$\boldsymbol{D}_{1d}$	4,5,...,62 (6,9,12,...,60)
	$\boldsymbol{T}(\sqrt{3})\boldsymbol{D}_{3d}$	$\boldsymbol{D}_{3d}$	6,9,12,...,60 (9,18,...54)
	$\boldsymbol{T}(\sqrt{3})\boldsymbol{D}_{9d}$	$\boldsymbol{D}_{9d}$	$18, 27, \ldots, 63$
	$\boldsymbol{T}_c(\sqrt{3})\boldsymbol{S}_{18}$		9
$(n,n)@(n+5,n+5)$	$\boldsymbol{T}(1)\boldsymbol{D}_{1d}$	$\boldsymbol{D}_{1d}$	2,3,...,36 (5,10, 15,20,25,30,35)
	$\boldsymbol{T}(1)\boldsymbol{D}_{5d}$	$\boldsymbol{D}_{5d}$	$10, 15, \ldots, 35$
	$\boldsymbol{T}_c(1)\boldsymbol{S}_{10}$		5
$(3n,2n)@(3n+6,2n+4)$ $(3n+6,2n+4)@(2n+4,3n+6)$	$\boldsymbol{T}^1_2(\sqrt{57})\boldsymbol{D}_1$	$\boldsymbol{D}_2$	$1, 3, \ldots, 13$
	$\boldsymbol{T}(\sqrt{57})\boldsymbol{D}_2$		$2, 4, \ldots, 14$
$(4n,n)@(4n+8,n+2)$ $(4n,n)@(n+2,4n+8)$	$\boldsymbol{T}^1_2(\sqrt{7})\boldsymbol{D}_1$	$\boldsymbol{D}_2$	1,3,...,13 1,3,...,11
	$\boldsymbol{T}(\sqrt{7})\boldsymbol{D}_2$		2,4,...,12 2,4,...,12 (6)
	$\boldsymbol{T}^3_{14}(\sqrt{7})\boldsymbol{D}_1$	$\boldsymbol{D}_{14}$	– 13
	$\boldsymbol{T}^{13}_{14}(\sqrt{7})\boldsymbol{D}_2$		– 6
$(7n,3n)@(7n+7,3n+3)$ $(7n,3n)@(3n+3,7n+7)$	$\boldsymbol{T}(\sqrt{237})\boldsymbol{D}_1$	$\boldsymbol{D}_1$	$1, 2, \ldots, 7$
$(8n,n)@(8n+8,n+1)$ $(8n,n)@(n+1,8n+8)$	$\boldsymbol{T}(\sqrt{219})\boldsymbol{D}_1$	$\boldsymbol{D}_1$	$1, 2, \ldots, 7$

2.3 Symmetry Based Band Calculations

Due to Wigner's pioneering work, the conceptual importance of symmetry in physics is well recognized. However, except in particle physics, it is commonly used *a posteriori*, to give a profound explanation of the results obtained. Particularly in solid state physics, crystal symmetries which have been known for more than a century, are applied in this way: description of phase transitions, selection rules for various processes, tensor shapes. Perhaps the only exception is Bloch's theorem, *a priori* giving the form of the electronic quantum states and ionic displacements; even this theorem treats the translational periodicity only, which significantly restricts possible symmetry based conclusions, although it reduces the calculations to the elementary cell. The elementary cells of SWNTs contain many ($2q$) atoms, which makes the Bloch theorem based calculations very robust, and frequently too demanding even for the best computers. Nevertheless, it is shown that nontrivial helical, rotational and U-axis symmetries suffice to generate a tube from a single atom. Here we briefly sketch the modified group projector technique, giving an algorithm to reduce the calculations to the full symmetry cell, or symcell, from which the structure is generated by a complete symmetry group. Besides the generalization of the Bloch theorem, it provides *a priori* assignation of the (electron/phonon) bands by all the conserved quantum numbers, enabling application of all possible selection rules for physical processes.

2.3.1 Modified Wigner Projectors

Within the quantum mechanical formalism, the dynamics of the system is determined by its Hamiltonian H, being an operator in the state space $\mathcal{H}$. Its eigenstates $\mid n\rangle$ and eigenvalues E_n solve the time independent Schrodinger equation $H \mid n\rangle = E_n \mid n\rangle$. Besides a few analytically solvable cases, this eigenproblem is hard even for a numerical approach, especially for complex systems. Below we describe the symmetry to simplify calculations.

Symmetries of a system, comprised in the group $\boldsymbol{L}$, are in the state space represented by the unitary operators $D(\ell)$, $\ell \in \boldsymbol{L}$. These are reduced to the block-diagonal matrices, with irreducible representations $D^{(\mu)}(\ell)$ (of the dimension $|\mu|$, each appearing f_μ times) on the diagonal in the symmetry adapted basis (SAB). Such a basis $|\mu t_\mu m\rangle$ ($m = 1, \dots, |u|$) satisfies:

$$\forall \ell \in \boldsymbol{L}: \quad D(\ell) \mid \mu t_\mu m\rangle = \sum_{m'=1}^{|\mu|} D^{(\mu)}_{m'm}(\ell) \mid \mu t_\mu m'\rangle, \quad t_\mu = 1, \dots, f_\mu. \quad (2.16)$$

As the Hamiltonian commutes with each $D(\ell)$ there is a SAB also being the eigenbasis of the Hamiltonian. Thus, the symmetry enables us to solve the Hamiltonian eigenequation by parts: each irreducible representation singles out one f_μ dimensional eigenequation. In the case of finite groups, this reduction is realized by the Wigner group projectors [16]. However, for crystalline

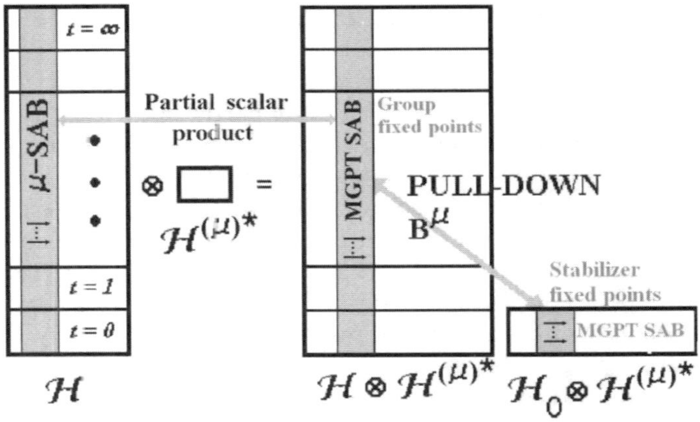

Fig. 2.4. Scheme of the modified group projector technique in the induced spaces

systems, due to the infinite symmetry group and the infinite dimensional state, this procedure, involving summation of $D(\ell)$ over group elements, is inappropriate at least numerically. These deficiencies are cured by the modified group projector technique (MGPT) [17,18]. It is fully implemented into the polymer symmetry simulation package *POLSym* devoted to the physics of the quasi-1D crystals [19], which is used in the forthcoming calculations.

The corner stone of MGPT is simple: in the space $\mathcal{H} \otimes \mathcal{H}^{(\mu)^*}$ (here $\mathcal{H}^{(\mu)}$ is the space of the irreducible representation $D^{(\mu)}$, and the asterisk denotes its dual) the basis $|\,\mu t_\mu\rangle$ of the fixed points of the representation $D \otimes D^{(\mu)^*}$ generates SAB $|\,\mu t_\mu m\rangle$ in $\mathcal{H}$ by the partial scalar product:

$$|\,\mu t_\mu m\rangle = \langle \mu m\,|\,\mu t_\mu\rangle \quad (m = 1, \ldots, |\mu|), \tag{2.17}$$

where $|\,\mu m\rangle$ are the standard vectors (2.16) of the irreducible representation. Thus, one needs to find the set of fixed points (Fig. 2.4): this is the range of the identity representation Wigner projector L^μ for $D \otimes D^{(\mu)^*}$ (and not D; thus the name modified projector). Since the fixed points of the whole group coincide with the common fixed points of the group generators, common eigenvectors for the eigenvalue 1 of the group generators are found; summation over the group is avoided.

The infinite dimension of the state space is overcome with the help of symcell and the convenient properties of L^μ. The action of the group $\boldsymbol{L}$ on a single atom gives a set of atoms called orbit. This way, the whole compound decomposes into disjoint orbits. Taking from each orbit O_P an arbitrary atom P (orbit representative), one gets the symcell. Further, all the transformations of $\boldsymbol{L}$ leaving P invariant form a *finite* subgroup $\boldsymbol{L}_P$, called stabilizer of P, which induces the partition of $\boldsymbol{L}$ onto the cosets: $\boldsymbol{L} = \boldsymbol{L}_P + z_1\boldsymbol{L}_P + \ldots$. The coset representatives form transversal $\boldsymbol{Z}_P = \{z_0 = e, z_1, \ldots\}$. Obviously, the complete orbit O_P is generated from the orbit representative by the action of $\boldsymbol{Z}_P$.

We consider the case when all the transversals are the same, $\boldsymbol{Z}_P = \boldsymbol{Z}$; otherwise, one can use the intersection of all the transversals; this enlarges the set of orbit representatives and the dimension of the eigenvalue problem. In the single-particle models, such as the tight-binding one, the total state space $\mathcal{H}$ is decomposed as the sum $\mathcal{H} = \sum_t D(z_t)\mathcal{H}_0$ of the finite dimensional isomorphic subspaces generated by the symmetry operations from the symcell space $\mathcal{H}_0$. In fact, in the electron (phonon) band calculations each symcell atom contributes by some number of orbitals (Cartesian basis). spanning the atomic space $\mathcal{H}_{0P}$, and the symcell space is the sum of these: $\mathcal{H}_0 = \sum_P \mathcal{H}_{0P}$. As the group action does not mix the orbits, the representation $D(\boldsymbol{L})$ is decomposed into the orbit subrepresentations $D(\boldsymbol{L}) = \sum_P D_P(\boldsymbol{L})$, each of them being *induced* from a representation of $\boldsymbol{L}_P$ by the transversal $\boldsymbol{Z}_P$. Therefore, in each of the spaces $\mathcal{H}_{0P}$ acts the representation $D^{\downarrow}_P$, the representation D_P of $\boldsymbol{L}$ being restricted to the stabilizer $\boldsymbol{L}_P$. Thus, for each irreducible representation of $\boldsymbol{L}$, one finds the representation of the stabilizer: $\gamma^{\mu}(s) = \sum_P D^{\downarrow}_P(s) \otimes D^{(\mu)^*}(s)$ (for each s from $\boldsymbol{L}_P$), defined in the *finite dimensional* product $\mathcal{H}_0 \otimes \mathcal{H}^{(\mu)^*}$. The main fact is that the fixed point subspace of $D \otimes D^{(\mu)^*}$ in $\mathcal{H} \otimes \mathcal{H}^{(\mu)^*}$ is isometric [17] to the fixed points of γ^{μ} in $\mathcal{H}_0 \otimes \mathcal{H}^{(\mu)^*}$. One has to find a modified Wigner projector for γ^{μ}, and any basis $|0; \mu t_{\mu}\rangle$ in its range. Only a finite number of symmetry elements (stabilizer or site symmetry group) and finite dimensional representations are used!

The basis $|\mu t_{\mu} m\rangle$ is achieved by (2.17), where the basis of the fixed points $|\mu t_{\mu}\rangle$ is generated from the symcell space basis $|0; \mu t_{\mu}\rangle$ found with help of the mentioned partial isometry B^{μ} (c is a positive number making this operator isometry on $\mathcal{H}_0 \otimes \mathcal{H}^{(\mu)^*}$):

$$|\mu t_{\mu}\rangle = B^{\mu}\,|0; \mu t_{\mu}\rangle, \quad B^{\mu} = c\sum_P B^{\mu}_P, \quad B^{\mu}_P = \sum_t \mathbf{1}_{0P} \otimes D^{(\mu)^*}(z_t). \tag{2.18}$$

Note that the vectors $|\mu t_{\mu}\rangle$ are given by the components in the atomic basis. Their components on atoms obtained from symcell by the action of the transversal element z_t are found as the terms (2.18):

$$|\mu t_{\mu}\rangle^t = c\sum_m D^{(\mu)^*}_{mm'}(z_t)\,|\mu t_{\mu} m\rangle^0, \quad |\mu t_{\mu} m\rangle^0 = \langle \mu m\,|0; \mu t_{\mu}\rangle. \tag{2.19}$$

Finally, to find SAB as the eigenbasis for H, in the same space $\mathcal{H}_0 \otimes \mathcal{H}^{(\mu)^*}$ the pulled-down Hamiltonian is used:

$$H^{\downarrow\mu} = \sum_P B^{\mu^{\dagger}}_P H B^{\mu}_P. \tag{2.20}$$

It commutes with all $\gamma^{\mu}(s)$, and its eigenvectors, which are simultaneously the fixed points of γ^{μ}, generate the μ-th multiplet of SAB in $\mathcal{H}$ through (2.18) and (2.17). The corresponding eigenenergies of H and $H^{\downarrow\mu}$ coincide [17].

The procedure must be performed for each irreducible representation of $\boldsymbol{L}$, and therefore the energies obtained are automatically assigned by the irreducible representation, i.e. by the complete set of the quantum numbers. The dimension of the eigenproblem in each step is the product of the dimension of the irreducible representation (the degeneracy of the eigenvalues obtained) and the dimension of the symcell space (the number of included orbitals for electrons and trice the number of atoms for phonons). When $\boldsymbol{L}$ describes pure translational symmetry, the algorithm reduces to the Bloch theorem, showing that MGPT generalizes it maximally, i.e. to the full symmetry group.

2.3.2 Symmetry and Band Topology

Some general properties of electron and phonon bands are deduced directly from the irreducible representations of the symmetry group, since Hamiltonian eigenvalues are neatly related to them.

Roto-translational elements of the line groups [20] form the abelian subgroup $\boldsymbol{T}_q^r(a)\boldsymbol{C}_n$, with one-dimensional representations

$$ {}_kA_m(\ell_{ts0}) = \mathrm{e}^{\mathrm{i}\psi_m^k(t,s)}, \quad \text{where } \psi_m^k(t,s) = \frac{kan + 2\pi mr}{q}t + \frac{2\pi m}{n}s \tag{2.21} $$

defined by the pair $\mu = (k, m)$ of quasi momentum (along the z-axis) $k \in (-\pi/a, \pi/a]$ and quasi-angular momentum (z component) $m \in (-q/2, q/2]$. Accordingly, the corresponding symmetry adapted Bloch states in the state space are $\mid kmt_{km}\rangle$. Besides these transformations, the NT symmetry group contains $U = \ell_{001}$, reversing both momenta; consequently, $U \mid kmt_{km}\rangle = \mid -k, -mt_{k,m}\rangle$. Only when $k = 0, \pi/a$ and simultaneously m takes very special values (e.g. for $k = 0$ and $m = 0, q/2$) vectors $|kmt_{km}\rangle$ and $|-k, -mt_{k,m}\rangle$ are either the same or of the opposite sign. Consequently, half $k \in [0, \pi/a]$ of the Brillouin zone suffices to count different representations, and the pairs ${}_kA_m$ and ${}_{-k}A_{-m}$ are joined into double degenerate representations ${}_kE_m$. Only the special pairs (k, m) are unaffected by U, giving even and odd pairs with respect to U one dimensional representations (e.g. ${}_0A_0^{\pm}$, ${}_0A_{q/2}^{\pm}$). The corresponding SAB becomes $\mid km\Pi_U; t_{km\Pi_U}\rangle$, where the parity $\Pi_U = \pm$ for the special representations. Achiral tube representations are four-dimensional ${}_kG_m$ in general, since additionally σ_v changes the sign of m and σ_h that of k; still, as σ_v leaves the invariant $m = 0, n$ bands, and σ_h energies at $k = 0, \pi/a$, these special representations are one- or two-dimensional, obtaining corresponding parities. The parity with respect to the horizontal (vertical) mirror plane is denoted as $\Pi_\mathrm{h} = \pm$ ($\Pi_\mathrm{v} = A/B$). Therefore, the label of representation has the general form ${}_k\Gamma_m^{\Pi}$, where $\Pi = \pm$ for U or σ_h even or odd representations (no superscript when neither of the two parities is defined), and Γ shows the dimension: A (or B for σ_v odd representations) E and G for one-, two- and four-dimensional representations. Frequently it is convenient to use parity quantum numbers $\Pi = \pm 1$ and $\Pi = 0$ for even/odd states and without specific parity.

This analysis shows that the energy bands of NTs are at least two-fold [21]. Indeed, ignoring at first U symmetry, one obtains bands assigned by the quantum numbers km of the subgroup $\boldsymbol{T}_q^r(a)\boldsymbol{C}_n$: one band $\epsilon_m(k)$ along the full Brillouin zone for each $t_{km} = 1, \ldots, f_{km}$. Then, *a posteriori* interrelating pairs of the Bloch states by U symmetry, one finds the degeneracy $\epsilon_m(k) = \epsilon_{-m}(-k)$ of the eigenenergies. In achiral cases bands are four-fold, except those with $m = 0, n$ which are double degenerate and the corresponding Bloch functions are either even or odd in σ_{v}.

Note that U and σ_{h} even and odd representations may appear only at the band edges, for $k = 0, \pi/a$. The corresponding states are Van Hove singularities. Indeed, for example for the band $\epsilon_0(k)$, either U or σ_{h} symmetry implies $\epsilon_0(k) = \epsilon_0(-k)$, and such an even function of k is extremal at $k = 0$. Therefore, in achiral SWNTs all bands end up by Van Hove singularities at $k = 0$, and in chiral ones only the bands with $m = 0, q/2$. For chiral SWNTs q is quite large (at least 14 for (4,1), but typically more than one hundred) and these systematic density of states (DOS) singularities are rare. On the contrary, in double-wall tubes Q is at most 14, making singularities much more frequent; in fact, Q is usually one or two when all the bands end up at $k = 0$ by DOS singularity.

Also, the dimension of the symcell space significantly differs in single- and double-wall tubes. The symcell of any single-wall carbon nanotube contains a single atom, making the eigenproblem for the phonon spectra at most 12-dimensional, while the electronic eigenproblem is at most four- and 16-dimensional in $p^{\perp}$- and sp^3-orbital model, respectively. On the other hand, the number of atoms in the symcell of double-wall carbon nanotubes varies from three (in the most symmetric tubes) to several hundreds when the translational symmetry and U-axis only are present.

2.3.3 Quantum Numbers and Selection Rules

Quantum mechanical description of the physical processes of NTs is inevitably related to the conservation involved laws. In general, a perturbation $Q_{km\Pi}$ with well defined quantum numbers k, m and Π (Π stands for all appearing parities) induces interband transitions of electrons and creation/annihilation of phonons in NT. The behavior of NT is determined by the transition probabilities, being the squares of absolute values of the matrix elements $\langle k_{\mathrm{f}} m_{\mathrm{f}} \Pi_{\mathrm{f}}; t_{\mathrm{i}} | \, Q_{km\Pi} \, | k_{\mathrm{i}} m_{\mathrm{i}} \Pi_{\mathrm{i}}; t_{\mathrm{i}} \rangle$ (initial and final states are labelled as i and f). Therefore, Clebsch-Gordan coefficients of NT symmetry groups [21], as proportional to these matrix elements (by the Wigner-Eckart theorem), give the most detailed information on the conservation laws. The selection rules, singling out the allowed transitions, are their comprehensive form. Note that assignation of bands and Bloch functions of NT by the complete set of quantum numbers is inevitable for application of the selection rules.

For the vertical interband transitions induced by a perturbation with $k = 0$, the quasi momentum is conserved, and the selection rules are:

$$\Delta k = k_\mathrm{f} - k_\mathrm{i} = 0, \quad \Delta m = m_\mathrm{f} - m_\mathrm{i} = m, \quad \Pi_\mathrm{f}\Pi\Pi_\mathrm{i} \neq -1. \tag{2.22}$$

In the non-vertical transitions, the rules are the same, except for the Umklapp processes, when $k_\mathrm{i} + k$ are out of the Brillouin zone and Δm obtains an additional term [21]. This manifests that m is not a conserved quantum number: it is related to the isogonal group (2.6), also containing rotations that are not isolated (but only combined with translations) symmetries of NTs. Therefore, sometimes it is plausible to use the alternative set of conserved quantum numbers [5, 18]. These helical quantum numbers [20] are the helical momentum $\tilde{k}$, including linear and part of the angular momentum, and the remaining angular momentum $\tilde{m}$, related to k and m as:

$$\tilde{k} = k + m\frac{2r\pi}{nc} + \tilde{\kappa}\frac{2q\pi}{na}, \qquad \tilde{m} = m \bmod n, \tag{2.23}$$

$$k = \tilde{k} - \tilde{m}\frac{2r\pi}{na} + \kappa\frac{2\pi}{a}, \qquad m = (\tilde{m} - \kappa p) \bmod q. \tag{2.24}$$

The integers $\tilde{\kappa}$ and κ are introduced to provide $\tilde{k} \in (-q\pi/na, q\pi/na]$ and $k \in (-\pi/a, \pi/a]$, respectively, while p is the helicity (2.4).

2.3.4 Electron Bands

In the graphene layer the bonding of the carbon atoms is realized by three in-plane hybridized sp^2 orbitals, while the forth $p^\perp$ orbital (p state perpendicular to the plane) is responsible for the electronic properties. As mentioned before, the curvature induced rehybridization in the tubes with a diameter larger than 7 Å is weak, and the $p^\perp$ approximation is sufficiently good. However, for narrow nanotubes, the influence of the folded geometry on the atomic orbitals is significant, and the correct dynamical model is sp^3 hybridization, including $2s$ and three $2p$ carbon orbitals. As already emphasized, the simply folded configuration is unstable, but the relaxed one retains the same symmetry.

In the tight-binding (TB) approach, each atom C_{tsu} contributes to the state space by a set of orbitals $\mid tsu; \phi\rangle$: In the sp^3 model four orbitals ($\phi = 1, \ldots, 4$) are included, while in the $p^\perp$-approximation $\phi = 1$ denotes the $p^\perp$-orbital. Therefore, in the view of the single atom SWNT symcell, the pulled down Hamiltonian (2.20) takes the simple form:

$$H^{\downarrow}_{km\Pi} = \sum_{tsu} (D^{km\Pi})^T(l_{tsu}) \sum_{\phi\phi'} h^{\phi\phi'}_{tsu} \mid tsu, \phi\rangle\langle 000, \phi' \mid \otimes \mathbf{1}_{km\Pi}. \tag{2.25}$$

Recall that according to MGPT the first sum only runs over the carbon atoms C_{tsu} interacting with the initial one C_{000}, while the second sum is the part of the Hamiltonian describing this interaction through the potential matrix elements $h^{\phi\phi'}_{tsu} = \langle 000; \phi \mid H \mid tsu; \phi'\rangle$. Also, the same expression describes the pulled down overlap matrix, when instead of $h^{\phi\phi'}_{tsu}$ pure overlaps $S^{\phi\phi'}_{tsu} = \langle 000; \phi \mid tsu; \phi'\rangle$ are used. To proceed, additional assumptions on the included orbitals and neighbors, as well as to the proposed interaction are needed.

Tight-Binding $p^{\perp}$-Orbital Model

Within the $p^{\perp}$ approximation with overlaps of atomic orbitals neglected, substituting in (2.25) the form of the irreducible representations of the line groups [18], for each m the pulled down Hamiltonian is two-dimensional [21]:

$$H_m(k) = \begin{pmatrix} h_m^0(k) & h_m^{1*}(k) \\ h_m^1(k) & h_m^0(k) \end{pmatrix}, \quad h_m^u(k) = \sum_{ts} h_{tsu}^{11} \mathrm{e}^{\mathrm{i}\psi_k^m(t,s)} \; (u = 0, 1). \quad (2.26)$$

This refers to the chiral tubes, while the achiral ones can be treated as a special case with $H_m(k)$ and $H_{-m}(k)$ joined into a single four dimensional matrix. Solving the eigenproblem of (2.26), for each m two double degenerate bands are found as discussed in Sect. 2.3.2:

$$\epsilon_m^{\pm}(k) = h_m^0(k) \pm |h_m^1(k)|. \quad (2.27)$$

The corresponding generalized Bloch eigenfunctions are $(h_m^k = \mathrm{Arg}(h_m^1(k)))$:

$$|km;\pm\rangle = \sum_{ts} \mathrm{e}^{-\mathrm{i}\psi_m^k(t,s)} \big(|ts0\rangle \pm \mathrm{e}^{\mathrm{i}h_m^k} |ts1\rangle \big),$$

$$|-k,-m;\pm\rangle = \sum_{ts} \mathrm{e}^{\mathrm{i}\psi_m^k(t,s)} \big(|ts1\rangle \pm \mathrm{e}^{\mathrm{i}h_m^k} |ts0\rangle \big).$$

Note again that for the achiral tubes four-fold degeneracy appears due to $\epsilon_m^{\pm}(k) = \epsilon_{-m}^{\pm}(k)$, except that the bands $\epsilon_m^{\pm}(k) = \epsilon_m^{A/B}(k)$ for $m = 0, n$ with the opposite vertical mirror parity remain double degenerate.

The simplest $p^{\perp}$ approximation completely neglects local distortions of graphene hexagons and includes only the first three neighbors pointed to by $u_1 = u_2 = u_3 = 1$ and

$$t_1 = -\frac{n_2}{n}, \qquad t_2 = \frac{n_1}{n}, \qquad t_3 = t_1 + t_2,$$

$$s_1 = \frac{2n_1 + (1 + r\mathcal{R})n_2}{q\mathcal{R}}, \quad s_2 = \frac{(1 - r\mathcal{R})n_1 + 2n_2}{q\mathcal{R}}, \quad s_3 = s_1 + s_2.$$

Therefore, all the interatomic matrix elements are equal, $h_{t_i,s_i}^{11} = V \approx -3\,\mathrm{eV}$, while $h_{000}^{11} = 0$. The model is analytically solvable: the pairs of m-assigned bands are symmetrical with respect to the Fermi level:

$$\epsilon_{E_m}^{\pm}(k) = \pm |V \sum_{i=1}^{3} \mathrm{e}^{\mathrm{i}\psi_m^k(t_i,s_i)}|. \quad (2.28)$$

According to (2.28), for the tubes with $n_1 - n_3$ divisible by 3, there are bands crossing at the Fermi level $E_{\mathrm{F}} = 0$. Precisely, if $\mathcal{R} = 3$, two symmetric bands with $m = nr \pmod q$ cross at $k = 2\pi/3a$, while for $\mathcal{R} = 1$ two pairs of bands with $m = \pm\frac{q}{3}$ come together in $k = 0$. Apparently this implies that

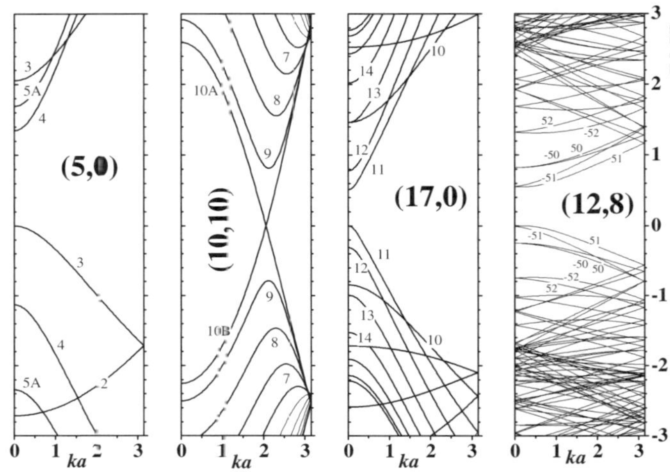

Fig. 2.5. Symmetry assigned $p^{\perp}$ orbital electronic bands of SWNTs. Note Van Hove singularities in $k = 0$ for the even or odd states of the U-axis, as predicted in Sect. 2.3.2, and a large gap in the case of (5,0). The bands are assigned by the quantum number m and in the case of the achiral tubes, by a vertical mirror parity if defined (A for even and B for odd). The bands of the chiral tubes are double degenerate (representations ${}_kE_m$); those of the achiral tubes are four-fold (representations ${}_kG_m$), except the ones with well defined vertical mirror parity (representations ${}_kE_m^A$ or ${}_kE_m^B$). The edge states are singlets or doublets

such tubes are conducting (Fig. 2.5). However, Landau's non-crossing rule forbids this: the crossing bands are assigned by the same quantum numbers, and any perturbation (such as neglected curvature effects) produces a small gap [5, 22]. The exceptions are the armchair tubes: for them $q = 2n$, $r = 1$, and $\mathcal{R} = 3$, giving $m = n$, and the crossed bands differ in vertical mirror parity [5, 23].

Thus, this naive model together with the non-crossing rule gives the classification of SWNTs according to their conducting properties: the armchair tubes are conducting, the other tubes with integer $(n_1 - n_2)/3$ are quasi-metallic with a small secondary gap less than 0.1 eV decreasing with the tube diameter [22] as D^{-2}. All other tubes are semiconducting, with the gap decreasing with the tube diameter as $1/D$; starting with more than 1 eV for narrow tubes, the gap tends to the zero graphene limit [4]. Of course, the whole model and therefore these conclusions are not applicable to ultra narrow tubes.

The $p^{\perp}$-orbital approximation may be refined by considering the interatomic overlaps, and for thick enough tubes such results give reasonably good predictions of optical properties [24]. Also, the approximation is improved by including higher level neighbors. A model with 12 neighbors (three first, six second and three third ones, Fig. 2.2), neglecting local distortions and overlaps effectively uses three different constants h_{tsu}^{11}, one for each level of neighbors. This model, with fitted constants, describes [25] satisfactorily the bands near the Fermi level, obtained by density functional techniques.

Density Functional Model

As has been stressed, $p^{\perp}$ approximation fails to describe electronic states of narrow SWNTs: large curvature substantially affects hybridization, and all sp^3 carbon orbitals have to be considered. The band structure may be satisfactorily studied only with the help of a density functional method. According to the exchange-correlation functional employed, several types of DF methods are distinguished. All of them give similar stable configurations (Table 2.1) and predict well the conductivity, while some other details of band structure may differ significantly. The most frequently used local density approximation [11, 26, 27] systematically overestimates the gap in semiconductors, making the predictions of optical properties not completely reliable. On the other hand, G. Seifert's [28] density functional tight-binding (DFTB) method applies the density functional technique to small molecules, and fits the Hamiltonian and overlap matrix elements of (2.25) as a functions of the interatomic distance $h^{\phi\phi'}(r)$ and $S^{\phi\phi'}(r)$. As these functions are well transferable to compounds with the same local structure, they can be used within the tight binding approach in studies of complex structures. Therefore, this method is more efficient (e.g. also thin SWNTs or inorganic tubes may be considered); possible deficiencies of the semiempirical foundation are at least compensated for to some extent controllable parameters (enabling to fit the experiments), and complete symmetry implementation (due to application of the DF potential through the tight-binding calculations). However, currently all DF methods fail to describe correctly Van der Waals systems; therefore, these methods are not reliable for multi-wall carbon nanotubes, despite some recent attempts to overcome this difficulty [29].

In comparison to the $p^{\perp}$ orbital results, DFTB bands of narrow tubes (see Fig. 2.6) differ substantially. In fact, the tubes (4,0), (5,0) and (4,1) are conductive, in contrast to the $p^{\perp}$ approximation based criterion. Of course, the symmetry based general properties of band topology, such as degeneracy and $k = 0, \pi/a$ Van Hove singularities, are manifested in both models.

2.3.5 Force Constants Phonon Dispersions

There are lattice dynamics studies of NTs based on various force field models. Since *ab initio* methods [30] are restricted to thin tubes, an efficient force constant method is necessary for massive calculations enlightening general NT properties [31]. To this end graphite force constants [32] are adjusted kinematically and dynamically to nanotube geometry. Each pair (α, β) of carbon atoms defines the stretching, out-of-plane and in-plane unit vectors

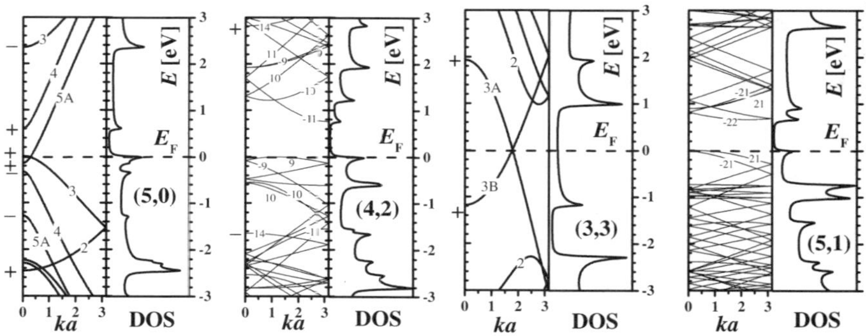

Fig. 2.6. DFTB bands and DOS of narrow tubes near Fermi level E_F (broken horizontal line at 0 eV). For the tubes (4,2) and (5,1) the bands relevant for the calculated optical interband transitions in Sect. 2.4.2 are in bold. If defined, the z-reversal parity at $k = 0$ and $k = \pi/a$ is indicated by $\pm$. Note that (5,0) is conducting

$\mid \alpha\beta, i\rangle$ $(i = 1, 2, 3)$: $\mid \alpha\beta, 1\rangle$ points from α to β, $\mid \alpha\beta, 2\rangle$ is perpendicular to $|\alpha\beta, 1\rangle$ and to the tube axis, while $|\alpha\beta, 3\rangle = |\alpha\beta, 1\rangle \times |\alpha\beta, 2\rangle$. The components of these vectors in the tube reference frame are $\mid \alpha\beta, 1\rangle = (S_{1i}^{\alpha\beta}, S_{2i}^{\alpha\beta}, S_{3i}^{\alpha\beta})$. These vectors and the force-constants $c_i^{\alpha\beta}$ are the eigenvectors and eigenvalues of the 3×3 matrix $\mathcal{D}_{\alpha\beta}$, describing the contribution of the pair (α, β) to the dynamical matrix $\mathcal{D}$ of nanotube [33]: $\mathcal{D}_{\alpha\beta} \mid \alpha\beta, i\rangle = c_i^{\alpha\beta} \mid \alpha\beta, i\rangle$. Therefore, $\mathcal{D}_{\alpha\beta} = \sum_i c_i^{\alpha\beta} \mid \alpha\beta, i\rangle\langle \alpha\beta, i \mid$ and its matrix elements are easily found: $\mathcal{D}_{\alpha\beta}^{ij} = \sum_p c_p^{\alpha\beta} S_{ip}^{\alpha\beta} S_{jp}^{\alpha\beta}$.

The kinematic modification provides the twisting mode exactly using the rotational sum rule [34]:

$$\sum_{\beta} (R_{\alpha\beta 1} \mathcal{D}_{\alpha\beta}^{2j} - R_{\alpha\beta 2} \mathcal{D}_{\alpha\beta}^{1j}) = 0, \quad \forall\, \alpha, \quad j = 1, 2, 3. \tag{2.29}$$

The sum runs over the relevant neighbors β of the atom α $(\beta \neq \alpha)$ and $R_{\alpha\beta i}$ is the Cartesian component of the vector $\boldsymbol{R}_{\alpha\beta} = R_{\alpha\beta} \,|\alpha\beta, 1\rangle$ from α to β in the equilibrium positions. Since $S_{32}^{\alpha\beta} = 0$ because $|\alpha\beta, 2\rangle$ is orthogonal to the tube axis, (2.29) reduces to:

$$\sum_{\beta} R_{\alpha\beta} (c_2^{\alpha\beta} S_{j2}^{\alpha\beta} S_{33}^{\alpha\beta} - c_3^{\alpha\beta} S_{j3}^{\alpha\beta} S_{32}^{\alpha\beta}) = 0, \quad j = 1, 2, 3;\ \forall\, \alpha.$$

Thus, each atom α gives two constraints on out-of-plane force constants $c_2^{\alpha\beta}$:

$$\sum_{\beta} R_{\alpha\beta} c_2^{\alpha\beta} S_{12}^{\alpha\beta} S_{33}^{\alpha\beta} = 0, \quad \sum_{\beta} R_{\alpha\beta} c_2^{\alpha\beta} S_{22}^{\alpha\beta} S_{33}^{\alpha\beta} = 0, \tag{2.30}$$

as SWNT is a single-orbit system, MGPT calculations are reduced to the orbit representative $\alpha = C_{000}$ only, and (2.30) imposes two conditions altogether. Particularly, the force constants model [32] involves eighteen up to the fourth level neighbors; collecting all the relevant coordinate factors and force constants into the vectors $S_k^\alpha = (R_{\alpha 1} S_{k2}^{\alpha 1} S_{33}^{\alpha 1}, \ldots, R_{\alpha 18} S_{k2}^{\alpha 18} S_{33}^{\alpha 18})$ and $c_2^\alpha = (c_2^{\alpha 1}, \ldots, c_2^{\alpha 18})$; (2.30) becomes a pair of the orthogonality conditions $c_2^\alpha \cdot S_k^\alpha = 0$ $(k = 1, 2)$. These quite general conditions for any dynamical model provide the twisting acoustic mode exactly. Thus, it is natural to correct minimally the graphene force constants by projecting c_2^α onto the orthocomplement of the plane defined by S_1^α and S_2^α.

The dynamical response to the configuration changes introduced by the folding of a graphene plane is described along the lines of [35]. In the α-frame, in which the z-axis coincides with the tube axis and the x-axis runs through the atom α, the eigenbasis of $\mathcal{D}_{\alpha\beta}$ is (φ and ψ are cylindrical coordinates of β and the angle between the horizontal xy-plane and the bond α-β):

$$\begin{aligned}
|\,\alpha\beta, 1\rangle &= (-\cos\psi \sin\frac{\varphi}{2}, \cos\psi\cos\frac{\varphi}{2}, \sin\psi),\\
|\,\alpha\beta, 2\rangle &= (\cos\frac{\varphi}{2}, \sin\frac{\varphi}{2}, 0),\\
|\,\alpha\beta, 3\rangle &= (-\sin\psi\sin\frac{\varphi}{2}, \sin\psi\cos\frac{\varphi}{2}, -\cos\psi).
\end{aligned} \tag{2.31}$$

Further, consider the graphene stretching, out-of-plane and in-plane unit vectors after the folding: $|\,\alpha_g\beta_g, 1\rangle$ is the tangent to the projection of the bond α-β to the tube's surface, $|\,\alpha_g\beta_g, 2\rangle$ is perpendicular to the tube at α, and $|\,\alpha_g\beta_g, 3\rangle = |\,\alpha_g\beta_g, 1\rangle \times |\,\alpha_g\beta_g, 2\rangle$. So, in the α-frame these vectors are:

$$\begin{aligned}
|\,\alpha_g\beta_g, 1\rangle &= \left(0, \frac{\cos\frac{\varphi}{2}\cos\psi}{\sqrt{1-\sin^2\frac{\varphi}{2}\cos^2\psi}}, \frac{\sin\psi}{\sqrt{1-\sin^2\frac{\varphi}{2}\cos^2\psi}}\right),\\
|\,\alpha_g\beta_g, 2\rangle &= (1, 0, 0),\\
|\,\alpha_g\beta_g, 3\rangle &= \left(0, \frac{\sin\psi}{\sqrt{1-\sin^2\frac{\varphi}{2}\cos^2\psi}}, -\frac{\cos\frac{\varphi}{2}\cos\psi}{\sqrt{1-\sin^2\frac{\varphi}{2}\cos^2\psi}}\right).
\end{aligned} \tag{2.32}$$

Assuming that the forces along $|\,\alpha_g\beta_g, i\rangle$ $(i = 1, 2, 3)$ are preserved in the course of rolling, one easily finds how the required modified constants $c_i^{\alpha\beta}$ are related to the original graphene ones $C_i^{\alpha_g\beta_g}$: $c_i^{\alpha\beta} = C_i^{\alpha_g\beta_g} / |\langle\alpha_g\beta_g, i\,|\,\alpha\beta, i\rangle|$. The expansion over sine and cosine of $\varphi/2$ yields:

$$\begin{aligned}
c_1^{\alpha\beta} &= C_1^{\alpha_g\beta_g}(2 - \cos\frac{\varphi}{2}),\\
c_2^{\alpha\beta} &= C_2^{\alpha_g\beta_g}(1 + \sin^2\psi(1 - \cos\frac{\varphi}{2})),\\
c_3^{\alpha\beta} &= C_3^{\alpha_g\beta_g}(1 + \cos^2\psi(1 - \cos\frac{\varphi}{2})).
\end{aligned} \tag{2.33}$$

It turns out that the dynamical and kinematical corrections each give similar SWNT force constants, and therefore the results [31] presented here, obtained by both corrections, are close to that of [35], but with exactly four acoustic modes.

2.4 Optical Absorption

Being both individual and bulk sensitive, the optical response measurements [36] are, together with Raman [37] and neutron scattering and electron and X-ray diffraction, widely used as a tool for diameter and chirality distribution analysis and for total SWNT yield estimations in macroscopic samples.

In this section, we evaluate numerically, using the line group theoretical methods, the optical conductivity tensor for individual SWNTs of arbitrary geometry. We use G. Seifert's [28] DFTB calculations as input for the *POLSym* code. The full line group symmetry of the SWNTs is taken into account (electron correlations and curvature effects are included) and the transition matrix elements of the momentum operator are, within the dipole approximation, exactly calculated from the generalized Bloch eigenfunctions.

2.4.1 Conventional Nanotubes

Prediction of the polarization dependent optical absorption [38] in NTs has been confirmed by optical ellipsometry [39], as well as by reflectivity measurements [40]. In addition, polarized Raman spectroscopy on fibers of aligned SWNTs, reflectance and absorption spectra and tensor invariant measurements of the Raman active modes [41] has given evidence of strongly polarized optical transitions. Anisotropy of the dielectric function has been studied in detail by S. Tasaki and coworkers [24]: a substantial optical rotatory power and circular dichroism as well as optical activity of the NT ensemble have been predicted. Also, polarized low-frequency optical spectra of SWNTs bundles, within the gradient approximation and the lowest-order isotropic model, has been calculated [42]. On the other hand, optical activity [7] and anisotropy of absorption [43] in the isolated NTs have been inferred on the line group symmetry arguments. Recently, polarized optical absorption in SWNTs has been thoroughly studied within the symmetry based approach [44].

In this subsection the polarized optical conductivity of SWNTs with diameters between 8 Å and 16 Å is studied. The calculations for bundled SWNTs with mean diameter 13.6 Å assuming a Gaussian diameter distribution with ΔD =0.5 Å are carried out as well.

In the relaxation-time approximation, the interband contribution (at absolute zero temperature) to the real part of the optical conductivity tensor is (we give here only the expression for the diagonal elements as, due to the symmetry of the tubes, the non-diagonal ones vanish [7]):

$$\mathbb{R}\mathrm{e}\,\sigma_{jj} = \frac{C}{\omega}\sum |\langle k_{\mathrm{f}} m_{\mathrm{f}} \Pi_{\mathrm{f}} |\, \nabla_j \,| k_{\mathrm{i}} m_{\mathrm{i}} \Pi_{\mathrm{i}} \rangle|^2 \delta(\Delta\epsilon - \hbar\omega), \tag{2.34}$$

where $C = 2\pi e^2 \hbar^2 / m_e^2$, $\Delta\epsilon = \epsilon_{\mathrm{f}} - \epsilon_{\mathrm{i}}$ is the energy absorbed, j denotes the direction of the electric field applied, while the subscripts i and f refer to the initial and the final state, respectively, and the summation runs over quantum numbers $k_{\mathrm{i}}, m_{\mathrm{i}}, \Pi_{\mathrm{i}}, k_{\mathrm{f}}, m_{\mathrm{f}}, \Pi_{\mathrm{f}}$ (Π represents all parities). As only two tensor components are independent [7], we introduce the following notation: $\sigma_{zz} = \sigma_{\|}$, $\sigma_{xx} = \sigma_{yy} = \sigma_{\perp}$. The form of the expression used is convenient for application of the selection rules. For any orientation of the electric field, the wave vector of the Bloch electron remains essentially unchanged in the optical transition: the crystal momentum conservation law reads $\Delta k = 0$.

Operator $\nabla_{\|}$ is reversed upon the U transformation and invariant under all the others (${}_0A_0^-$ tensor of the group $\boldsymbol{L}_{\mathcal{C}}$ and $\boldsymbol{L}_{\mathcal{ZA}}$). As for ∇_x and ∇_y, it is convenient to switch to the momentum standard components $\nabla_{\pm} = \nabla_x \mp \mathrm{i}\nabla_y$, as these carry 2D space of the representation ${}_0E_1$ for $\mathcal{C}$ and ${}_0E_1^+$ for the achiral tubes. The dipole optical transition selection rules are now easy to deduce.

The quasi-angular momentum m selection rules depend on the direction of the perturbing electric vector: if it is parallel to the tube axis ($\|$ polarization) the rule imposes $\Delta m = 0$, while if it is orthogonal onto the z-axis ($\perp$ polarization) Δm should be 1 (left circular polarization) or -1 (right circular polarization) in order for the dipole optical transition to take place.

As for the parities (if defined), the one with respect to σ_{v} is to be preserved while the one with respect to the U axis or to σ_{h} is to be reversed in the case of $\|$ polarization. For $\perp$ polarization it is *vice versa* concerning the σ_{h}-parity while no restriction is imposed onto the σ_{v}-parity.

Taking into account the selection rules and substituting $\sum_k$ by $\frac{1}{\pi}\int_k \mathrm{d}k$ in (2.34) one finds:

$$\mathbb{R}\mathrm{e}\,\sigma_{\|} = \frac{C}{\omega}\sum_m \int_0^{\pi} \mathrm{Im}\Big[\frac{|\langle km;+|\, \nabla_{\|} \,|km;-\rangle|^2}{\epsilon_m^+(k) - \epsilon_m^-(k) - \hbar\omega - i\eta}\Big]\,\mathrm{d}k, \tag{2.35}$$

where $|km;\pm\rangle$ represent the Bloch orbitals above and below the Fermi level, respectively, while η =0.04 eV is a phenomenological broadening that suppresses the height of the resonant peaks. Analogously, starting from (2.34) it is straightforward to obtain the relation for the perpendicularly polarized field:

$$\begin{aligned}\mathbb{R}\mathrm{e}\,\sigma_{\perp} = \frac{C}{\omega}\sum_m \int_0^{\pi} \mathrm{Im}\Bigg[&\frac{|\langle k, m+1;+|\, \nabla_+ \,|km;-\rangle|^2}{\epsilon_{m+1}^+(k) - \epsilon_m^-(k) - \hbar\omega - i\eta} \\ &+ \frac{|\langle k, m-1;+|\, \nabla_- \,|km;-\rangle|^2}{\epsilon_{m-1}^+(k) - \epsilon_m^-(k) - \hbar\omega - i\eta}\Bigg]\,\mathrm{d}k.\end{aligned} \tag{2.36}$$

Note that for $\mathcal{A}$ tubes $\|$ transitions for $m = 0, n$ are to be excluded as doubly degenerate electronic bands have different vertical mirror symmetry.

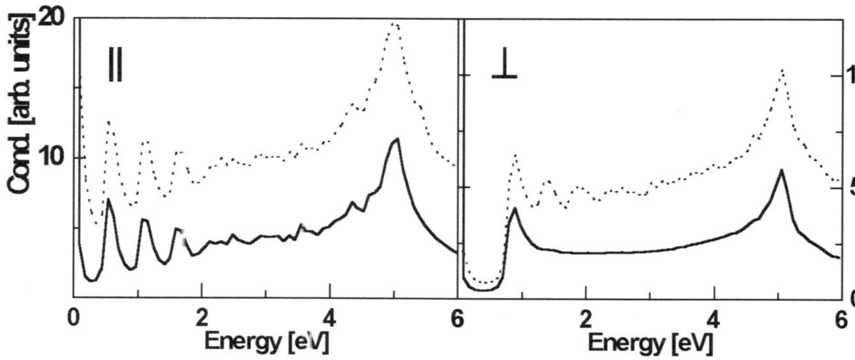

Fig. 2.7. Components of the optical conductivity tensor (*solid lines*) and their JDOS approximation (*dotted lines*) for the (10,10)-bundle. The intensity of the optical conductivity of the bundle is normalized to a single tube

Concerning the +/− parities, they are implicitly taken into account as symmetry adapted state functions are used. Namely, in spite of the systematic strong Van Hove singularities, the $\|$ transitions between the states with non-vanishing k (but close to $k = 0$) are highly suppressed. This is due to the continuity principle, since the transitions are forbidden at $k = 0$ by the z-reversal symmetry. As for the $\perp$ transitions, the parity selection rules are of no importance for the $\mathcal{A}$ tubes.

By contrast, for $\mathcal{Z}$ tubes the $\|$ transitions are not affected by the parities, while the $\perp$ transitions are almost entirely suppressed by them. For this type of tubes there are no DOS peaks outside the $k = 0$ vicinity while the z-reversal symmetry forbids the $\perp$ absorption at $k = 0$. Owing to the continuity of the matrix elements, the corresponding transition probabilities are small and the $\perp$ optical spectra features are highly suppressed despite the strong Van Hove singularities.

Concerning the $\mathcal{C}$ tubes, although the parities practically do not influence the optical absorption, we find that the transition probabilities are strongly helicity dependent: the results obtained differ considerably from those approximated by the joint density of states (JDOS).

A typical $\|$ polarized absorption spectrum of the bundled SWNTs (Gaussian distribution centered at D =13.6 Å with ΔD = 0.5 Å of 29 tubes comprising the (10,10)-bundle) is depicted in the left panel of Fig. 2.7. Note that each tube in the bundle is weighted with a Gaussian factor, i.e. the absorption intensity is normalized to a single tube. The first two peaks at 0.55 eV and 1.1 eV are the well-known semi-conducting tube response that scale inverse to the tube diameter. The third feature, at 1.65 eV, scaling up also with $1/D$, comes from $\mathcal{A}$ tubes and from quasi-metallic $\mathcal{Z}$ and $\mathcal{C}$ ones.

The collective excitation of the π-electron system (at zero-momentum transfer) in the tubes with the (10,10)-bundle typical diameters, we find (for

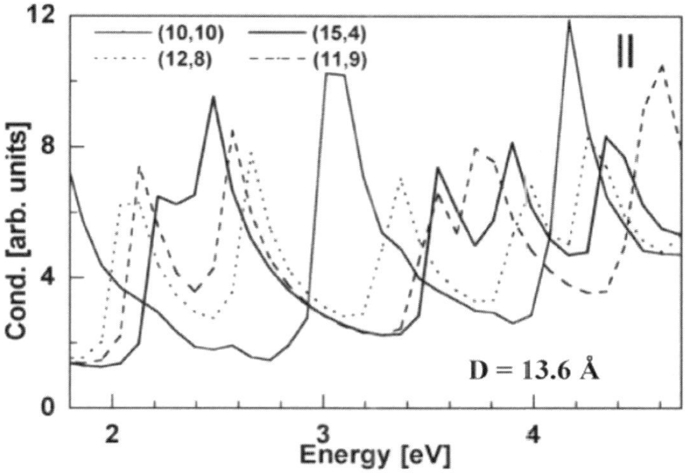

Fig. 2.8. Highly helicity sensitive || polarized absorption spectra features (in the energy region between the well pronounced low-energy peaks and the onset of the π-plasmon) for SWNTs (10,10), (15,4), (12,8) and (11,9) having the same diameter $D = 13.6$ Å but different wrapping angle

both polarizations) at 5 eV, Fig. 2.7. Considerable anisotropy is evident although the excitation is well pronounced for the $\perp$ polarized electric field as well. In the range between the low-energy inter-band transitions and the π-plasmon excitation energy few inter-band optical absorption features appear. As they are highly sensitive to the helicity of the particular tube (regarding the peak positions and absorption intensities), after averaging over a bundle they result in rather broad and weakly pronounced peaks. Their wrapping-angle sensitiveness is illustrated in Fig. 2.8 by tubes with diameter $D =13.6$ Å.

The main features of the $\perp$ polarized optical spectra are given in the right panel of Fig. 2.7. These are responses from all the types of tubes within the bundle apart from the tubes (15,3), (16,4), (15,5), (14,5), (13,5), (14,6) and (13,7), which are practically transparent throughout the entire range for such a polarization of the electric field.

We now turn to the inspection of the differences between the results presented here and the widely used JDOS approximated ones based on the lowest-order isotropic TB model. For || polarization the JDOS only based results are in reasonably good agreement with more accurate calculations (Fig. 2.9), aside from the overall absorption intensity enhancement and slight differences regarding positions, relative intensities and shapes of peaks. These differences seem to be more pronounced for tubes with small wrapping angle ($\mathcal{Z}$ tubes and tubes close to the $\mathcal{Z}$-direction) but can scarcely be noticed when averaged over a bundle. This is illustrated in the left panel of Fig. 2.7. On the contrary, in the case of the $\perp$ polarization, the JDOS approxima-

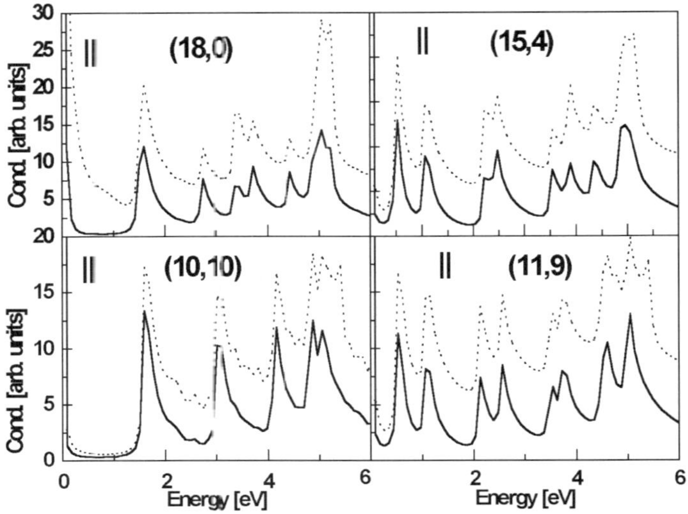

Fig. 2.9. $\|$ component of the optical conductivity (*bold line*) and the JDOS approximation to it (*dotted line*) for SWNTs (18,0), (15,4), (10,10), (11,9)

tion leads to incorrect predictions. As is evident from the calculated spectra for the (10,10)-bundle (Fig. 2.7, right panel), many of the absorption peaks resulting from the JDOS approximation of the $\perp$ component of the optical conductivity tensor, do not appear in the more accurately calculated optical spectrum. For isolated SWNTs these discrepancies are even more noticeable. For comparison see Fig. 2.10.

The results obtained reinforce the JDOS approximation of the optical conductivity for $\|$ polarized electric field, but lead to quite different conclusions as far as the $\perp$ component of the conductivity tensor is concerned. The discrepancies are most prominent for thin $\mathcal{Z}$ tubes, but also quite substantial even when the JDOS only based results are averaged over a bundle. We explain this by the influence of the parity selection rules on the transition probabilities, elucidating that the JDOS approximation is conceptually incompatible with the horizontal mirror parity selection rules for dipole absorption processes in achiral SWNTs. Namely, although very little weight is associated with the horizontal mirror parity states (as this parity is well defined only for a finite number of high symmetry states) due to the continuity principle (which cannot be included within the JDOS approximation), if the transition is not permitted at $k = 0$ the transitions between the nearby states with general, nonvanishing k are not very likely to occur. As all the bands of the achiral tubes are zero sloped at $k = 0$, this means that the JDOS approximation [43] to the optical absorption spectrum of the achiral NT cannot be entirely reliable. Especially not for the $\mathcal{Z}$ tubes as all their DOS Van Hove singularities show up in the $k = 0$ vicinity. In general, horizontal mirror

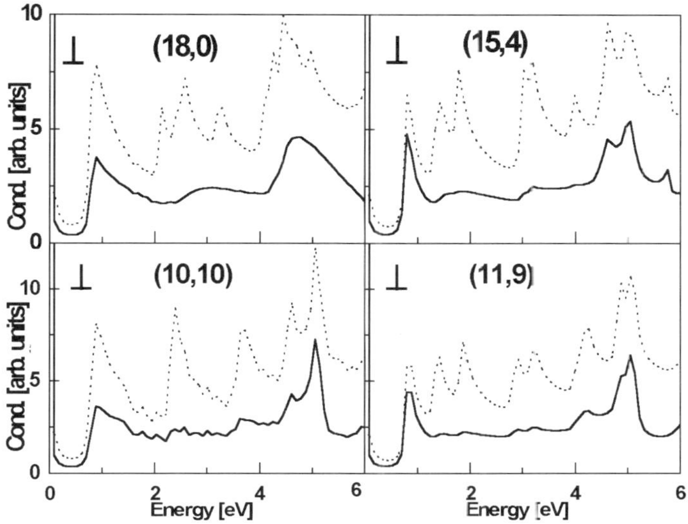

Fig. 2.10. ⊥ component of the optical conductivity (*bold line*) and the JDOS approximation to it (*dotted line*) for SWNTs (18,0), (15,4), (10,10), (11,9)

symmetry suppresses a vast number (all but two) of the perpendicularly polarized optical transitions in the $\mathcal{Z}$ tubes and all the parallel transitions close to $k = 0$ in the $\mathcal{A}$ tubes. Since the number of bands enlarges with diameter and they become zero sloped over larger $k = 0$ vicinity, the horizontal mirror parity influence on the interband transitions is diminished for thicker tubes. In contrast, the U parity does not affect the optical transitions in the chiral SWNTs.

2.4.2 Template Grown Nanotubes

Soon after the discovery of carbon nanotubes, the existence of narrow tubes was predicted [45]. Recent calculations confirmed [11] that SWNTs that have been synthesized inside channels of zeolite $AlPO_4$ (AFI) crystal [46], if isolated, should be energetically stable. The diameter of the zeolite-grown tubes is determined to be 4.2 ± 0.2 Å, while 7.3 Å is the inner diameter of the channels of the AFI crystal, adjacent centers of which are 13.7 Å distant [47, 48]. Such a highly defined geometry of the sample makes it easier to carry out the measurements of the electro-optical properties of the zeolite-grown tubes and gives measurements closer to the theoretical models. On the basis of the optical absorption measurements and *ab initio* electronic band structure calculations, the NTs (5,0), (3,3) and (4,2) were suggested to be zeolite-grown [47]. More recently, several full DFLD calculations of the optical properties of these three tubes have been reported [26, 27, 49]. Quite recently symmetry based calculations of the optical absorption in narrow NTs has been performed [50].

Here we study the polarized optical absorption of NTs (5,0), (4,2), (3,3) and (5,1), in energy region 0-6 eV. The method applied is conceptually the same as the one described in the previous subsection where the SWNTs that are thicker than icosahedral C_{60} are treated. Only here, instead a single $2p^{\perp}$ orbital (per carbon atom), we take the basis set $\{2s, 2p_x, 2p_y, 2p^{\perp}\}$, thus taking into account $\sigma - \pi$ hybridization. Also, we adopt the symmetry maintaining DFTB optimized configuration given in Table 2.1. An analogous optimization procedure was originally suggested by Cabria et al. [11], and we have used those results in [50]. This gives us an opportunity to compare the results of the full DFLD [11, 26, 27, 49] and the combined DFLD/DFTB [50] approaches to the full DFTB results here presented.

The considered SWNTs are of uniform thickness while their symmetry and chirality are quite diverse. As for the achiral NTs, symmetry of the zig-zag one, (5,0), is described by the line group $\boldsymbol{L}10_5/mcm$, which is a product of the screw axis $\boldsymbol{T}^1_{10}$ and the point group $\boldsymbol{D}_{5\mathrm{h}}$. The isogonal point group is $\boldsymbol{D}_{10\mathrm{h}}$ and it is not a subgroup of the symmetry group $\boldsymbol{L}10_5/mcm$. (Recall that SWNT symmetry groups are non-symmorphic [7].) Line group $\boldsymbol{L}6_3/mcm = \boldsymbol{T}^1_6\,\boldsymbol{D}_{3\mathrm{h}}$ describes the symmetry of the armchair tube (3,3), while $\boldsymbol{D}_{6\mathrm{h}}$ is the isogonal point group. Concerning the remaining two chiral tubes, (4,2) and (5,1), the screw axis group of each tube is $\boldsymbol{T}^9_{28}$ and $\boldsymbol{T}^{51}_{62}$. Thus, the isogonal point groups are $\boldsymbol{D}_{28}$ and $\boldsymbol{D}_{62}$, while the symmetry groups are $\boldsymbol{L}28_{22}22 = \boldsymbol{T}^9_{28}\,\boldsymbol{D}_2$ and $\boldsymbol{L}62_{45}22 = \boldsymbol{T}^{51}_{62}\,\boldsymbol{D}_1$.

In Fig. 2.6 the calculated electronic dispersions and the corresponding state density are depicted. The achiral tubes are found to be metallic, while (4,2) and (5,1) are indirect gap semiconductors (~ 0.76 eV and ~ 0.66 eV, respectively). Due to the σ^*-π^* hybridization the double-degenerate $m = 5$ electronic band with well defined vertical mirror parity (Fig. 2.6a) is lowered below the Fermi level, converting the (5,0) tube from a semiconductor (within a simple TB method that includes nearest-neighbor $2p^{\perp}$-$2p^{\perp}$ interaction only) to a metal. Interestingly, the electronic DOS at the Fermi level in the (5,0) turns out to be larger than that in the (3,3) NT.

In Fig. 2.11, the optical conductivity spectra for the light polarized parallelly to the NT axis is depicted. There are two strong peaks at 1 eV and 2.4 eV and a weaker one at 1.5 eV for the tube (5,0). These peaks are due to the inter-band transitions (Fig. 2.6a): ${}_kG_4 \to {}_kG_4$, ${}_kE^A_5 \to {}_kE^A_5$ and ${}_kG_3 \to {}_kG_3$. Here ${}_kG_m$ and ${}_kE^A_m$ denote four-fold and double degenerate electronic bands (below and above the Fermi level) which, at $k = 0$, end in double-degenerate states (transforming as ${}_0E^{\pm}_m$) and non-degenerate states (${}_0A^{\pm}_m$), respectively. Thus, the transitions at the Γ point are ${}_0E^-_4 \to {}_0E^+_4$ and ${}_0E^+_3 \to {}_0E^-_3$ in the line group notation or $E_{4\mathrm{u}} \to E_{4\mathrm{g}}$ and $E_{3\mathrm{u}} \to E_{3\mathrm{g}}$ in the isogonal point group notation. Since the polar vector along the tube axis transforms as the ${}_0A^-_0$ ($A_{2\mathrm{u}}$) tensor, the parities of the edge states favor these transitions [44]. Consequently, intense absorption at these particular energies is expected. On the other hand, edge states of both ${}_kE^A_5$ bands, ${}_0A^-_5$ and ${}_0A^+_5$ (i.e. B_2 and

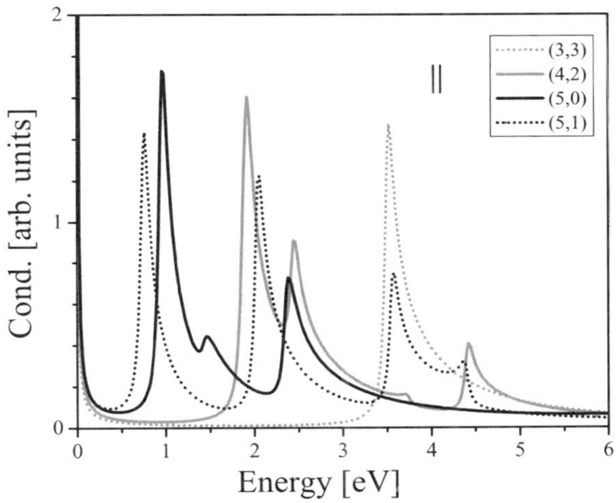

Fig. 2.11. Real part of the optical conductivity for incoming light polarized parallel to the NT axis

B_1), fall below the Fermi level, Fig. 2.6. The peak at 2 eV is due to the (5,1) tube inter-band transitions ${}_kE_{21} \rightarrow {}_kE_{21}$ (Fig. 2.6). The peaks beyond the visible absorptions, at 3.5 eV, at 4.5 eV and at 4.4 eV come from the tubes (3,3), (4,2) and (5,1) due to the dipole transitions ${}_kG_2 \rightarrow {}_kG_2$, ${}_kE_3 \rightarrow {}_kE_3$, and ${}_kE_{31} \rightarrow {}_kE_{31}$, respectively. The tube (4,2) also absorbs at 1.9 eV and at 2.5 eV. These peaks are attributed to the transitions: ${}_kE_{-10} \rightarrow {}_kE_{-10}$ and ${}_kE_{-11} \rightarrow {}_kE_{-11}$ (Fig. 2.6b).

The absorption measurements of the tubes encased in a porous zeolite [47] show a sharp peak at 1.37 eV with a shoulder at 1.19 eV and two broad bands centered at 2.1 eV and 3.1 eV. Although zeolite $AlPO_4$ is transparent from the near infrared to the ultraviolet, it might affect the electro-optical properties of the NTs inside its channels. Recent full DFT calculations [49] show that generally, the AFI crystal has a real effect on the electronic structure and optical absorption spectra of the NTs inside its channels (some of the peaks were blue-shifted and some red-shifted after tube-crystal interaction was included). Also, many-electron effects are included only through the DFTB input data, thus not taking into account influence of the excitonic effects [51] on the optical properties of NTs. Hence, only a qualitative comparison of our results of the experiment can be made. We find the calculated intense lowest energy optical transition, due to (5,0), at 1 eV, while the shoulder below it can be attributed to the transition in (5,1) (Note that the diameter of the (5,1) tube in the ideal cylindrical structure is 4.36 Å.). The first broad band can be interpreted by a convolution of the calculated absorption peaks: 1.9 eV, 2.5 eV of (4,2), 2 eV of (5,1) and 2.4 eV of (5,0). The second broad band might correspond to the convolution of the (3,3) and (5,1) tube optical response at 3.5 eV.

Table 2.3. Representations of the acoustic (T, L and W for transversal, longitudinal and twisting), radial-breathing (R), high-energy (H), IR and Raman active modes of SWNTs. IR and Raman activities are indicated by R and I, with the superscripts showing corresponding polarization of incident (and scattered) light. Contribution to the symmetric and antisymmetric part of the Raman tensor is emphasized by [R] and {R}. The total number of IR and Raman active-modes assigned by the representation is on the right.

Chiral					Zig-zag					Armchair				
${}_0A_0^+$	R H		$[\mathrm{R}]^{\|\|}\,[\mathrm{R}]^{\perp\perp}$	3	${}_0A_0^+$	R H		$[\mathrm{R}]^{\|\|}\,[\mathrm{R}]^{\perp\perp}$	2	${}_0A_0^+$	R H		$[\mathrm{R}]^{\|\|}\,[\mathrm{R}]^{\perp\perp}$	2
${}_0A_0^-$	L W	$I^{\|}$	$\{\mathrm{R}\}^{\perp\perp}$	1	${}_0A_0^-$	L	$I^{\|}$		1	${}_0A_0^-$	L	$I^{\|}$		1
					${}_0B_0^+$	W				${}_0B_0^+$	W		$\{\mathrm{R}\}^{\perp\perp}$	1
${}_0E_1$	T	$I^{\perp}$	$\mathrm{R}^{\|\perp}\,\mathrm{R}^{\perp\|}$	5	${}_0E_1^+$	T	$I^{\perp}$		2	${}_0E_1^+$	T	$I^{\perp}$		3
					${}_0E_1^-$			$\mathrm{R}^{\|\perp}\,\mathrm{R}^{\perp\|}$	3	${}_0E_1^-$			$\mathrm{R}^{\|\perp}\,\mathrm{R}^{\perp\|}$	2
${}_0E_2$			$[\mathrm{R}]^{\perp\perp}$	6	${}_0E_2^+$			$[\mathrm{R}]^{\perp\perp}$	3	${}_0E_2^+$			$[\mathrm{R}]^{\perp\perp}$	4

2.5 Phonons

Along the lines prescribed in Sect. 2.3.1 the full symmetry implemented calculation of phonon dispersions within described dynamical models is carried out for single- and double-wall NTs of various chiralities and diameters. This enables one to study various related properties, among which sound velocity, specific heat and optical scattering are selected for their comparability with standard experiments. In general, in each of these analyses the underlying physical process singles out some characteristic vibrational modes.

2.5.1 Infinite SWNTs

The dynamical matrix $\mathcal{D}$ is constructed from the adjusted graphene force constants as it discussed in Sect. 2.3.5. It takes the role of the Hamiltonian, and within MGPT it is easily diagonalized, due to the fact that the symcell contains a single atom, making the dimension of the eigenproblem at most 12. The symmetry is in this case represented by the dynamical representation D; its decomposition into irreducible components [7] shows the number of modes assigned by any set of quantum numbers. This gives a quick, although very rough, insight to Raman [53], infra red (IR) or other properties. In Table 2.3 the quantum numbers of the relevant modes are given as the labels of the corresponding irreducible representations.

Acoustic Branches: Sound Velocity and Heat Capacity

The acoustic modes, being rigid body translations and coaxial twisting of SWNT, transform like polar vector and axial angular momentum under the symmetry operations. Longitudinal acoustic (LA) and twisting (TW) modes are nondegenerate, with the same momenta quantum numbers $k = m = 0$

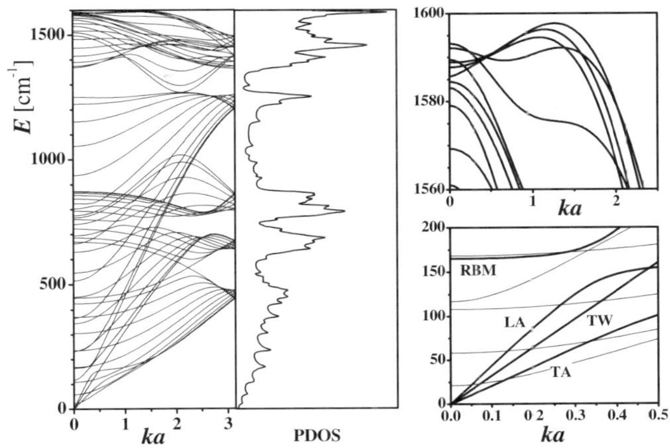

Fig. 2.12. Left: phonon dispersions and the corresponding density of states of the (5,5) tube. Right: high energy bands with overbending (upper panel) and, in the lower panel, acoustic and radial breathing branches in the Γ point region

and U parity $\Pi_U = -1$; they differ only in the achiral tubes due to the mirror planes: $\Pi_\mathrm{h}^\mathrm{LA} = -\Pi_\mathrm{h}^\mathrm{TW} = -1$ and $\Pi_\mathrm{v}^\mathrm{LA} = -\Pi_\mathrm{v}^\mathrm{TW} = +1$. On the other hand, the x and y components, combined by the rotations of SWNT, give two degenerate transversal acoustic (TA) modes, characterized by $k = 0$. Instead of the Cartesian components, SAB is formed by the linear combinations corresponding to $m = \pm 1$: e.g. for linear momentum one gets $p_\pm = (p_x \mp \mathrm{i}\, p_y)/2$. As for the parities, $\Pi_U = \Pi_\mathrm{v} = 0$ and $\Pi_\mathrm{h} = +1$.

All the acoustic branches are linear in k near the Γ point (Fig. 2.12), with the almost tube independent slopes equal to the sound velocities [31, 32, 35]: $v_\mathrm{TA} = 9.41\,\mathrm{km/s}$, $v_\mathrm{LA} = 20.37\,\mathrm{km/s}$, $v_\mathrm{TW} = 14.98\,\mathrm{km/s}$. The analogy with the graphene TA branch is used in the literature [54, 55] to propose quadratic k-dependence of the nanotube TA branch. To clarify the question connected to the specific heat, the graphene and SWNT acoustic branches are interrelated. Close to the Γ point the TA graphene acoustic branch is quadratic, while the other two (degenerate) are linear in the wave vector, as clearly explained in [56]. In the folding procedure, the degenerated in-plane translations (graphene LA modes along and perpendicular to the chiral vector) become TW and LA acoustic modes of a SWNT. Simultaneously, the graphene TA mode obviously turns into the homogeneous oscillations of the tube diameter, i.e. the radial breathing (RB) mode. Therefore, by analogy with the graphene, in the nanotube one expects linear LA and TW and a quadratic RB branch. Indeed, as discussed in Sect. 2.3.2, the totally symmetric RB mode (thus $\Pi_U = +1$) must be of Van Hove type singularity for all the SWNTs types, which implies that the corresponding band is quadratic in k near $k = 0$ (see Fig. 2.12). Note that two degenerate nanotube TA modes are not related to the graphene acoustic modes.

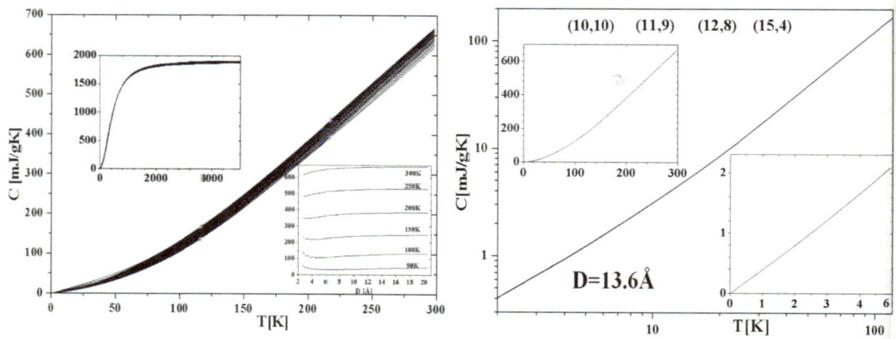

Fig. 2.13. Left: heat capacity C_{ph} dependence on temperature for tubes with diameters D from 2.8 Å to 20.5 Å (225 tubes from (2,2) to (24,4)). The broadening of the curve is due to the variations in C_{ph} with the tube diameters being depicted for various temperatures in the upper inset. The high temperature limit is visible in the lower inset. Right: heat capacities for four tubes with the same diameter $D = 13.6$ Å coincide at all temperatures: the central (log-log plot) and upper panels show that at low temperatures C_{ph} is linear in T, the overall behavior is in the lower panel

Using phonon dispersions and densities of phonon states [57, 58], the specific heat of SWNTs may be calculated. Theoretical predictions of the specific heat of some particular isolated SWNT [54, 57, 59] are in reasonably good agreement with the heat capacity measurements on nanotube bundles [58, 60, 61]. On the other hand, a fine dependence of this quantity on the nanotube parameters has been found [31] by calculations on many SWNTs in a wide range of diameters and chiralities, Fig. 2.13.

Below the temperature $T_{\text{o}} = \hbar\omega^{\text{o}}_{\text{min}}/6k_B$, where $\omega^{\text{o}}_{\text{min}}$ is the minimal frequency of optical modes, only the acoustic modes contribute to the SWNT specific heat [54, 58]. It is found that T_{o} depends on the tube diameter (in Å) as $T_{\text{o}} = (7.2 + 0.05D + 1.045D^2)^{-1}10^3$ K. The lowest optical branch is assigned by $m = 2$. It starts at $k = 0$ with Raman active modes and attains its minimum $\omega^{\text{c}}_{\text{min}}$ at k_{o}, which rapidly decreases with the tube diameter and depends on chirality (e.g. for achiral tubes $k_{\text{o}} = 0$). At low temperatures specific heat is linear in T; the temperature range of the linear regime gets narrower when the tube diameter increases, yielding continuous crossover to the graphene T^2 dependence.

In Fig. 2.13 the calculated specific heat temperature dependence up to 300 K for 225 SWNTs is presented. It nicely matches the measured values [60]. Notable broadening of the curve with the temperature indicates slight differences in C_{ph} of various tubes at higher temperatures. However, this weak diameter dependence is completely saturated for considerably thin tubes (upper inset). The high-temperature limit of $3k_B/m$ agrees reasonably well with the expected value for the carbon systems [58, 60]. The chirality shows no signature in the specific heat of the tubes, as clearly illustrated in the right panel: the tubes (10,10), (12,8), (15,4) and (11,9), with the same diameter $D = 13.6\,\text{Å}$ but different chiral angle, have the same specific heat.

Infra Red Activity

Infra red activity is determined by the exciting phonons in the electrical field. Within the dipole approximation [62], the perturbation operator is reduced to the projection $\boldsymbol{e} \cdot \boldsymbol{p}$ of the total ionic momentum to the polarization vector $\boldsymbol{e}$ of the incoming electric field. Consequently, linear momenta quantum numbers single out IR active modes. As discussed above $k = 0$, while other quantum numbers depend on the field polarization: parallel to the tube axis field (involving p_z momentum) $m = 0$, $\Pi_U = \Pi_\mathrm{h} = -1$, $\Pi_\mathrm{v} = +1$, and for the perpendicular linear and circular polarizations (relating p_x and p_y and the standard components $p_\pm$) $m = 1$, $\Pi_U = \Pi_\mathrm{v} = 0$, $\Pi_\mathrm{h} = +1$. Excluding acoustic modes, for chiral tubes there are six IR active modes, out of which only one with $m = 0$ is active under the parallel polarization of the light and all the others ($m = 1$) under the perpendicular (either linear or circular) polarization of the light. It is important to note that the armchair configuration has no active modes for the parallel polarization, unlike the zig-zag case where there is one such mode. Perpendicularly (or xy-circularly) polarized light may activate three (two) modes in the armchair (zig-zag) tubes.

Raman Spectra

In the first order Raman scattering in the incoming field $\boldsymbol{E}_\mathrm{i}$, exciting electrons induces in the target polarization $\boldsymbol{P}$ described by the Raman (polarizability) tensor R as $\boldsymbol{P} = \mathsf{R}\boldsymbol{E}_\mathrm{i}$. This way the ionic system is affected indirectly, i.e. phonons are created at the cost of the partial deexcitation of the electrons. The residual excitation energy is emitted as the scattered field $\boldsymbol{E}_\mathrm{s}$. The transition probabilities are determined by the component $\mathsf{R}_\mathrm{is} = \boldsymbol{e}_\mathrm{i}\mathsf{R}\boldsymbol{e}_\mathrm{s}$ of the Raman tensor corresponding to the incoming and scattered light polarizations. In general, as a second rank polar tensor, these components carry the quantum numbers $k = 0$, $m = 0, \pm 1, \pm 2$, and only the corresponding phonons participate in the scattering. Which of them will be excited depends on the polarizations $\boldsymbol{e}_\mathrm{i}$ and $\boldsymbol{e}_\mathrm{s}$, as discussed below and outlined in Table 2.3.

The most relevant symmetric Raman tensor $[\mathsf{R}]$ besides the isotropic (totally symmetric) component with $m = 0$, $\Pi_U = \Pi_\mathrm{v} = \Pi_\mathrm{h} = +1$ includes anisotropic components with $m = 1$, $\Pi_U = \Pi_\mathrm{v} = 0$, $\Pi_\mathrm{h} = -1$ and $m = 2$, $\Pi_U = \Pi_\mathrm{v} = 0$, $\Pi_\mathrm{h} = +1$. Totally symmetric phonons are activated if both polarizations are parallel, $\boldsymbol{e}_\mathrm{i} = \boldsymbol{e}_\mathrm{s}$. These modes are one radial breathing, and two (chiral tubes) or one (achiral) high energy (HE) modes. Anisotropic $m = 1$ modes are excited for the crossed polarizations, $\boldsymbol{e}_\mathrm{i} \perp \boldsymbol{e}_\mathrm{s}$, and $m = 2$ phonons are created if the polarizations are perpendicular mutually (or left-

and right-hand circular) and to the tube axis. Altogether, this gives 14 for chiral tubes and 8 for achiral tubes with Raman active modes [7,53].

For the chiral systems [63], as well as in the resonant scattering, which is known to be important in SWNTs [64], the antisymmetric part {R} of the Raman tensor may be important. Only one of its components is specifically antisymmetric (Table 2.3), with quantum numbers differing from those listed for [R]; even this component is absent for zig-zag tubes. It is characterized by $m = 0$, $\Pi_U = -1$ for chiral tubes, and additionally by $\Pi_\mathrm{v} = -1$, $\Pi_\mathrm{h} = +1$ for armchair tubes. The modes are activated by mutually parallel polarizations perpendicular to the tube. The components with $m = 1$, $\Pi_U = \Pi_\mathrm{v} = 0$, $\Pi_\mathrm{h} = -1$ are excited for the crossed polarizations, as well as for the symmetric components with the same assignation.

In the Raman measurements [52] on carbon nanotubes the most emphasized peaks are observed at the frequencies of the totally symmetric vibrations: besides the low energy (up to 700 cm^{-1}) radial breathing mode, also a high energy (below 1600 cm^{-1}) one is activated. Except for ultra narrow tubes, RB frequency is chirality independent, but directly reflects the diameter of the tube, which is widely used in the sample characterization by Raman measurements. The fit of the results for 1280 tubes gives (for ω in cm^{-1} and D in Å):

$$\omega_{\mathrm{RB}} = 2243/D; \tag{2.37}$$

Some deviations from this law are noted for ultra narrow tubes. Two HE modes of chiral tubes have frequencies depending both on the chirality and diameter:

$$\omega_{\mathrm{HE}}^{\mathcal{Z}} = 1588 + \frac{59.8\cos 6\theta}{D^2}, \quad \omega_{\mathrm{HE}}^{\mathcal{A}} = 1588 - \frac{757.6\cos\theta}{D^2} - \frac{1069.9\cos\theta}{D^4}. \tag{2.38}$$

In achiral tubes there is a single totally symmetric high-energy mode (Table 2.3). Their frequencies are also given by (2.38): while $\omega_{\mathrm{HE}}^{\mathcal{Z}}$ gives the zig-zag HE mode for the chiral angle $\theta = 0$, the armchair HE mode is obtained from $\omega_{\mathrm{HE}}^{\mathcal{A}}$ for $\theta = 30°$.

Usually it is assumed that the RB mode has only radial components, while the HE mode is circumferal in armchair tubes and longitudinal in zig-zag tubes (thus, in chiral tubes one is considered as circumferal and the other one as longitudinal). However, symmetry in general forbids only circumferal components in zig-zag tubes and longitudinal ones in the armchair cases. This is confirmed by precise numerical calculations [31]: the RB small longitudinal component depends on the chiral angle θ and decreases with the tube diameter D. Indeed, in the armchair tubes, as the atoms are in the horizontal mirror planes, σ_h even (parity +) and odd modes (−) have vanishing longitudinal and horizontal components, respectively, i.e. vibrations are either perpendicular or parallel to the tube axis. Analogously, in zig-zag tubes the

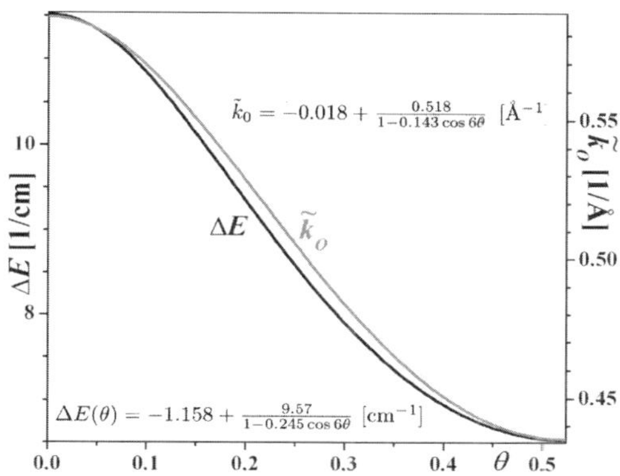

Fig. 2.14. Overbending ΔE and its maximum position $\tilde{k}_0$ as a function of the chiral angle θ for the 754 SWNTs (among 1280 tubes with diameters within 2.8–50.0 Å, those with maximal D for given θ are selected). The data are precisely fitted by the given expressions

atoms are in the vertical mirror planes; thus the σ_v-even (A) displacements are in these planes, while the odd ones (B) are circumferal. In addition to these symmetry constraints, HE mode(s) are perpendicular to the RB modes. Note that $\Pi_\mathrm{h} = \pm 1$ characterizes all $k = 0$ modes, while $\Pi_\mathrm{v} = \pm 1$ is realized along the whole bands with $m = 0, n$.

In contrast to almost all covalent solids, the graphite phonon branch corresponding to the longitudinal high energy optic mode at the Γ point has a local minimum. This feature is usually referred to as "overbending", since the local maxima appear outside the Γ point inside Brillouin zone. The analogous property of the SWNT phonon dispersions [31] is essential for the theoretical interpretation [64] of the dependence of the first-order Raman peak's position and double-peak structure of the high-energy mode ($\approx 1600\,\mathrm{cm}^{-1}$) on the laser excitation energy. Namely, these features are attributed to the double-resonant process coming from the phonon modes out of the Γ point. In the view of the generally quite weak dispersion around $k = 0$, the excitation energy dependence is expected to be very sensitive to the vibrational spectra details. In particular, the accurate theoretical investigation may help the experimental sample characterization of the SWNTs. Indeed, from the Raman measurements the slope of the excitation energy dependence of the high-energy mode frequency can be obtained, while the quantum theory of the double resonant scattering relates this dependence to the overbending position k_0, its absolute value $\Delta E = E(k_0) - E(0)$ and slope, and the phonon eigenvectors (which can be easily calculated for any SWNT).

The overbending is analyzed (Fig. 2.14) for a large number of SWNTs of all chiralities and diameters up to 50 Å. The overbending maximum is found for the band assigned by $\tilde{m} = 0$, at the helical wave vector $\tilde{k}_0$ depending on the tube chirality. Quantum numbers k and m are readily found by (2.23) for each SWNT, despite somewhat cumbersome general expression; e.g. only for the achiral tubes $m = 0$. The absolute value ΔE is strongly chirality dependent, while for a fixed chiral angle, although it decreases with the diameter, it rapidly saturates for the very thin tubes, showing no further diameter dependence. Thus, the maximal overbending of $11.5\,\mathrm{cm}^{-1}$ is in the zig-zag tubes and the minimal ($6.50\,\mathrm{cm}^{-1}$) in the armchair tubes.

2.5.2 Commensurate Double-Wall Nanotubes

The Dynamical Matrix and Its Reduction

Counting firstly the atoms of the wall W, and after that the atoms of W′, the dynamical matrix obtains the block structure

$$\mathcal{D} = \mathcal{D}^0_{\mathrm{WW}'} + \mathcal{D}'_{\mathrm{WW}'}, \quad \mathcal{D}^0_{\mathrm{WW}'} = \begin{pmatrix} \mathcal{D}^0_{\mathrm{W}} & 0 \\ 0 & \mathcal{D}^0_{\mathrm{W}'} \end{pmatrix}, \quad \mathcal{D}'_{\mathrm{WW}'} = \begin{pmatrix} d_{\mathrm{WW}'} & \mathcal{D}_{\mathrm{WW}'} \\ \mathcal{D}_{\mathrm{W}'\mathrm{W}} & d_{\mathrm{WW}'} \end{pmatrix}. \tag{2.39}$$

Here, $\mathcal{D}_{\mathrm{WW}'}$ consists of the submatrices $\mathcal{D}_{\alpha\beta'}$ comprising the intralayer coupling, as well as the transposed matrix $\mathcal{D}_{\mathrm{W}'\mathrm{W}}$; $\mathcal{D}^0_{\mathrm{W}}$ and $\mathcal{D}^0_{\mathrm{W}'}$ are the dynamical matrices of the isolated walls. Finally, $d_{\mathrm{WW}'}$ (and analogously $d_{\mathrm{W}'\mathrm{W}}$) is the interlayer coupling dependent diagonal matrix; in the view of the translational sum rules for isolated and coupled walls, $\mathcal{D}^0_{\alpha\alpha} = -\sum_{\beta(\neq\alpha)} \mathcal{D}_{\alpha\beta}$ and $\mathcal{D}_{\alpha\alpha} = \mathcal{D}^0_{\alpha\alpha} - d_{\alpha\alpha}$, its elements are $d_{\alpha\alpha} = -\sum_{\beta'} \mathcal{D}_{\alpha\beta'}$.

Assuming the intralayer interaction unaffected, the same force constants model (Sect. 2.3.5) as for SWNTs is used. The interlayer submatrices are derived from the potential (2.14) in the Hessian form:

$$\mathcal{D}_{\alpha\beta'} = \frac{\partial^2 v(\boldsymbol{r}_\alpha - \boldsymbol{r}'_\beta)}{\partial x_i^\alpha \partial x_j^{\beta'}}. \tag{2.40}$$

The dynamical matrix is invariant under the symmetries of DWCN, which is used to facilitate the diagonalization. If DWCN is incommensurate, then it has point group symmetry, and its finiteness prevents efficient reduction; therefore, only finite tubes, with at most a few hundred atoms, can be numerically studied. On the contrary, infinite line groups enable consideration of perfect infinite commensurate DWCNs. Therefore, the results presented obtained by the methods described in Sect. 2.3.1 refer to CDWNTs. Recall that the symcell of the DWNT contains many atoms, in contrast to SWNTs generated by a single atom. This is a consequence of greatly reduced roto-translational symmetry (in comparison to SWNTs), and therefore the range of m usually consists only of the special values 0 and $Q/2$; the corresponding states are even or odd in z-reversal parities. As discussed in Sect. 2.3.2, this is manifested as a large number of singularities in the density of states (Fig. 2.15).

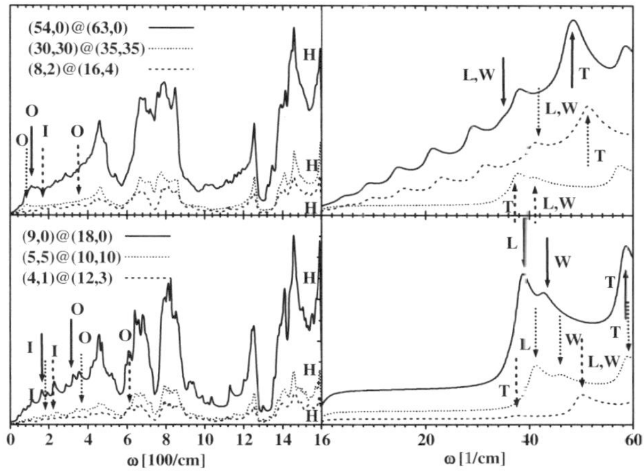

Fig. 2.15. Phonon density of states of CDWNTs with characteristic peaks L, T, W, I, O and H, denoting LRL, TRL, WRL, in- and out-of-phase BL and HE modes, respectively. Low frequency region is on the right

Perturbative Interpretation

One of the main common characteristics of CDWCN phonon bands is nicely illustrated by the example of (5,5)@(10,10) in Fig. 2.16; it is notable that the branches of W@W′ resemble the union of branches of W and W′, which would be obtained in the case of noninteracting walls This is a consequence of the considerably lower interlayer interaction (2.14), in comparison to the intralayer force constants field. Accordingly, the elements of submatrices $\mathcal{D}_{\alpha\beta'}$ are much less than those of $\mathcal{D}_{\alpha\beta}$ (unless $\alpha = \beta$), and may be considered as a perturbation to the dynamical matrix of the noninteracting walls. This invokes fruitful interpretation of CDWNT modes in terms of the modes of the isolated layers [54, 65].

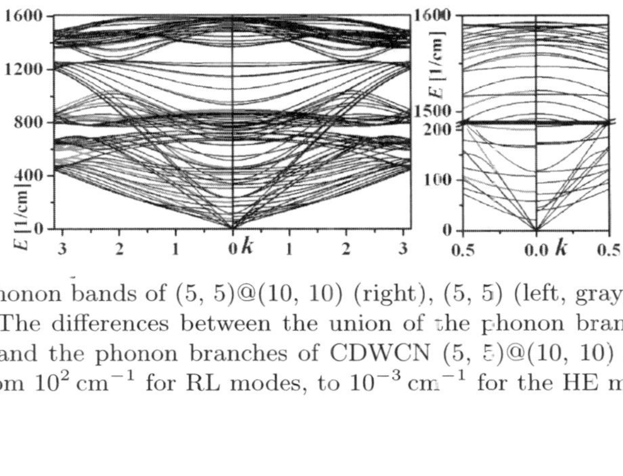

Fig. 2.16. Phonon bands of (5, 5)@(10, 10) (right), (5, 5) (left, gray) and (10, 10) (left, black). The differences between the union of the phonon branches of (5, 5) and (10, 10) and the phonon branches of CDWCN (5, 5)@(10, 10) decrease with frequency (from $10^2\,\mathrm{cm}^{-1}$ for RL modes, to $10^{-3}\,\mathrm{cm}^{-1}$ for the HE modes)

Obviously, in view of (2.39) the interlayer interaction matrix $\mathcal{D}'_{\mathrm{WW}'}$ is a perturbation of the noninteracting wall dynamical matrix $\mathcal{D}^0_{\mathrm{WW}'}$. In the basis $\{|\,a\rangle, |\,a'\rangle\}$ (a and a' enumerate modes of W and W′) of the normal modes of noninteracting walls $\mathcal{D}^0_{\mathrm{WW}'}$ is diagonal, with eigen energies ω_a^2 and $\omega_{a'}^2$, while the perturbation retains the same structure. However, the off-diagonal matrix elements $\kappa_{aa'} = \sqrt{m}\langle a\,|\,\mathcal{D}'_{\mathrm{WW}'}\,|\,a'\rangle\sqrt{m'}$ take the role of the coupling of the two oscillatory modes $|\,a\rangle$ and $|\,a'\rangle$. Recall that the normal mode basis is weighted by the square roots of the length mass densities (2.7) of the walls: $m = cD$ and $m' = cD'$. If modes $|\,a\rangle$ and $|\,a'\rangle$ are not coupled to the others, the corresponding two-dimensional submatrix of $\mathcal{D}_{\mathrm{WW}'}$ can be diagonalized independently of the rest of $\mathcal{D}_{\mathrm{WW}'}$. This submatrix is the dynamical matrix [65] of the two oscillators with frequencies ω_a^2 and $\omega_{a'}^2$, and masses m' and m, coupled by the oscillator with force constant $\kappa_{aa'}$:

$$\mathcal{D}_{aa'} = \begin{pmatrix} \omega_a^2 - \kappa_{aa'}/m & \kappa_{aa'}/\sqrt{mm'} \\ \kappa_{aa'}/\sqrt{mm'} & \omega_{a'}^2 - \kappa_{aa'}/m' \end{pmatrix} . \tag{2.41}$$

Hence, the modes of the two coupled isolated walls become two DWNT modes with frequencies

$$\Omega_\pm^2 = \frac{1}{2}\left[\omega_a^2 + \omega_{a'}^2 + \frac{\kappa_{a'a}}{\mu_+} \pm \sqrt{(\omega_a^2 - \omega_{a'}^2)^2 + 2\frac{\kappa_{a'a}}{\mu_-}(\omega_a^2 - \omega_{a'}^2) + \frac{\kappa_{a'a}^2}{\mu_+^2}}\right] , \tag{2.42}$$

where $\mu_\pm = m'm/(m \pm m')$. The higher frequency Ω_+ corresponds to the out-of-phase oscillations of the walls and Ω_- to the in-phase ones. Note again that this simple approach is plausible only for the pairs of modes negligibly mixed with other modes, and such pairs may be singled out by symmetry. In fact, the interaction potential is invariant under all the symmetries of both walls, as obvious from (2.11), and further elucidated in Sect. 2.6. Therefore, the modes with the same quantum numbers are coupled much more strongly than differently assigned modes. Accordingly, the perturbative interpretation is plausible for the pairs of equally assigned modes, with frequencies well separated from the other modes with the same quantum numbers. Further, as the interaction field is effectively averaged in the rapid oscillations and thus partly cancelled, the coupling of low frequency modes is in general more intensive.

Rigid Layer Modes

The acoustic modes are equally assigned in all achiral SWNTs by mutually different quantum numbers. Since their zero frequency is much less than the other modes of same assignation, the pairs of LA, TA and TW modes are suitable for perturbative analysis. Hence, substituting $\omega_a = \omega_{a'} = 0$ in (2.42) one finds

$$\Omega_- = 0, \quad \Omega_+(D) = \sqrt{\frac{2\kappa}{c}\frac{D+\Delta/2}{D(D+\Delta)}}, \tag{2.43}$$

i.e. one acoustic and one low frequency optic DWNT mode for each SWNT acoustic pair. In both modes the layers vibrate as rigid bodies, in-phase in the acoustic and out-of-phase in the optic mode. Therefore, the out-of-phase vibrations are rigid-layer modes (Fig. 2.15), transversal (TRL), longitudinal (LRL) and twisting (TWRL), in correspondence to their in-phase TA, LA and TW counterparts.

The branches starting at $k = 0$ with acoustic modes are twofold, except that the TA branch is fourfold for $\mathcal{ZZ}$ and $\mathcal{AA}$ CDWNTs (other tubes have two close TA branches). These bands are linear in k near $k = 0$, with the slope almost independent of the tube (the differences are less than 1%) and close to that for SWNTs, giving the sound velocities $v_{\mathrm{TA}} = 9.54\,\mathrm{km\ s}^{-1}$, $v_{\mathrm{LA}} = 20.64\,\mathrm{km\ s}^{-1}$ and $v_{\mathrm{TW}} = 15.18\,\mathrm{km\ s}^{-1}$.

The coupling constant κ is itself a function of the tube diameter and the interlayer distance. The number of interacting atoms per unit length increases with D, yielding an increase in κ as well. Proposing a polynomial function $\kappa(D)$, it turns out that for LRL and TWRL modes the first order polynomial fits the numerical data well [66], while for TRL modes a quadratic term must be added. Thus, with the convenient form $\kappa(D) = c\omega_\infty^2(\delta_0 + D + \gamma D^2)/2$, the rigid layer frequencies are:

$$\Omega_+(D) = \omega_\infty \sqrt{\frac{(D+\delta_0+\gamma D^2)(D+\Delta/2)}{D(D+\Delta)}}. \tag{2.44}$$

The model is clearly justified by Fig. 2.17 comparing the numerical results to the analytic form (2.44) with the parameters given in Table 2.4 (assuming

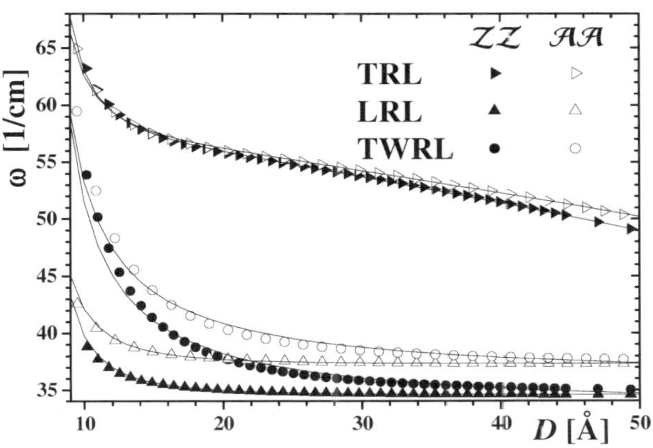

Fig. 2.17. Rigid layer mode frequencies vs. DWNT diameter. The solid lines are fitted curves

Table 2.4. Fitting parameters for RL modes (see (2.44) and Fig. 2.17).

	$(n,0)@(n+9,0)$, $\Delta=7\,\text{Å}$			$(n,n)@(n+5,n+5)$, $\Delta=6.78\,\text{Å}$		
	ω_∞	δ_0	γ	ω_∞	δ_0	γ
LRL	34.7	-4.00	0	37.4	-4.00	0
TWRL	33.8	1.33	0	36.0	1.33	0
TRL	62.5	-4.67	-0.0068	61.7	-4.39	-0.006

frequency in cm^{-1} and the interior-wall diameter in Å). Additional verification may be found in the large D limit, when the graphene values should be reproduced. Then LRL and TWRL modes become two degenerate A point ($k_z=\pi/c$, $k_x=k_y=0$) modes of graphite [67], describing rigid out-of-phase vibrations (in two perpendicular directions) of the adjacent graphite layers. Indeed, as $\gamma=0$, for both modes and for both types of achiral DWNTs $\Omega_+(\infty)=\omega_\infty=35\,\text{cm}^{-1}$, matching the singled out graphite frequency. Note that this matching is dominantly due to symmetry. As discussed in detail in Sect. 2.6, symmetry completely determines the patterns of wall-wall interaction, particularly the periodicity and this way its slopes in Φ and Z around stable configuration. This yields correct experimental frequencies for graphite as the large D DWNT limit, despite the fact that the Lenard-Jones potential used is assumed to be too smooth to describe directly out-of-phase rigid-layer modes of the adjacent graphite layers.

These results may be of experimental interest, since TRL and LRL modes are IR active (TWRL is also IR active for tubes with at least one chiral wall), and correspond to the density of state singularities (Fig. 2.15).

The rigid layer modes influence the low temperature heat capacity. In fact, if there were no interlayer interaction, the DWNT heat capacity would be equal to that of the walls, as an average of almost the same quantities. However, the coupling of the walls' acoustic modes (altogether eight modes) produces only four DWNT acoustic modes, while another four become optical. Therefore, the heat capacity is close to that of the layers, except in the low temperature region, contributed by the acoustic branches only, where it is significantly lowered (Fig. 2.18). This agrees nicely with the available experimental data [58, 68].

Breathing-Like and High-Energy Modes

The same arguments as for the acoustic modes of the layers justify perturbative interpretation for the totally symmetric ($k=m=0$, even in all parities) radial breathing and high-energy modes (Fig. 2.19), being mutually well separated by frequency. The frequencies of the resulting [65] in-phase and out-of-phase breathing-like (BL) modes are found by substituting frequencies (2.37) in (2.42). The numerical results [66], well fitted by $\kappa_{\text{BL}}=-246404+38799D$, are in accordance with earlier predictions [54] and match the experimen-

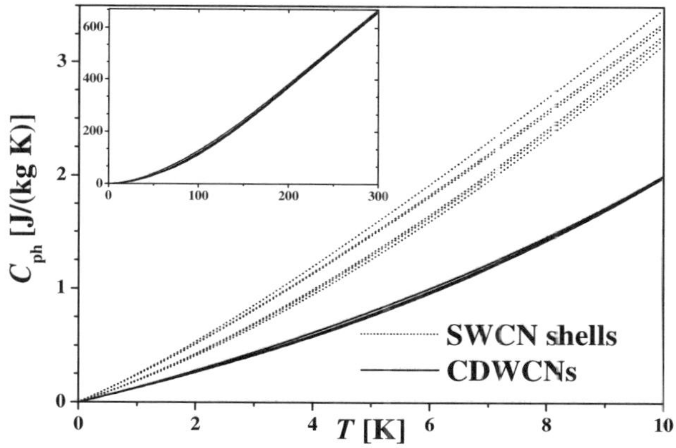

Fig. 2.18. Heat capacity of double wall tubes $\mathcal{AA}_5$, $\mathcal{ZZ}_9$ (8,2)@(16,4) and their SWNT constituents in the low temperature regime, and up to 300 K (inset)

tal results [69–71]. As for the high energy branches, it is known from measurements that they hardly differ from those of SWNT. Also, the difference not greater than $10^{-3}\,\mathrm{cm}^{-1}$ is found numerically. Fit by (2.42) with isolated wall frequencies (2.38) give coupling constants $\kappa_{\mathrm{HE}}^{\mathcal{ZZ}} = -25875 + 6528D$ and $\kappa_{\mathrm{HE}}^{\mathcal{AA}} = 9329 + 6307D$.

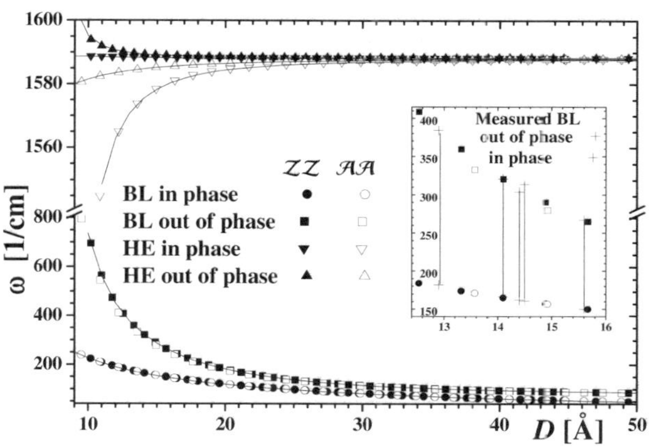

Fig. 2.19. Rigid layer (left) and breathing-like and high-energy (right) frequencies vs. DWNT diameter. The solid lines are fitted curves. Inset: comparison of the calculated and the experimental results [71]

2.6 Symmetry Breaks Friction: Super-Slippery Walls

The symmetry of double-wall nanotubes gives a profound insight to the interlayer interaction, as well as an efficient way to estimate it. Such analysis clarifies the main characteristics of this potential: while in general it is quite weak, a variety of the potential shapes may be created by the appropriate choice of the chirality and length of the walls. Therefore, DWNTs are perfect bearings, and attractive candidates for nano-electromechanical devices [14, 72].

2.6.1 Symmetry and Interaction

To enlighten symmetry properties of the interwall interaction potential $V(\Phi, Z)$, at first Eq. (2.11) is rewritten as the sum

$$V(\Phi, Z) = \sum_{t's'u'} V_{\text{in}}(\boldsymbol{r}'_{t's'u'}) \qquad (2.45)$$

of the energy of the outer wall atoms in the potential created by the inner wall:

$$V_{\text{in}}(\boldsymbol{r}) = \sum_{tsu} v(\boldsymbol{r}_{tsu}, \boldsymbol{r}). \qquad (2.46)$$

This form manifests that V_{in} is invariant under the inner tube symmetry group. Therefore it can be expanded over the basis of invariant functions:

$$V_{\text{in}}(\boldsymbol{r}) = \sum_{M \geq 0, \omega} \alpha_{\omega}^{M}(D') C_{\omega}^{M}(\varphi, z). \qquad (2.47)$$

Here, the basis consists of the functions $C_{\omega}^{M}(\varphi, z) = \cos(M\varphi + 2\pi\omega z)$; rotational and screw axis invariance, respectively, restricts [73] nonnegative integers M and real numbers ω by:

$$M = 0 \pmod{n}, \qquad Mr + \omega na = 0 \pmod{q}. \qquad (2.48)$$

Due to the summation in (2.45), the total potential $V(\Phi, Z)$ is also invariant under the outer wall symmetries, and therefore under all the transformations generated by the symmetries of both walls. Consequently, to $V(\Phi, Z)$ contribute only the terms of (2.47) subdued to the additional restrictions of the form (2.48), but with the outer wall parameters q', r', n' and a'. Taking this into account, the summation in (2.45) is easily performed; the total potential per atom of an infinitely long DWNT is:

$$v_{\infty}(\Phi, Z) = 2 \sum_{M \geq 0, \omega}'' \alpha_{|\omega}^{M} \cos(M\Phi + 2\pi\omega Z) \cos(M\varphi_{000} + 2\pi\omega z_{000}). \qquad (2.49)$$

The double prime indicates two pairs of restrictions (2.48), imposed by the roto-translational symmetries of the walls on M and ω.

Mutually commuting roto-translational transformations of the walls together generate the breaking group $\boldsymbol{L}_{\mathrm{B}}^{\mathrm{R}}$ [13]. Note that if walls did not interact, the system would be invariant under any pair of independently performed symmetries of the walls, i.e. under the product group $\boldsymbol{L}_{\mathrm{W}}^{\mathrm{R}} \times \boldsymbol{L}_{\mathrm{W}'}^{\mathrm{R}}$. Hence, the interaction itself imposes the symmetry breaking from the product to the intersection (2.12). Denoting by $|\boldsymbol{L}|$ the number of transformations in $\boldsymbol{L}$, the symmetries of the noninteracting and interacting walls are related by:

$$|\boldsymbol{L}_{\mathrm{WW}'}^{\mathrm{R}}||\boldsymbol{L}_{\mathrm{B}}^{\mathrm{R}}| = |\boldsymbol{L}_{\mathrm{W}}^{\mathrm{R}}||\boldsymbol{L}_{\mathrm{W}'}^{\mathrm{R}}|. \tag{2.50}$$

Thus, while the roto-translational symmetry of the DWNT is highly reduced in comparison to that of the walls, the breaking group, being the symmetry group of the interaction potential (2.49), is much greater. Particularly, the periods A_{B} and $2\pi/N_{\mathrm{B}}$ (along Z and Φ, respectively) of the interaction potential decrease with the periods A and $2\pi/N$ of DWNT. For the commensurate DWNTs, the breaking group is the line group [13]:

$$\boldsymbol{L}_{\mathrm{B}}^{\mathrm{R}} = \boldsymbol{T}_{Q_{\mathrm{B}}}^{R_{\mathrm{B}}}(A_{\mathrm{B}})\boldsymbol{C}_{N_{\mathrm{B}}}, \tag{2.51}$$

with the parameters (compare to (2.13) for Q, N and A):

$$N_{\mathrm{B}} = \frac{nn'}{N}\frac{\sqrt{\tilde{q}\tilde{q}'}}{\mathrm{GCD}(\frac{r'\hat{a}n - r\hat{a}'n'}{N}, \sqrt{\tilde{q}\tilde{q}'})}, \quad Q_{\mathrm{B}} = \frac{qq'}{\mathrm{GCD}(q,q')}, \quad A_{\mathrm{B}} = \frac{aa'}{AN}\frac{Q_{\mathrm{B}}Q}{qq'},$$

$$\frac{R_{\mathrm{B}}}{\mathrm{GCD}(q,q')} = \frac{(r'\hat{a}q - r\hat{a}'q')\hat{a'}^{\phi(\hat{a})} + rq'\hat{a'}}{\hat{a}\hat{a}'} \quad (\mathrm{mod}\ \mathrm{GCD}(N\tilde{q}\tilde{q}', r'\hat{a}q - r\hat{a'}q')).$$

By convention, R_{B} is the (unique) solution of the last equation, which is coprime with Q_{B} and less than Q_{B}. In the incommensurate case the breaking group is:

$$\boldsymbol{L}_{\mathrm{B}}^{\mathrm{R}} = \boldsymbol{C}_{Q_{\mathrm{B}}}\boldsymbol{T}(0). \tag{2.52}$$

Here, $\boldsymbol{T}(0)$ is the translational group generated by the incommensurate periods a and a', i.e. including the translations for $ta + t'a'$ for all the integers t and t'. This is the quasi-continual group, as for any real x and arbitrary small ε, there are t and t' such that $|ta + t'a' - x| \leq \varepsilon$. Therefore, the invariance of $V(\Phi, Z)$ under the group $\boldsymbol{T}(0)$, in view with its physically indispensable continuity, implies that it is constant along Z. In fact, incommensurate tubes can be viewed as the limiting case: $A = \infty$, thus $A_{\mathrm{B}} = 0$, while due to the Z independence, the (fractional) translations do not affect the potential and Q_{B} takes the role of N_{B}.

It is clear from the definition of the breaking group that two pairs (2.48) (one for each wall) of conditions on M and ω are equivalent to the single such pair with the breaking group parameters Q_{B}, R_{B}, N_{B} and A_{B}. As the breaking group is quite large with respect to the single-wall symmetry, this implies that most of the terms in (2.46) do not contribute to the interaction $V(\Phi, Z)$, meaning that the inter-shell friction is in general quite low. In addition, the rotational and translational breaking independently give rough insight to the interaction potential: large N_B leaves only high M terms in (2.49), as well as small A_B allows only large ω terms. Together with the natural assumption that the amplitudes $\alpha^M_{|\omega|}$ rapidly decrease with M and ω (being justified numerically [73]), this means that large rotational and translational breaking are manifested as slow varying potential along Φ and Z, respectively. Numerical calculations [14, 15, 74, 75] and recent experiments [76, 77] confirm this theoretical prediction; particularly, the static friction has been estimated [77] to be lower than $6 \cdot 10^{-15}\,\mathrm{N/Å^2}$.

As remarked in Sect. 2.2.2, most DWNTs are incommensurate (even 99.5% of the ones studied), with Z independent interaction potential. Also in commensurate cases breaking is large, and despite several highly symmetric CDWNTs, potential is almost constant along one of the coordinates. Therefore, for nanomechanical applications it is interesting to consider DWNTs with one finite layer, providing additional variety of potential patterns. The summation over the atoms in (2.11) is slightly modified to include only m outer wall monomers (the outer wall length is $n'a'(m-1)/q'$): $t' = 0, 1, \ldots, m-1$, while all possible values of u' and s' are allowed as before. After performing this summation, the total interaction per atom is:

$$v_m(\Phi, Z) = 2 \sum_{M \geq 0, \omega}'' \alpha^M_\omega \cos(M\Phi + 2\pi\omega Z)$$

$$\times \cos(M\varphi_0 + 2\pi\omega z_0) \frac{\sin(\pi m \frac{r'M + n'a'\omega}{q'})}{m \sin(\pi \frac{r'M + n'a'\omega}{q'})}. \qquad (2.53)$$

Now the double prime restricts the summation to the solutions of the system:

$$M = 0 \pmod{n}; \quad M = 0 \pmod{n'}; \quad Mr + \omega n a = 0 \pmod{q}. \qquad (2.54)$$

The horizontal U axis of the receiver wall is in the middle of the ring with $\Phi_m = \Phi + \frac{2\pi r'}{q'}\frac{m-1}{2}$ and $Z_m = Z + \frac{n'a'}{q'}\frac{m-1}{2}$ (the infinite tube U axis at (Φ, Z) is no longer a symmetry, as the monomers are above it only). In comparison to the infinite case, there is no restriction imposed on M and ω by the other wall screw-axis. Consequently, the potential $v_m(\Phi, Z)$ can be viewed as $v_\infty(\Phi, Z)$ superposed to $v^{\mathrm{fin}}_m(\Phi, Z)$ comprising the damped (factor $1/m$) oscillations. Also, a resonant effect is observed from (2.53): whenever m is a multiple of $q'A/n'a'$ the dumping term vanishes.

2.6.2 Numerical Results

Several pairwise potentials are used in the literature for calculating the interlayer interaction [14, 15, 74, 75]. This has caused quantitative differences of the results: e.g. for the energy barrier of the relative sliding of the walls in (5,5)@(10,10) Charlier and Michenaud got 0.52 meV/atom (longitudinal) and 0.23 meV/atom (circumferal), and Palser 0.295 meV/atom and 0.085 meV/atom. Nevertheless, there is agreement that the relative sliding of two nested tubes is easy.

Here we discuss potential barriers calculated with the pairwise potential of the Van der Waals type (2.14). Except for a few highly symmetric achiral DWNTs, qualitatively the same results are obtained with another potential [15, 73] taking care of π-bonding. The amplitudes α_ω^M are derived by fast Fourier transform of the potential V_{in}. The interior tube is, long at least 100 Å (at least 40 elementary cells); the potential is scanned on the grid of 41×41 points for φ and z at the radius of the outer-wall. The constant term α_0^0 is set to zero while the numerical error of the calculated interaction potential is estimated to be 10^{-10} meV.

The patterns shown in Fig. 2.20 illustrate the variety of potential shapes of DWNTs with infinite walls. There are basically two different patterns of the interaction potential. Firstly, "horizontal stripes" appear in all the studied commensurate DWNT with achiral walls, with the exception of (5,5)@(10,10); in fact, due to the high incompatibility of the rotational symmetries of the layers with larger diameters, the calculated potential is almost Φ independent. Secondly, "vertical stripes" enabling super-slippery sliding along Z are typical for the incommensurate tubes. A large number of commensurate tubes interact negligibly within the numerical precision, due to the symmetries of the highly incompatible layers, which is manifested as a perfectly flat potential surface.

Additional potential shapes appear in DWNTs with a finite outer wall (Fig. 2.21). Note that, depending on the chirality of the layers, the easiest sliding is allowed along helices with angles varying from $-90°$ to $90°$. Also, there are sharply defined minima or maxima. From (2.53) it is clear that a change in the length of the outer wall may result in a different pattern, due to m-dependance of $v_m^{\text{fin}}(\Phi, Z)$. Also, the larger barrier reported [14] for the tubes (5,5)@(15,4) and (9,0)@(15,4) can be easily explained: the tube

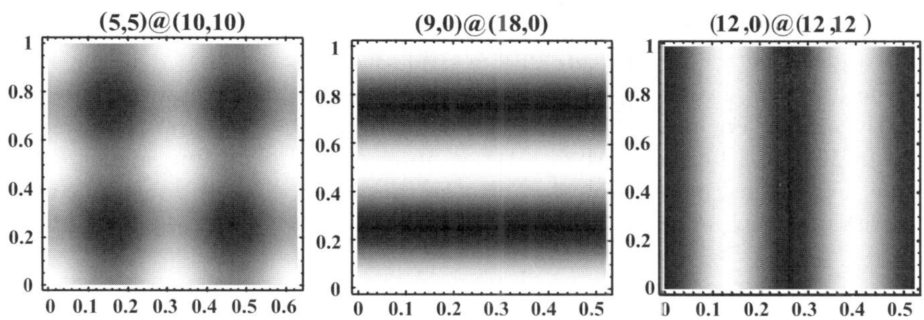

Fig. 2.20. Density plot of $V(\Phi, Z)$ for infinite (5,5)@(10,10), (9,0)@(18,0) and (12,0)@(12,12) (Φ and Z/a are along the vertical horizontal and vertical axes)

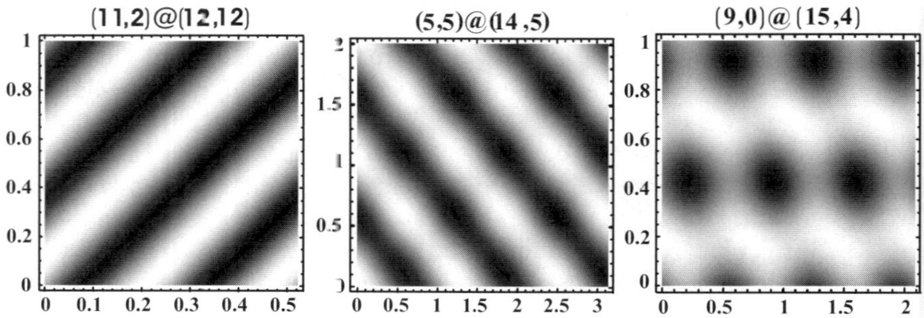

Fig. 2.21. Density plot of $V(\Phi, Z)$ for (11,2)@(12,12), (5,5)@(14,5) and (9.0)@(14,5), with the outer walls of 40, 194 and 301 monomers, respectively

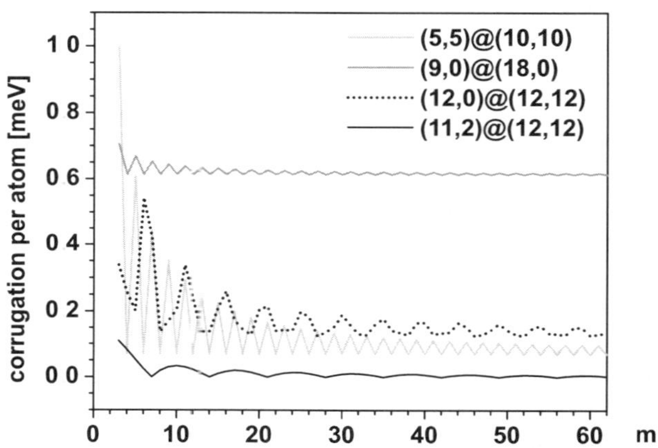

Fig. 2.22. Corrugation per atom as a function of m for several DWNTs with the outer wall made of m monomers

(15, 4), without rotational symmetry, does not restrict the values of M, and consequently $v_m^{\mathrm{fir}}(\Phi, Z)$ contains all the $\omega \neq 0$ terms, including those with significant amplitudes α_ω^M.

To get a better insight into the possible "low energy cost motions" the corrugation for several tubes is calculated (Fig. 2.22). The resonant effect is clearly visible. For the (11, 2)@(12, 12) tube $q'A/n'a' = 14$, and whenever m is a multiple of 14 the corrugation per atom drops to its limiting $m = \infty$ value. For achiral commensurate DWNTs this effect is more pronounced, as it occurs for each even m.

References

1. S. Iijima: Nature, **354**, 56 (1991)
2. A. Thess, R. Lee, P. Nikolaev, H. Dai, P.Petit, J. Robert, C.H. Xu, Y.H. Lee, S.G. Kim, A.G. Rinzler, D.T. Colbert, G.E. Scuseria, D. Tomanek, J.E. Fischer and R.E. Smally: Science **273**, 483 (1996)
   C. Journet, W.K. Maser, P. Bernier and A. Loiseau: Nature **388**, 756 (1997)
3. M.S. Dresselhaus, G. Dresselhaus and R. Saito: Phys. Rev. B **45**, 6234 (1992)
4. N. Hamada, S. Sawada and A. Oshiyama: Phys. Rev. Lett. **68**, 1579 (1992)
5. C.T. White, D.H. Robertson and J.W. Mintmire: Phys. Rev. B **47**, 5485 (1993)
   J.W. Mintmire and C.T. White, Phys. Rev. Lett. **81**, 2506 (1998)
6. R.A. Jishi, L. Venkataraman, M.S. Dresselhaus and G. Dresselhaus: Phys. Rev. B **16**, 11176 (1995)
7. M. Damnjanović, I. Milošević, T. Vuković and R. Sredanović: Phys. Rev. B **60**, 2728–2739 (1999)
8. M. Vujičić, I. Božović and F. Herbut: J. Phys. A **10**, 1271 (1997)
9. I. Milošević and M. Damnjanović: Phys. Rev. B **47**, 7805 (1993)
10. H. Abud and G. Sartori: Ann. Phys. **150**, 307 (1983)
11. I. Cabria, W. Mintmire and C.T. White: Phys. Rev. B **67**, 121406(R) (2003)
12. Y.E. Lozovik, A.M. Popov and A.V. Belikov: Phys. Sol. State **45**, 1396 (2003)
13. M. Damnjanović, I. Milošević, E. Dobardžić, T. Vuković and B. Nikolić: J. Phys. A **36**, 10349–10360 (2003)
14. R. Saito, R. Matsuo, T. Kimura, G. Dresselhaus G and M.S. Dresselhaus: Chem. Phys. Lett. **348**, 187 (2001)
15. A. Kolmogorov, V. Crespi: Phys. Rev. Lett. **85**, 4727 (2000)
16. E.P. Wigner: *Group Theory and its Applications to the Quantum Mechanics of Atomic Spectra* (Academic Press, New York 1959)
17. M. Damnjanović and I. Milošević: J. Phys. A **28**, 1669–79 (1995)
18. M. Damnjanović, T. Vuković and I. Milošević: J. Phys A **33**, 6561–72 (2000)
19. I. Milošević, A. Damjanović, and M. Damnjanović: Ch. XIV. In: *Quantum Mechanical Simulation Methods in Studying Biological Systems*, ed by D. Bicout and M. Field (Springer-Verlag, Berlin, Heidelberg, New York 1996)
20. M. Damnjanović, I. Milošević, T. Vuković and J. Maultzsch: J. Phys. A **36**, 5707–17 (2003)
21. T. Vuković, I. Milošević and M. Damnjanović: Phys. Rev. B **65**, 045418 (2002)
22. C. Kane and E.J. Mele: Phys. Rev. Lett. **78**, 1932 (1997)
23. M. Damnjanović, T. Vuković and I. Milošević: Solid State Comm. **116**, 265 (2000)
24. S. Tasaki, K. Maekawa and T. Yamabe: Phys. Rev. B **57**, 9301 (1998)
25. S. Reich, J. Maultzsch and C. Thomsen: Phys. Rev. B **66**, 035412 (2002)
26. M. Machón, S. Reich, C. Thomsen, D. Sánchez-Portal, P. Ordejón: Phys. Rev. B **66**, 155410 (2002)
27. H.J. Liu and C.T. Chan: Phys. Rev. B **66**, 115416 (2002)
28. D. Porezag, Th. Frauenheim, Th. Köhler, G. Seifert and R. Kaschner: Phys. Rev. B **51**, 12947 (1995)
29. E. Schröder and P. Hyldgaard: Surface Sci. **532-5**, 880 (2003)
30. O. Dubay and G. Kresse: Phys. Rev. B **67**, 035401 (2003)
31. E. Dobardžić, I. Milošević, B. Nikolić, T. Vuković and M. Damnjanović: Phys. Rev. B **68**, 045408 (2003)

32. R.A. Jishi, L. Venkataraman, M.S. Dresselhaus and G. Dresselhaus: Chem. Phys. Lett. **209**, 77 (1993)
33. M. Born and K. Huang: *Dynamical Theory of Crystal Lattices* (Oxford University Press, Oxford, 1954)
34. O. Madelung: *Solid State Theory*, (Springer-Verlag, Berlin, Heidelberg, New York 1978)
35. R. Saito, T. Takeya, T. Kimura, G. Dresselhaus and M.S. Dresselhaus: Phys. Rev. B **57**, 4145 (1998)
36. O. Jost, A.A. Gorbunov W. Pompe, T. Pichler, R. Friedlein, M. Knupfer, M. Reibold, H.-D. Bauer, L Dunsch, M.S. Golden and J. Fink: Appl. Phys. Lett. **75**, 2217 (1999);
S. Kazaoui, N. Minami, R. Jacquemin, H. Kataura and Y. Achiba: Phys. Rev. B **60**, 13339 (1999);
X. Liu, T. Pichler, M. Knupfer, M.S. Golden, J. Fink, H. Kataura and Y. Achiba: Phys. Rev. B **66**, 045411 (2002)
37. H. Kuzmany, W. Plank, M. Hulman, Ch. Kramberger, A. Grüneis, Th. Pichler, H. Peterlik, H. Kataura and Y. Achiba: Eur. Phys. J. B **22**, 307 (2001)
38. H. Ajiki and T. Ando: Physica B **201**, 349 (1994)
39. W.A. de Heer, W.S. Bacsa, A. Chatelain, T. Gerfin, R. Humphrey-Baker, L. Forro and D. Ugarte: Science **268**, 845 (1995)
40. F. Bommeli, L. Degioergi, P. Wachter, W.S. Bacsa, W.A. de Heer and L. Forro: Solid State Commun. **99**, 513 (1996)
41. H.H. Gommans, J.W. Alldredge, H. Tashiro, J. Park, J. Magnuson and A.G. Rinzler: J. Appl. Phys. **88**, 2509 (2000);
J. Hwang, H.H. Gommans, A. Ugawa, H. Tashiro, R. Haggenmueller, K.I. Winey, J.E. Fischer, D.B. Tanner and A.G. Rinzler: Phys. Rev. B **62**, R13310 (2000);
S. Reich, C. Thomsen, G.S. Duesberg and S.Roth: Phys. Rev. B **63**, 041401(R) (2001)
42. M.F. Lin: Phys. Rev. B **62**, 13153 (2000)
43. I. Božović, N. Božović and M. Damnjanović: Phys. Rev. B **62**, 6971 (2000)
44. I. Milošević, T. Vuković, S. Dmitrović and M. Damnjanović: Phys. Rev. B **67**, 165418 (2003)
45. S. Sawada and N. Hamada: Solid State Commun. **83**, 917 (1992)
46. Z.K. Tang, H.D. Sun, J. Wang, J. Chen and G. Li: Appl. Phys. Lett. **73**, 2287 (1998)
N. Wang, Z.K. Tang, G.D. Li and J.S. Chen: Nature (London) **408**, 50 (2000)
47. Z.M. Li, Z.K. Tang, H.J. Liu, N. Wang, C.T. Chan, R. Saito, S. Okada, G.D. Li, J.S. Chen, N. Nagasawa and S. Tsuda: Phys. Rev. Lett. **87**, 127401 (2001)
48. Y.F. Chan, H.Y. Peng, Z.K. Tang and N. Wang: Chem. Phys. Lett. **369**, 541 (2003)
49. X.P. Yang, H.M. Weng and J. Dong: Eur. Phys. J. B **32**, 345 (2003)
50. I. Milošević, B. Nikolić and M. Damnjanović: Phys. Rev. B **69**, 113408 (2004)
51. S.M. Bachilo, M.S. Strano, C. Kittrell, R.H. Hauge, R.E. Smalley, R.B. Weisman: Science **298** 2361 (2002)
C.L. Kane and E.J. Mele: Phys. Rev. Lett. **90** 207401 (2003)
C.D. Spataru, S. Ismail-Beigi, L.X. Benedict and S.G. Louie: cond-mat/0310220v1

52. L.X. Benedict, S.G. Louie and M.I. Cohen: Phys. Rev. B **52**, 8541 (1995); G.S. Duesberg, I. Loa, M. Burghard, K. Syassen, S. Roth: Phys. Rev. Lett. **85**, 5436 (2000);
A. Jorio, A.G. Souza Filho, V.W. Brar, A.K. Swann, M.S. Ünlü, B.B. Goldberg, A. Right, J.H. Hafner, C.M. Lieber, R. Saito, G. Dresselhaus, M.S. Dresselhaus: Phys. Rev. B **65**, 121402(R) (2002)
53. O.E. Alon: Phys. Rev. B. **63**, 201403(R) (2001)
54. V.N. Popov and L. Henrard: Phys. Rev. B **65**, 235415 (2002)
55. G.D. Mahan: Phys. Rev. B. **65**, 235402 (2002)
56. R. Saito, G. Dresselhaus, M. Dresselhaus: *Physical Properties of Carbon Nanotubes* (Imp. College Press, London 1998)
57. L.X. Benedict, S.G. Louie and M.L. Kohen: Solid State Commun. **100**, 177 (1996)
58. A. Mizel, L.X. Benedict, L.M. Cohen, S.G. Louie, A. Zettl, N.K. Budraa and W.P. Beyermann: Phys. Rev. B **60**, 3264 (1999)
59. A. Charlier, E. McRae, M.-F. Charlier, A. Spire and S. Forster: Phys. Rev. B **57**, 6689 (1998)
60. J. Hone, B. Batlogg, Z. Benes, A.T. Johnson, J.E. Fischer: Sciense **289**, 1730 (2000)
61. J.C. Lasjaunias, K. Biljaković, Z. Benes, J.E. Fischer and P. Monceau: Phys. Rev. B **65**, 113409 (2002)
62. J.L. Birman: *Theory of Crystal Space Groups and Infra-Red and Raman Lattice Processes of Insulating Crystals*, (Springer-Verlag, Berlin, Heidelberg, New York 1974)
63. D.A. Long: *Raman Spectroscopy* (McGRaw-Hill, New York 1977)
64. C. Thomsen and S. Reich: Phys. Rev. Lett. **85**, 5214 (2000)
J. Maultzsch, S. Reich and C. Thomsen: Phys. Rev. B **64** 121407 (2001); **65**, 233402 (2002)
65. E. Dobardžić, J. Maultzsch, I. Milošević, C. Thomsen and M. Damnjanović: Phys. Status Solidi b **237**, R7 (2003)
66. E. Dobardžić, I. Milošević, T. Vuković, B. Nikolić and M. Damnjanović: Eur. Phys. J. B **34**, 409 (2003)
M. Damnjanović, E. Dobardžić, I. Milošević, T. Vuković, B. Nikolić: New Journal of Physics **5**, (1) 1–15 (2003)
M. Damnjanović, I. Milošević, E. Dobardžić, T. Vuković, B. Nikolić, Phys. Rev. B **69** 153401 (2004)
67. A. Jishi and G. Dresselhaus: Phys. Rev. B **26**, 4514 (1982)
68. V.N. Popov: Phys. Rev. B **66**, 153408 (2002)
69. S. Bandow, M. Takizawa, K. Hirahara, M. Yudasaka and S. Iijima: Chem. Phys. Lett. **337**, 48 (2001)
70. J.M. Benoit, J.P. Buisson, O. Chauvet, C. Godon and S. Lefrant: Phys. Rev. B **66**, 073417 (2002)
71. S. Bandow, G. Chen, G.U. Sumanasekera, R. Gupta, M. Yudasaka, S. Iijima and P.C. Eklund: Phys. Rev. B **66**, 075416 (2002)
72. Y.E. Lozovik, A.V. Minogin, A.M. Popov: Phys. Lett. A **313**, 112 (2003)
73. M. Damnjanović, T. Vuković and I. Milošević: Eur. Phys. J. B **25**, 131 (2002)
T. Vuković, M. Damnjanović and I. Milošević: Physica E **16**, 259 (2003)
74. J.C. Charlier and J.P. Michenaud: Phys. Rev. Lett. **70**, 1858 (1993)
75. A.H.R. Palser: Phys. Chem. Chem. Phys. **1** 4459 (1999)

76. J. Cumings and A. Zettl: Science **289**, 602 (2000)
77. M. Yu, B. Yakobson and R. Ruoff: J. Phys. Chem. B **104**, 8764 (2000)

3 Elastic Continuum Models of Phonons in Carbon Nanotubes

A. Raichura, M. Dutta, and M.A. Stroscio

In this chapter, elastic continuum models are used to describe phonons in carbon nanotubes and vibrational modes in biological structures. Based on elastic continuum theory, acoustic vibrational modes are modeled for both zigzag and armchair nanotubes of finite length using a variational solution of Donnell's equation. The acoustic phonon modes in these calculations are determined for both even and odd modes of the acoustic displacement. The dispersion relations vary with the length of the tube. The displacement field of the nanotube is used to calculate the deformation potential interaction Hamiltonian. In addition, the optical vibrational modes are derived for finite length nanotubes in the elastic continuum approximation. A quantum mechanical normalization prescription is applied to facilitate the determination of the optical phonon modes. The dispersion relation is calculated based on the continuum approach and the quantum normalized amplitude is used to calculate the optical deformation potential. These fully three-dimensional elastic continuum models are compared with other approaches to modeling phonons in carbon nanotubes such as the popular zone-folding technique. In a related underlying topic, the applicability of continuum models for the analysis of nanoscale structures is demonstrated for the case fullerenes. It is shown that the b_2 elongation mode of C_{60} may be described within the continuum approximation. Indeed, for these fullerene structures, the frequencies of selected vibrational modes are predicted to within a few percent.

3.1 Introduction

Since their discovery by Iijima in 1991, carbon nanotubes have been shown to have numerous electronic and optical applications [1–4]. Electronic transport in nanotubes has acquired a lot of attention recently, the quantitative understanding of which requires a detailed knowledge of both the structure and the vibrations of the tube. The phonon modes of finite-length nanotubes are of considerable interest due to their potential applications in carbon nanotube (CNT) metal-insulator-semiconductor (MIS) capacitors [5], and carbon nanotube field effect transistors [6, 7]. In all these devices, the ends of the nanotubes are in contact with the leads creating electrical and mechanical constraints. The typical length of the nanotubes used in these applications is about 2 nm to 20 nm. These nanotubes can be modeled as finite length nanotubes with suitable boundary conditions at their ends. Herein, analytical continuum modes are presented for both acoustic and optical phonon modes in single-wall nanotubes (SWNTs) and multi-wall nanotubes (MWNTs). Results presented in this article include the dispersion relations, mode amplitudes and interaction Hamiltonians for CNTs of finite length.

3.2 Acoustic Modes in Single Wall Nanotubes

Various models have been used to study the mechanics of single-wall nanotubes (SWNTs) The strong similarity of the chemistry of carbon nanotubes to graphite allows theoretical analyses to be done based on the empirical methodologies imported from studies on graphite. They range from direct zone-folding [8] of the results for graphene to the quantum-mechanical studies based on the tight binding Hamiltonians fitted to graphene properties [9,10]. Effective interatomic potentials [11], force-constant models [12] and *ab initio* calculations [13] have also been performed. The results obtained using these different techniques vary from the qualitative picture offered by zone folding, with intrinsic deficiencies for low frequencies, to the very quantitative results of the *ab intio* calculations.

For small systems such as nano spheres or tubes, continuum models are known to yield a useful approximate description and provide valuable insight. Continuum models have been applied successfully to study embedded nanoscale structures in one and two dimensions such as heterostructures [14]. They have also been applied to nonembedded nanoscale structures. For example, the continuum model has been used in the study of the radial breathing mode frequencies of nanospheres, including the effects of damping due to environmental interactions [15–17]. Also, the model has been applied to the study of both the normal modes and the energetics of fullerenes [18,19]. In such continuum models, the nanoscale properties of a material are frequently described with adequate accuracy in terms of bulk material properties and the elastic properties of a dimensionally confined structure.

In the elastic continuum model of a SWNT, the SWNT is represented as an elastic cylindrical membrane of finite thickness. The vibrational equations for the SWNT are solved using Donnell's equation of motion of a cylindrical membrane and using a variational solution of Rayleigh-Ritz method to solve Donnell's equation [20].

3.2.1 Model

We begin by taking the Donnell's equation of motion for a thin elastic cylindrical shell to be given by:

$$\frac{\partial^2 u_x}{\partial x^2} + \frac{1-v}{2a^2}\frac{\partial^2 u_x}{\partial \theta^2} + \frac{1+v}{2a}\frac{\partial^2 u_\theta}{\partial x \partial \theta} + \frac{v}{a}\frac{\partial w}{\partial x} - \frac{1-v^2}{E}\rho\frac{\partial^2 u_x}{\partial t^2} = 0 \quad (3.1)$$

$$\frac{1+v}{2a}\frac{\partial^2 u_x}{\partial x \partial \theta} + \frac{1-v}{2a}\frac{\partial^2 u_x}{\partial \theta^2} + \frac{1}{a^2}\frac{\partial w}{\partial x} - \frac{1-v^2}{E}\rho\frac{\partial^2 u_\theta}{\partial t^2} = 0$$

$$\frac{v}{a}\frac{\partial u_x}{\partial x} + \frac{1}{a^2}\frac{\partial u_\theta}{\partial \theta} + \frac{w}{a^2} + \frac{h^2}{12}\nabla^4 w + \frac{1-v^2}{E}\rho\frac{\partial^2 v}{\partial t^2} = 0$$

where

u_x = displacement in the axial direction
u_θ = displacement in the circumferential direction
w = displacement in the radial direction
v = Poisson ratio (for graphene, $v = 0.145$)
ρ = density (for graphene, $= 2.27\,\mathrm{gm/cm^3}$)
E = Young's modulus (for graphene, E = 1.060 TPa)
a = radius of the nanotube
h = thickness of the nanotube.

We divide our solution into two parts – odd and even. For the odd part, the phonon displacement in the three directions can be assumed to be:

$$u_x = A\left[-\sin\frac{\mu}{a}\left(\frac{l}{2}-x\right) + k\sinh\frac{\mu}{a}\left(\frac{l}{2}-x\right)\right]\cos(n\theta)\cos(\omega t) \quad (3.2)$$

$$u_\theta = B\left[\cos\frac{\mu}{a}\left(\frac{l}{2}-x\right) + k\cosh\frac{\mu}{a}\left(\frac{l}{2}-x\right)\right]\sin(n\theta)\cos(\omega t)$$

$$w = C\left[\cos\frac{\mu}{a}\left(\frac{l}{2}-x\right) + k\cosh\frac{\mu}{a}\left(\frac{l}{2}-x\right)\right]\cos(n\theta)\cos(\omega t)$$

where A, B and C are constants.

As mentioned earlier, the tubes being clamped, hence the equivalent boundary conditions are taken to be

$$u_x = u_\theta = w = \frac{\partial w}{\partial x} = 0 \text{ at } x = 0 \text{ and } x = l \quad (3.3)$$

Substituting the above boundary condition in the assumed displacements of (3.2), we get

$$k = \frac{\sin\left(\dfrac{\mu l}{2a}\right)}{\sinh\left(\dfrac{\mu l}{2a}\right)} \quad (3.4)$$

$$\tan\left(\frac{\mu l}{2a}\right) + \tanh\left(\frac{\mu l}{2a}\right) = 0 \quad (3.5)$$

which has the roots $\frac{\mu l}{a} = 1.506\pi, 3.5\pi, 5.5\pi, \ldots$ corresponding to 1, 3, 5, ... axial half waves.

Substituting the assumed displacements of (3.2) into the Donnell's equation and carrying out the Rayleigh Ritz analysis [20], we get

$$\frac{A}{C} = \frac{v\mu\left(n^2 + \frac{1-v}{2}\mu^2\frac{\theta_2}{\theta_1} - \Delta\frac{\theta_2}{\theta_1}\right) - \left(\frac{1+v}{2}\right)^2\mu^2 n^2}{\left(\mu^2\frac{\theta_1}{\theta 2} + \frac{1-v}{2}n^2 - \Delta\right)\left(n^2 + \frac{1-v}{2}\mu^2\frac{\theta_2}{\theta_1} - \Delta\right) - \left(\frac{1+v}{2}\right)^2\mu^2 n^2\frac{\theta_2}{\theta_1}} \tag{3.6}$$

$$\frac{B}{C} = \frac{v\mu^2 n\frac{1-v}{2}\frac{\theta_2}{\theta_1} + n\left(\mu^2\frac{\theta_1}{\theta 2} + \frac{1-v}{2}n^2 - \Delta\right)}{\left(n^2 + \frac{1-v}{2}\mu^2\frac{\theta_2}{\theta_1} - \Delta\right)\left(\mu^2\frac{\theta_1}{\theta 2} + \frac{1-v}{2}n^2 - \Delta\right) - \left(\frac{1+v}{2}\right)^2\mu^2 n^2\frac{\theta_2}{\theta_1}} \tag{3.7}$$

These classical modes can be quantized by making the amplitude of the phonon displacement equal to that given in [21]

$$\frac{1}{V}\int (u_x u_x^* + u_\theta u_\theta^* + ww^*)\, dV = \frac{\hbar}{2M\omega} \tag{3.8}$$

where M is the mass of the two carbon atoms in the unit cell of the graphene sheet which constitutes the wall of the carbon nanotube. Substituting the assumed displacements into this quantization condition, it follows that,

$$C = \frac{\pi l\hbar}{2\pi M\omega}\frac{1}{\left(\frac{A}{C}\right)^2 I_+ + \left[\left(\frac{B}{C}\right)^2 + 1\right] I_-} \tag{3.9}$$

$$\begin{aligned} I_\pm = \mp\frac{\pi\exp\left(-\frac{\mu l}{a}\right)}{4\pi}&[4\cos\left(\frac{\mu l}{2a}\right)\sin\left(\frac{\mu l}{2a}\right)a\exp\left(\frac{\mu l}{a}\right) \\ &+4ka\cos\left(\frac{\mu l}{2a}\right)\exp\left(\frac{\mu l}{2a}\right) \mp k^2 a\exp\left(\frac{2\mu l}{a}\right) \mp 2\mu l\exp\left(\frac{\mu l}{a}\right) \\ &+2k^2\mu l\exp\left(\frac{\mu l}{a}\right) \mp 4ka\cos\left(\frac{\mu l}{2a}\right)\exp\left(\frac{3\mu l}{2a}\right) \\ &\pm 4ka\sin\left(\frac{\mu l}{2a}\right)\exp\left(\frac{\mu l}{2a}\right) \pm k^2 a \pm 4ka\sin\left(\frac{3\mu l}{2a}\right)] \end{aligned} \tag{3.10}$$

Similarly, for the even modes, we assume the displacements to be

$$u_x = A\left[-\cos\frac{\mu}{a}\left(\frac{l}{2}-x\right)+k\cosh\frac{\mu}{a}\left(\frac{l}{2}-x\right)\right]\cos(n\theta)\cos(\omega t) \quad (3.11)$$

$$u_\theta = B\left[\sin\frac{\mu}{a}\left(\frac{l}{2}-x\right)+k\sinh\frac{\mu}{a}\left(\frac{l}{2}-x\right)\right]\sin(n\theta)\cos(\omega t)$$

$$w = C\left[\sin\frac{\mu}{a}\left(\frac{l}{2}-x\right)-k\sinh\frac{\mu}{a}\left(\frac{l}{2}-x\right)\right]\cos(n\theta)\cos(\omega t)$$

where A, B and C are constants.

Substituting the above displacements into the boundary conditions of (3.3) we get,

$$k = \frac{\cos\left(\frac{\mu l}{2a}\right)}{\cosh\left(\frac{\mu l}{2a}\right)} \quad (3.12)$$

$$\tan\left(\frac{\mu l}{2a}\right) - \tanh\left(\frac{\mu l}{2a}\right) = 0 \quad (3.13)$$

which has the roots $\frac{\mu l}{a} = 2.5\pi, 4.5\pi, 6.5\pi, \ldots$ corresponding to 2, 4, 6, ... axial half waves.

Following the same procedure used for odd modes we obtain,

$$\frac{A}{C} = \frac{v\mu\left(n^2+\frac{1-v}{2}\mu^2\frac{\theta_2}{\theta_1}-\Delta\frac{\theta_2}{\theta_1}\right)-\left(\frac{1+v}{2}\right)^2\mu^2n^2}{\left(\mu^2\frac{\theta_1}{\theta 2}+\frac{1-v}{2}n^2-\Delta\right)\left(n^2+\frac{1-v}{2}\mu^2\frac{\theta_2}{\theta_1}-\Delta\right)-\left(\frac{1+v}{2}\right)^2\mu^2n^2\frac{\theta_2}{\theta_1}} \quad (3.14)$$

$$\frac{B}{C} = \frac{v\mu^2 n\frac{1+v}{2}\frac{\theta_2}{\theta_1}-n\left(\mu^2\frac{\theta_1}{\theta 2}+\frac{1-v}{2}n^2-\Delta\right)}{\left(n^2+\frac{1-v}{2}\mu^2\frac{\theta_2}{\theta_1}-\Delta\right)\left(\mu^2\frac{\theta_1}{\theta 2}+\frac{1-v}{2}n^2-\Delta\right)-\left(\frac{1+v}{2}\right)^2\mu^2n^2\frac{\theta_2}{\theta_1}} \quad (3.15)$$

and

$$C = \frac{\pi l h}{2\pi M\omega}\frac{1}{\left(\frac{A}{C}\right)^2 I_+ + \left[\left(\frac{B}{C}\right)^2+1\right]I_-} \quad (3.16)$$

$$
\begin{aligned}
I_{\pm} = \pm \frac{\pi \exp\left(-\frac{\mu l}{a}\right)}{4\pi} & [4 \cos\left(\frac{\mu l}{2a}\right) \sin\left(\frac{\mu l}{2a}\right) a \exp\left(\frac{\mu l}{a}\right) \\
& + 4ka \cos\left(\frac{\mu l}{2a}\right) \exp\left(\frac{\mu l}{2a}\right) \mp k^2 a \exp\left(\frac{2\mu l}{a}\right) \mp 2\mu l \exp\left(\frac{\mu l}{a}\right) \\
& + 2k^2 \mu l \exp\left(\frac{\mu l}{a}\right) \mp 4ka \cos\left(\frac{\mu l}{2a}\right) \exp\left(\frac{3\mu l}{2a}\right) \\
& \pm 4ka \sin\left(\frac{\mu l}{2a}\right) \exp\left(\frac{\mu l}{2a}\right) \pm k^2 a \pm 4ka \sin\left(\frac{3\mu l}{2a}\right)]
\end{aligned}
\tag{3.17}
$$

3.2.2 Dispersion Curves

The frequencies of the lowest breathing modes for the (10,0), (10,10) and (40,0) carbon nanotubes are shown in Figs. 3.1–3.6 below. As expected, the dispersion relation is discrete as a result of axial dimensional confinement. The values obtained for the (10,10) nanotube agree well those given by Suzurra and Ando [22] for small values of the wave vector. The effects of finite nanotube length are evident in the breathing modes at small wave vectors.

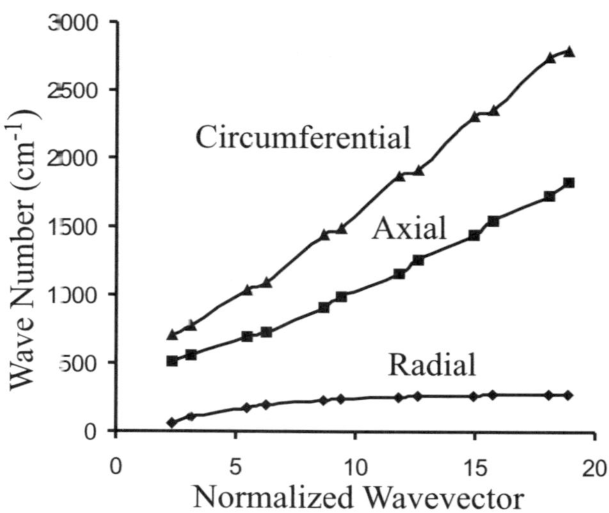

Fig. 3.1. Dispersion curve for (10,0) nanotube with a length of 16 Angstroms

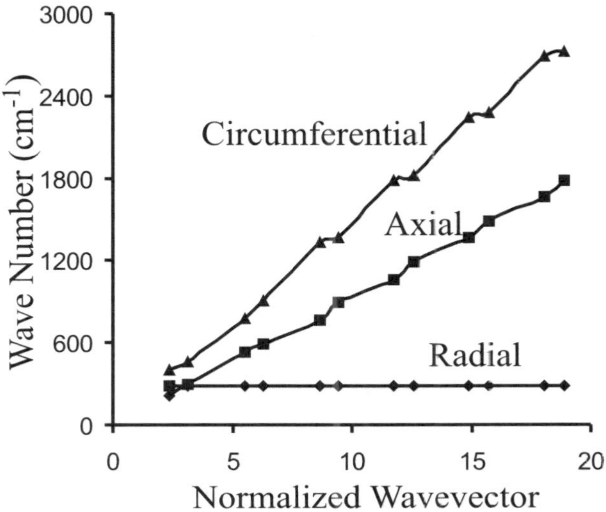

Fig. 3.2. Dispersion curve for (10,0) nanotube with a length of 80 Angstroms

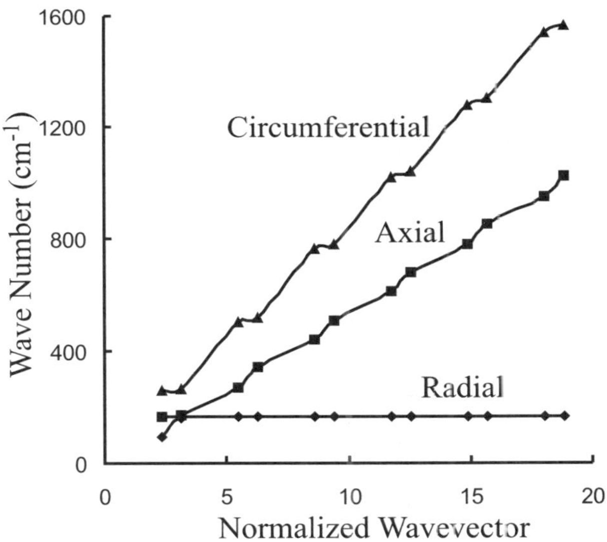

Fig. 3.3. Dispersion curve for (10,10) nanotube with a length of 28 Angstroms

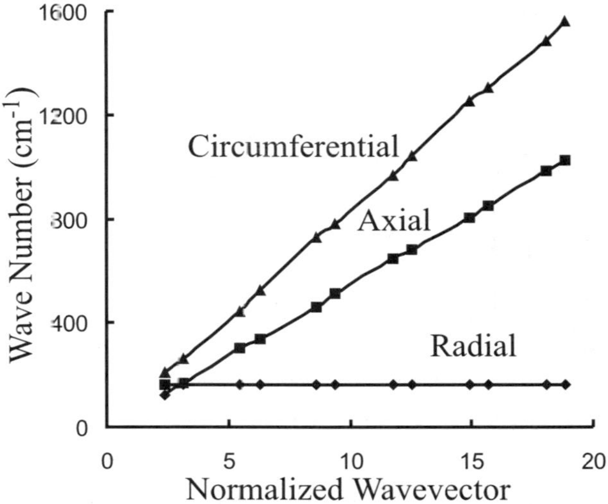

Fig. 3.4. Dispersion curve for (10,10) nanotube with a length of 70 Angstroms

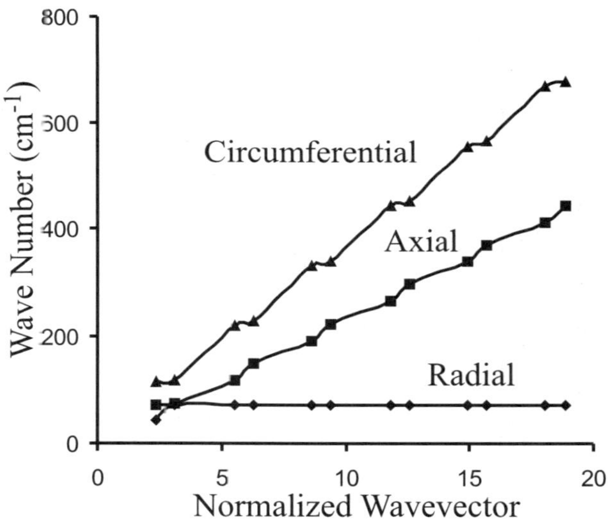

Fig. 3.5. Dispersion curve for (40,0) nanotube with a length of 64 Angstroms

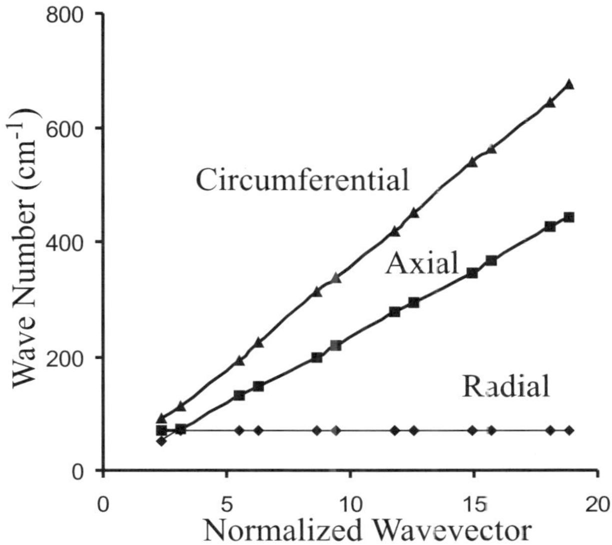

Fig. 3.6. Dispersion curve for (40,0) nanotube with a length of 160 Angstroms

3.2.3 Deformation Potential

The divergence of the displacement may be calculated as

1. Odd Case

$$\begin{aligned}\nabla.u = &\frac{\mu A}{a}\left[\cos\frac{\mu}{a}\left(\frac{l}{2}-x\right)-k\cosh\frac{\mu}{a}\left(\frac{l}{2}-x\right)\right]\cos(n\theta)\\ &+\frac{nB}{r}\left[\cos\frac{\mu}{a}\left(\frac{l}{2}-x\right)+k\cosh\frac{\mu}{a}\left(\frac{l}{2}-x\right)\right]\cos(n\theta)\\ &+\frac{C}{r}\left[\cos\frac{\mu}{a}\left(\frac{l}{2}-x\right)+k\cosh\frac{\mu}{a}\left(\frac{l}{2}-x\right)\right]\cos(n\theta)\end{aligned} \tag{3.18}$$

2. Even Case

$$\begin{aligned}\nabla.u = &\frac{\mu A}{a}\left[-\sin\frac{\mu}{a}\left(\frac{l}{2}-x\right)-k\sinh\frac{\mu}{a}\left(\frac{l}{2}-x\right)\cos(n\theta)\right]\\ &+\frac{nB}{r}\left[\sin\frac{\mu}{a}\left(\frac{l}{2}-x\right)-k\sinh\frac{\mu}{a}\left(\frac{l}{2}-x\right)\cos(n\theta)\right]\\ &+\frac{C}{r}\left[\sin\frac{\mu}{a}\left(\frac{l}{2}-x\right)-k\sinh\frac{\mu}{a}\left(\frac{l}{2}-x\right)\cos(n\theta)\right]\end{aligned} \tag{3.19}$$

The divergence relation can be used to calculate the deformation potential of the tube. The deformation potential interaction Hamiltonian is then given by the Bardeen ansatz:

$$H_{def} = D_{ac} \nabla \cdot u \tag{3.20}$$

where D_{ac} is the deformation potential constant whose value is 24 eV [22,23].

Figures 3.7–3.10 represent the deformation potential corresponding to a tube of radius 0.7 nm and having one half, two halves, three halves and four halves axial waves respectively. As seen, the deformation potential is maximum at the center of the tube for tubes having odd number of axial waves. The relatively small values of the deformation potential at the ends of the CNT may be one of the reasons for the observed failure of the CNTs near the center of the CNT [24]. Figures 3.11–3.14 depict the deformation potential of the tube of radius 1.6 nm. The carrier interactions are determined by the deformation potential and it is clear from the present model of finite-length CNTs that the mechanical perturbations along the length of the nanotube play a major role in determining the relative strength of the interaction at different axial locations along the length of the CNT.

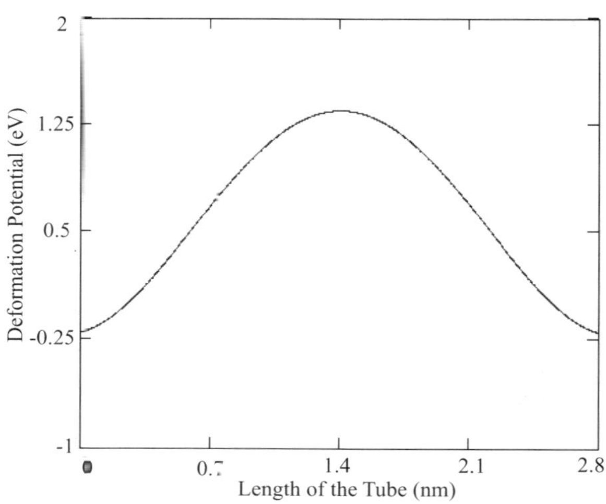

Fig. 3.7. Deformation potential for (10,10) nanotube with a length of 28 Angstroms and m = 1

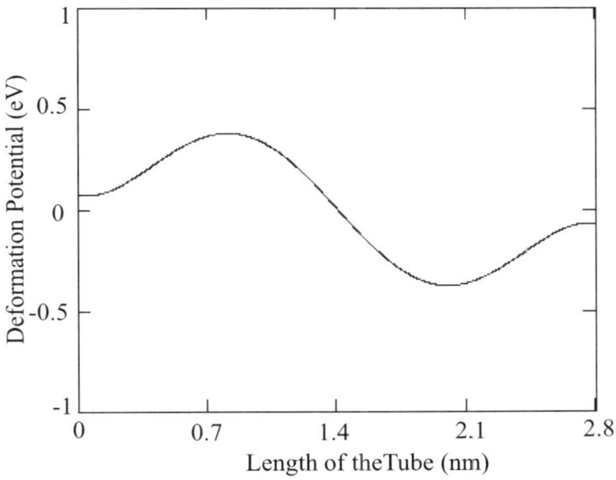

Fig. 3.8. Deformation potential for (10,10) nanotube with a length of 28 Angstroms and m = 2

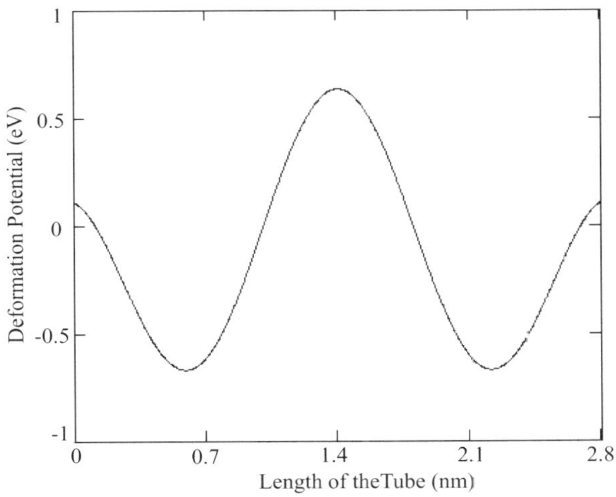

Fig. 3.9. Deformation potential for (10,10) nanotube with a length of 28 Angstroms and m = 3

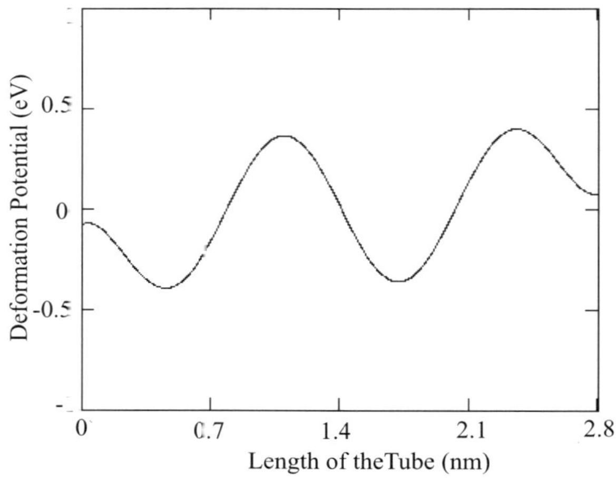

Fig. 3.10. Deformation potential for (10,10) nanotube with a length of 28 Angstroms and m = 4

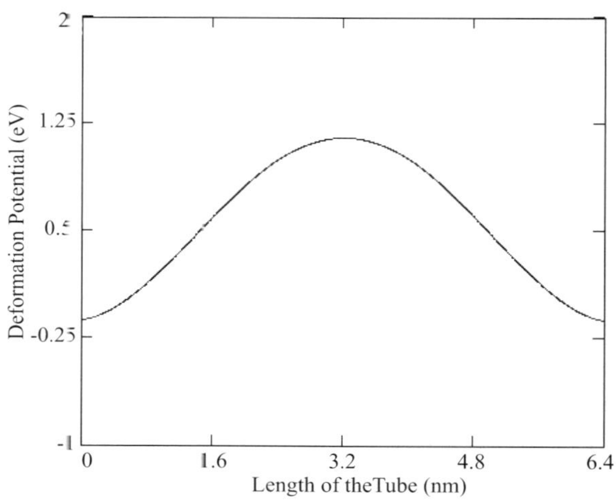

Fig. 3.11. Deformation potential for (40,0) nanotube with a length of 64 Angstroms and m = 1

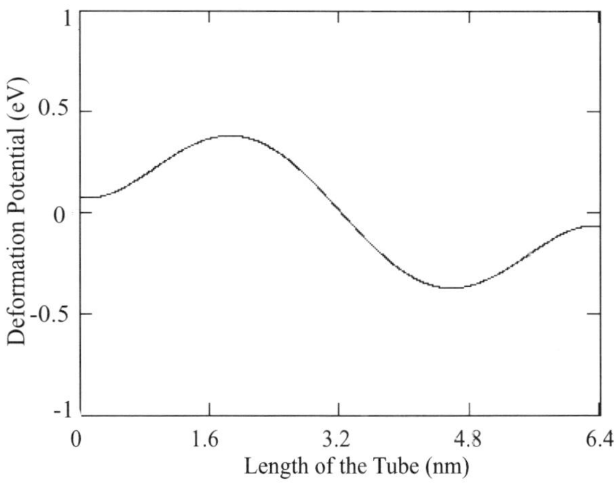

Fig. 3.12. Deformation potential for (40,0) nanotube with a length of 64 Angstroms and m = 2

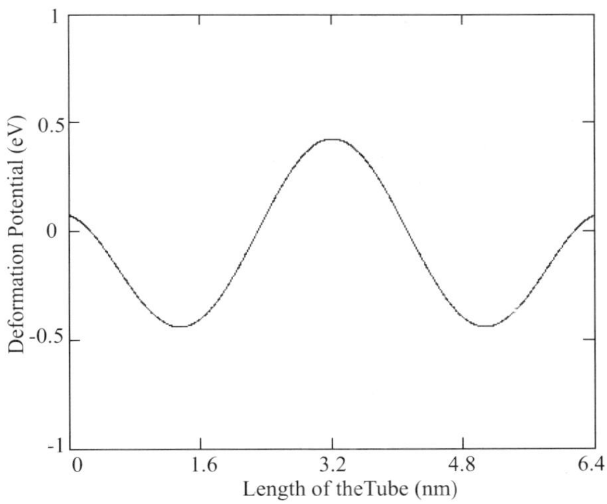

Fig. 3.13. Deformation potential for (40,0) nanotube with a length of 64 Angstroms and m = 3

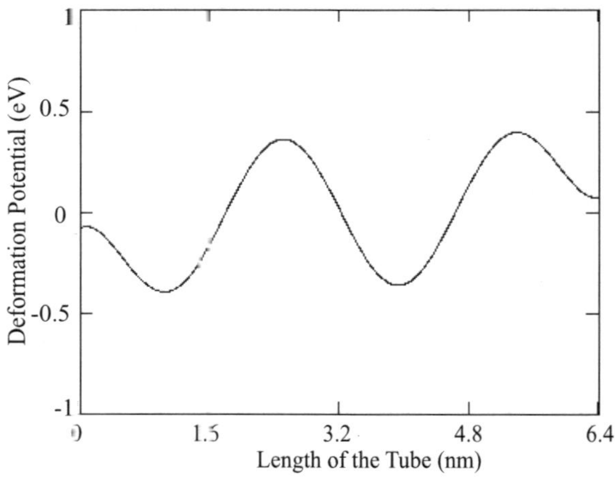

Fig. 3.14. Deformation potential for (40,0) nanotube with a length of 64 Angstroms and m = 4

3.3 Optical Modes in Multi-wall Nanotubes

The longitudinal optical modes for infinite length multi-wall nanotubes (MWNTs) has been calculated by Stroscio et al. [25] using the elastic continuum model. Here, using the same approach, we have calculated the optical modes for finite length nanotubes.

3.3.1 Model

We start our analysis by taking the modified ionic displacements as discussed by Born and Huang [26] and as used by Stroscio et al. [25] to model optical phonons in MWNT. The ionic displacement is related to the actual displacement by

$$u = \left(\frac{M_r}{V_o}\right)^{\frac{1}{2}} w = (\rho)^{\frac{1}{2}} w \tag{3.21}$$

where M_r is the reduced mass, ρ is the mass density and V_o is the volume of the unit cell.

The displacement u satisfies the Helmholtz equation,

$$\left(\nabla^2 + k_i^2\right) u^{(i)} = 0 \tag{3.22}$$

where,

$$k_i^2 = \left(\omega_i^2 - \omega^2\right) \beta_i^{-2} \tag{3.23}$$

where ω_i is the zone-center LO-phonon frequency for material region, i, and β_i is the acoustic phonon velocity.

The phonon displacements satisfying the Helmholtz equation may be taken as

$$\begin{aligned} u_z &= C_{mn} \exp(im\Phi) \left[N'_m (q_{mn}a) J_m (q_{mn}r) - J'_m (q_{mn}a) N_m (q_{mn}r)\right] \\ &\quad \times \sin(q_z z) \\ u_\Phi &= \frac{-C_{mn} q_{mn} \exp(im\Phi)}{r q_z} \left[N'_m (q_{mn}a) J_m (q_{mn}r) - J'_m (q_{mn}a) N_m (q_{mn}r)\right] \\ &\quad \times \cos(q_z z) \\ u_r &= \frac{-C_{mn} im \exp(im\Phi)}{q_z} \left[N'_m (q_{mn}a) J_m (q_{mn}r) - J'_m (q_{mn}a) N_m (q_{mn}r)\right] \\ &\quad \times \cos(q_z z) \end{aligned} \tag{3.24}$$

Applying the boundary conditions

$$u_z = u_r = \frac{du_r}{dz} = 0 \text{ at } z = 0, l \text{ and } r = a, b \tag{3.25}$$

we have

$$q_z = \frac{k\pi}{l} \tag{3.26}$$

where k is an integer, and

$$N'_m (q_{mn}a) J'_m (q_{mn}b) - J'_m (q_{mn}a) N'_m (q_{mn}b) = 0 \tag{3.27}$$

leading to 'double infinite' set of eigenvalues,

$$q = q_{mn}, \quad m = 0, 1, 2 \ldots \quad n = 0, 1, 2 \ldots \tag{3.28}$$

given by the positive roots of (3.27).

The dispersion relation is then given by

$$\omega_{mn}^2 = \omega_{LO}^2 - \beta_i^2 \left(q_{mn}^2 + q_z^2\right) \tag{3.29}$$

where ω_{LO} is the zone-center LO phonon frequency of the material composing the nanotube.

3.3.2 Normalization of LO Phonon Modes

The phonon amplitude is normalized using (3.8) giving rise to

$$C_{mn} = \frac{l^{\frac{1}{2}} \sqrt{a^2 - b^2}}{2^{\frac{1}{2}}} \left[\frac{\hbar}{M_r \omega_{mn}}\right]^{\frac{1}{2}} \frac{1}{\left(I_r^{(m)}\right)^{\frac{1}{2}}} \tag{3.30}$$

where

$$I_r^{(m)} = \frac{l}{2}\left\{I_{r1}^{(m)} + \left(\frac{q_{mn}}{q_z}\right)^2 I_{r2}^{(m)} + \left(\frac{m}{q_z}\right)^2 I_{r3}^{(m)}\right\} + \frac{\sin(2q_z l)}{4q_z}\left\{-I_{r1}^{(m)} + \left(\frac{q_{mn}}{q_z}\right)^2 I_{r2}^{(m)} + \left(\frac{m}{q_z}\right)^2 I_{r3}^{(m)}\right\} \tag{3.31}$$

$$I_{r1}^{(m)} = N_m'^2(q_{mn}a)\,I_1^{(m)} + J_m'^2(q_{mn}a)\,I_2^{(m)} - 2N_m'(q_{mn}a)\,J_m'(q_{mn}a)\,I_3^{(m)} \tag{3.32}$$

$$I_{r2}^{(m)} = N_m'^2(q_{mn}a)\,I_4^{(m)} + J_m'^2(q_{mn}a)\,I_5^{(m)} - 2N_m'(q_{mn}a)\,J_m'(q_{mn}a)\,I_6^{(m)} \tag{3.33}$$

$$I_{r3}^{(m)} = N_m'^2(q_{mn}a)\,I_7^{(m)} + J_m'^2(q_{mn}a)\,I_8^{(m)} - 2N_m'(q_{mn}a)\,J_m'(q_{mn}a)\,I_9^{(m)} \tag{3.34}$$

$$I_1^{(m)} = \frac{b^2}{2}\left[J_m^2(q_{mn}b) - J_{m-1}(q_{mn}b)\,J_{m+1}(q_{mn}b)\right] - \frac{a^2}{2}\left[J_m^2(q_{mn}a) - J_{m-1}(q_{mn}a)\,J_{m+1}(q_{mn}a)\right] \tag{3.35}$$

$$I_2^{(m)} = \frac{b^2}{2}\left[N_m^2(q_{mn}b) - N_{m-1}(q_{mn}b)\,N_{m+1}(q_{mn}b)\right] - \frac{a^2}{2}\left[N_m^2(q_{mn}a) - N_{m-1}(q_{mn}a)\,N_{m+1}(q_{mn}a)\right] \tag{3.36}$$

$$I_3^{(m)} = \frac{b^2}{4}\left[2J_m(q_{mn}b)\,N_m(q_{mn}b) - J_{m+1}(q_{mn}b)\,N_{m-1}(q_{mn}b)\right] - \frac{b^2}{4}J_{m-1}(q_{mn}b)\,N_{m+1}(q_{mn}b) - \frac{a^2}{2}J_m(q_{mn}a)\,N_m(q_{mn}a) + \frac{a^2}{4}\left[J_{m+1}(q_{mn}a)\,N_{m-1}(q_{mn}a) + J_{m-1}(q_{mn}a)\,N_{m+1}(q_{mn}a)\right] \tag{3.37}$$

$$
\begin{aligned}
I_4^{(m)} = & \frac{b^2}{2}\left[J_{m+1}^2\left(q_{mn}b\right) - J_m\left(q_{mn}b\right)N_{m-1}\left(q_{mn}b\right)\right] \\
& -\frac{a^2}{2}\left[J_{m+1}^2\left(q_{mn}a\right) - J_m\left(q_{mn}a\right)N_{m+1}\left(q_{mn}a\right)\right] \\
& +\frac{m}{2q_{mn}^2}\left[-J_0^2\left(q_{mn}b\right) - J_m^2\left(q_{mn}b\right) - 2\sum_{k=1}^{m-1}J_k^2\left(q_{mn}b\right)\right] \\
& +\frac{m}{2q_{mn}^2}\left[J_0^2\left(q_{mn}a\right) - J_m^2\left(q_{mn}a\right) - 2\sum_{k=1}^{m-1}J_k^2\left(q_{mn}a\right)\right] \\
& +\frac{2m}{q_{mn}^2}\left[J_0^2\left(q_{mn}b\right) + \sum_{k=1}^{m}J_k^2\left(q_{mn}b\right)\right] \\
& -\frac{2m}{q_{mn}^2}\left[J_0^2\left(q_{mn}a\right) + \sum_{k=1}^{m}J_k^2\left(q_{mn}a\right)\right]
\end{aligned} \tag{3.38}
$$

$$
\begin{aligned}
I_5^{(m)} = & \frac{b^2}{2}\left[N_m^2\left(q_{mn}b\right) - N_m\left(q_{mn}b\right)N_{m+1}\left(q_{mn}b\right)\right] \\
& -\frac{a^2}{2}\left[N_m^2\left(q_{mn}a\right) - N_m\left(q_{mn}a\right)N_{m+1}\left(q_{mn}a\right)\right] \\
& +\frac{m}{2q_{mn}^2}\left[-N_0^2\left(q_{mn}b\right) - N_m^2\left(q_{mn}b\right) - 2\sum_{k=1}^{m-1}N_k^2\left(q_{mn}b\right)\right] \\
& +\frac{m}{2q_{mn}^2}\left[N_0^2\left(q_{mn}a\right) - N_m^2\left(q_{mn}a\right) - 2\sum_{k=1}^{m-1}N_k^2\left(q_{mn}a\right)\right] \\
& -\frac{2m}{q_{mn}}\int_a^b N_{m+1}\left(q_{mn}r\right)N_m\left(q_{mn}r\right)dr
\end{aligned} \tag{3.39}
$$

$$
\begin{aligned}
I_6^{(m)} = & \frac{b^2}{4}\left[2J_{m+1}\left(q_{mn}b\right)N_{m+1}\left(q_{mn}b\right) - J_{m+1}\left(q_{mn}b\right)N_{m-1}\left(q_{mn}b\right)\right] \\
& -\frac{b^2}{4}\left[J_{m-1}\left(q_{mn}b\right)N_{m+1}\left(q_{mn}b\right)\right] - \frac{a^2}{4}\left[2J_{m+1}\left(q_{mn}a\right)N_{m+1}\left(q_{mn}a\right)\right] \\
& +\frac{a^2}{4}\left[J_{m+1}\left(q_{mn}a\right)N_{m-1}\left(q_{mn}a\right) + J_{m-1}\left(q_{mn}a\right)N_{m+1}\left(q_{mn}a\right)\right] \\
& -\frac{m}{2q_{mn}^2}\left[J_0\left(q_{mn}b\right)N_0\left(q_{mn}b\right) + J_m\left(q_{mn}b\right)N_m\left(q_{mn}b\right)\right] \\
& -\frac{m}{2q_{mn}^2}\left[2\sum_{k=1}^{m-1}J_k\left(q_{mn}b\right)N_k\left(q_{mn}b\right) - J_0\left(q_{mn}a\right)N_0\left(q_{mn}a\right)\right] \\
& +\frac{m}{2q_{mn}^2}\left[J_m\left(q_{mn}a\right)N_m\left(q_{mn}a\right) + 2\sum_{k=1}^{m-1}J_k\left(q_{mn}a\right)N_k\left(q_{mn}a\right)\right] \\
& -\frac{m}{q_{mn}}\left[\int_a^b N_{m+1}\left(q_{mn}r\right)J_m\left(q_{mn}r\right)dr + \int_a^b N_{m+1}\left(q_{mn}r\right)J_m\left(q_{mn}r\right)dr\right]
\end{aligned} \tag{3.40}
$$

$$I_7^{(m)} = -\frac{1}{2m}\left[J_0^2(q_{mn}b) + J_m^2(q_{mn}b) + 2\sum_{k=1}^{m-1} J_k^2(q_{mn}b)\right] + \frac{1}{2m}\left[J_0^2(q_{mn}a) + J_m^2(q_{mn}a) + 2\sum_{k=1}^{m-1} J_k^2(q_{mn}a)\right] \quad (3.41)$$

$$I_8^{(m)} = -\frac{1}{2m}\left[N_0^2(q_{mn}b) + N_m^2(q_{mn}b) + 2\sum_{k=1}^{m-1} N_k^2(q_{mn}b)\right] + \frac{1}{2m}\left[N_0^2(q_{mn}a) + N_m^2(q_{mn}a) + 2\sum_{k=1}^{m-1} N_k^2(q_{mn}a)\right] \quad (3.42)$$

$$I_9^{(m)} = -\frac{1}{2m}\left[J_0(q_{mn}b)N_0(q_{mn}b) + J_m(q_{mn}b)N_m(q_{mn}b)\right] - \frac{1}{2m}\left[2\sum_{k=1}^{m-1} J_k(q_{mn}b)N_k(q_{mn}b) - J_0(q_{mn}a)N_0(q_{mn}a)\right] + \frac{1}{2m}\left[J_m(q_{mn}a)N_m(q_{mn}a) + 2\sum_{k=1}^{m-1} J_k(q_{mn}a)N_k(q_{mn}a)\right] \quad (3.43)$$

For the lowest azimuthal mode with m=0,

$$C_{0n} = \frac{l\sqrt{a^2+b^2}}{2^{\frac{1}{2}}}\left[\frac{\hbar}{M_r\omega_{0n}}\right]^{\frac{1}{2}}\frac{1}{\left(I_r^{(0)}\right)^{\frac{1}{2}}} \quad (3.44)$$

where,

$$I_r^{(0)} = \frac{l}{2}\left\{I_{r1}^{(0)} + \left(\frac{q_{0n}}{q_z}\right)^2 I_{r2}^{(0)}\right\} + \frac{\sin(2q_z l)}{4q_z}\left\{-I_{r1}^{(0)} + \left(\frac{q_{0n}}{q_z}\right)^2 I_{r2}^{(0)}\right\} \quad (3.45)$$

$$I_{r1}^{(0)} = N_0'^2(q_{0n}a)\,I_1^{(0)} + J_0'^2(q_{0n}a)\,I_2^{(0)} - 2N_0'(q_{0n}a)\,J_0'(q_{0n}a)\,I_3^{(0)} \quad (3.46)$$

$$I_{r2}^{(0)} = N_0'^2(q_{0n}a)\,I_4^{(0)} + J_0'^2(q_{0n}a)\,I_5^{(0)} - 2N_0'(q_{0n}a)\,J_0'(q_{0n}a)\,I_6^{(0)} \quad (3.47)$$

$$I_1^{(0)} = \frac{b^2}{2}\left[J_0^2(q_{0n}b) + J_1^2(q_{0n}b)\right] - \frac{a^2}{2}\left[J_0^2(q_{0n}a) + J_1^2(q_{0n}a)\right] \quad (3.48)$$

$$I_2^{(0)} = \frac{b^2}{2}\left[N_0^2(q_{0n}b) + N_1^2(q_{0n}b)\right] - \frac{a^2}{2}\left[N_0^2(q_{0n}a) + N_1^2(q_{0n}a)\right] \quad (3.49)$$

$$I_3^{(0)} = \frac{b^2}{2}\left[J_0(q_{0n}b)N_0(q_{0n}b) + J_1(q_{0n}b)N_1(q_{0n}b)\right] - \frac{a^2}{2}\left[J_0(q_{0n}a)N_0(q_{0n}a) + J_1(q_{0n}a)N_1(q_{0n}a)\right] \quad (3.50)$$

$$\begin{aligned} I_4^{(0)} &= \frac{b^2}{2}\left[J_1^2\left(q_{0n}b\right) - J_0\left(q_{0n}b\right)J_2\left(q_{0n}b\right)\right] \\ &-\frac{a^2}{2}\left[J_1^2\left(q_{0n}a\right) - J_0\left(q_{0n}a\right)J_2\left(q_{0n}a\right)\right] \end{aligned} \tag{3.51}$$

$$\begin{aligned} I_5^{(0)} &= \frac{b^2}{2}\left[N_1^2\left(q_{0n}b\right) - N_0\left(q_{0n}b\right)N_2\left(q_{0n}b\right)\right] \\ &-\frac{a^2}{2}\left[N_1^2\left(q_{0n}a\right) + -N_0\left(q_{0n}a\right)N_2\left(q_{0n}a\right)\right] \end{aligned} \tag{3.52}$$

$$\begin{aligned} I_6^{(0)} &= \frac{b^2}{4}\left[2J_1\left(q_{0n}b\right)N_1\left(q_{0n}b\right) - J_2\left(q_{0n}b\right)N_0\left(q_{0n}b\right)\right] \\ &-\frac{b^2}{4}J_0\left(q_{0n}b\right)N_2\left(q_{0n}b\right) - \frac{a^2}{2}J_1\left(q_{0n}a\right)N_1\left(q_{0n}a\right) \\ &+\frac{a^2}{4}\left[J_2\left(q_{0n}a\right)N_0\left(q_{0n}a\right) + J_0\left(q_{0n}a\right)N_2\left(q_{0n}a\right)\right] \end{aligned} \tag{3.53}$$

3.3.3 Optical Deformation Potential

The optical deformation is given by

$$H_{opt-def} = \vec{D}_{op} \cdot \vec{u} \tag{3.54}$$

where, D_{op} is the optical-phonon deformation potential constant [27].

Since the axial phonons are important in determining the transport along the length of the tube, the optical deformation potential can be approximated by

$$H_{opt-def} = |D_{op}|\,|u_z| \tag{3.55}$$

The carrier interactions are determined by the optical deformation potential interaction. The optical-phonon and the acoustic-phonon deformation potentials have different origins; indeed, the optical deformation potential has its origins in the vibrations of two sublattices against each other whereas the acoustic deformation potential arises from a volume-changing dilatation of the crystal. Accordingly, the optical deformation potential strongly depends on the symmetry of band structure, and may, in fact, vanish [27] for high symmetry situations such as those existing close to selected band minima. The tight binding calculations of [28] yield an optical deformation potential mean free path of about 100 nm for CNTs suggesting that effective optical deformation potential constants have approximate magnitude of several times 10^8 eV/cm.

3.4 Quantized Vibrational Modes in Hollow Spheres

The study of the hollow spheres has gained more importance of late due to the similarity between icosahedral C_{60} molecule and the spherical shell. The quantized vibrational modes for the spherical shell have been calculated by Kahn et al. [29] using the classical modes [30].

The displacement for an axisymmetric mode n can be written as

$$w_n = R_n P_n \left[\cos(\theta)\right] \tag{3.56}$$

$$w_n = Q_n P_n^1 \left[\cos(\theta)\right] \tag{3.57}$$

where the functions P_n are the Legendre polynomials, the functions P_n^1 are the associated Legendre polynomials of the first kind and the first order and θ is the polar angle. These, along with the coefficients R_n and Q_n, determine the shape of the mode.The ratio of coefficients has been determined to be [30]

$$\frac{R_n}{Q_n} = \frac{1}{2}\left(\frac{\left[n(n+1)-3-v\right]}{1+v} \pm \left\{\frac{\left[3+v-n(n+1)\right]^2}{(1+v)^2} + 4n(n+1)\right\}^{\frac{1}{2}}\right) \tag{3.58}$$

where the plus and minus sign refer to the upper and lower branches of the dispersion relations respectively.

Using the normalization condition of (3.8),

$$Q_n^2 = \left\{ \begin{array}{c} \dfrac{(2n+1)\hbar}{4M\omega_\pm} \\ -\dfrac{1}{4}\left[\begin{array}{c} \dfrac{2n(n+1)-3-v}{(1+v)^2} \\ \times\left(1 \pm \left[3+v-n(n+1)+4n(n+1)(1+v)^2\right]^{\frac{1}{2}}\right) \\ +4n(n+1) \end{array} \right] \end{array} \right\} \tag{3.59}$$
$$\times \frac{1}{n(n+1)}$$

The angular frequencies $\omega_\pm$ are

$$\omega_\pm = \frac{E}{\rho 2R^2(1-v^2)}\left[n(n+1)+1+3v\right] \pm \sqrt{\left[n(n+1)+1+3v\right]^2 - 4(1-v^2)\left[n(n+1)-2\right]} \tag{3.60}$$

The modes obtained through this calculations is within 2%-6% of the experimental values [31]. This outstanding agreement may be surprising in view of the fact that the wall of the buckyball is only monolayer thick. However, since the elastic properties of the buckyball depend primarily on those of the two-dimensional graphene-related sheet used to form the buckyball surface, such agreement is reasonable.

3.5 Conclusions

The elastic continuum model is used for calculating the quantized vibrational modes of nanostructures such as nanotubes and fullerenes. The displacement amplitude of the nanostructures should be quantized so that each mode has proper quantum mechanical energy. The analytical continuum models presented in this article provide expressions for acoustic and optical mode amplitudes. These amplitudes are related to both optical and acoustic deformation interaction Hamiltonians that provide the basis for scattering rate calculations for a variety of CNTs. These results describe the mode amplitude variations due to the finite length of the nanotube and they manifest the role of boundary conditions on the deformation potential interactions. For the case of acoustic modes, the use of membrane theory leads to tractable mode solutions that yield analytic expressions for the deformation potential interaction. The Hamiltonians given in the article provide the basis for determining carrier-phonon scattering rates for CNTs of finite length. These results provide a framework for the general computation of carrier transport phenomena as they are determined by the deformation potential.

Acknowledgements

This research was supported by the AFOSR under the direction of Dr. Todd Steiner.

References

1. M.R. Falvo et al: Nature (London) **389**, 582 (1997)
2. Z. Yao, H.W.C. Postma, L. Balents, and C. Dekker: Nature (London) **402**, 273 (1999)
3. Z.K. Tang et al: Science **292**, 2462 (2001)
4. T.W. Odom, J.L. Huang, P. Kim, and C.M. Leiber: Nature (London) **391**, 62 (1998)
5. Jing Guo, Sebastian Goasguen, Mark Lundstrom, and Supriyo Datta: Appl. Phys. Lett. **81**, 1486 (2002)
6. Jing Guo, Mark Lundstrom, and Supriyo Datta: Appl. Phys. Lett. **80**, 3192 (2002)
7. J.A. Misewich, R. Martel, Ph. Avouris, J.C. Tsang, S. Heinze, J, Tersoff: Science **300**,783 (2003)

8. R.A. Jishi, L Venkataraman, M.S. Dresselhaus and G. Dresselhaus: Chem. Phys. Lett. **209**,77 (1993)
9. J. Yu, R.K. Kalia and P.Vashista: J. Chem. Phys. **103**, 6697 (1995)
10. M. Menon, E. Ritcher and K.R. Subbaswamy: J. Chem. Phys. **104**, 5875 (1996)
11. D.H Robertson, D.W. Brenner and J.W. Mintmire: Phys Rev. B. **45**, 12592 (1992)
12. J.P. Lu, Phys Rev. Lett. **79**: 1297 (1997)
13. Daniel Sanchez-Portal, Emilio Artacho, Jose M. Soler, Angel Rubio, and Pablo Oedejon: Phys. Rev. B **59**, 12678 (1999)
14. C. Trallero-Giner, F. Garcya-Moliner, V. Velasco, and M. Cardona: Phys. Rev. B **45**, 11944 (1992)
15. T.D. Krauss and F.W. Wise: Phys. Rev. Lett. **79**, 5102 (1997)
16. J.H. Hodak, A. Henglien, and G.V. Hartland: J. Chem. Phys. **111**, 8613 (1999)
17. M.A. Stroscio and M. Dutta: Phys. Rev. B **60**, 7722 (1999); M.A. Stroscio and Mitra Dutta: *Phonons in Nanostructures* (Cambridge University Press, Cambridge 2001)
18. L.T. Chadderton: J. Phys Chem. Solids **54**, 1027 (1993)
19. X. Xiaoyu, L. Ji-Xing, and O.Y. Zhong-Can: Mod. Phys. Lett. **9**, 1649 (1995)
20. H. Kraus: *Thin Elastic Shells* (Wiley, New York 1967)
21. Daniel Kahn, K.W.Kim and Michael A. Stroscio: Journal of Appl. Phys. **89**, 5107 (2001)
22. Hidekatsu Suzurra and Tsuneya Ando: Phys. Rev. B **65**, 235412 (2002)
23. Amit Raichura, Mitra Dutta and Michael A. Stroscio: Journal of Appl. Phys. **94**, 4060 (2003)
24. Philip G. Collins, M. Hersam, M. Arnold, R. Martel, and Ph. Avouris: Phys. Rev. Lett. **86**, 3128 (2001)
25. M.A. Stroscio, Mitra Dutta, Daniel Kahn, and Ki Wook Kim: Superlatt. Microstruct. **29**, 405–409 (2001)
26. Max Born and Kun Huang: *Dynamical Theory of Crystal Lattices* (Oxford University Press, Oxford 1954)
27. Karl Hess: *Advanced Theory of Semiconductor Devices* (Prentice Hall, NJ 1988)
28. Zhen Yao, Charles L. Kane, and Cees Dekker: Phys. Rev. Lett. **84**, 2941 (2000)
29. Daniel Kahn, K.W. Kim, and Michael A. Stroscio: Journal of Appl. Phys. **89**, 5107 (2001)
30. W.E. Baker: J. Acoust. Soc. Am. **33**, 1749 (1961)
31. L.T. Chadderton: J. Phys. Chem. Solids **54**, 1027 (1993)

Part II

Synthesis and Characterization

4 Direct Growth of Single Walled Carbon Nanotubes on Flat Substrates for Nanoscale Electronic Applications

Shaoming Huang and Jie Liu

This chapter describes the growth of single walled carbon nanotubes on suitable substrates that can be directly used for device fabrication. The chapter focuses on the growth of nanotubes using chemical vapor deposition (CVD) methods. The control of diameters and orientation of the produced nanotubes is discussed. More importantly, a new "fast-heating" CVD method that can produce long and well aligned of nanotubes is described in details. The control of location and orientation offered by this method offer a great advantage for device fabrication, representing a significant advance in controlling the structures of nanoscaled materials through synthesis.

4.1 Introduction

Over the last decade, many nanoelectronic devices based on individual single walled carbon nanotubes (SWNTs), such as quantum wires [1], logic gates [2–7], field emitters [8, 9], field-effect transistors [2, 4, 7, 10–16], diodes [17], inverters [15] and sensors [15, 18–24] etc., have been demonstrated. The normal way to make these nanotube devices is to first suspend purified carbon nanotubes in solvents, then deposit them on a suitable substrate followed by evaporating electrodes (metals) onto them by photolithography or e-beam lithography. These methods used to fabricate nanotube devices are time consuming and unreliable. Furthermore, there remain many problems in this solution-based device fabrication approach. First, the nanotubes have to be purified. Although researchers have succeeded in producing SWNTs with high yield, the as-produced SWNTs always contain some impurities including other carbonaceous materials and catalysts etc. Purification normally involves strong acid treatment and sonication, which may introduce defects into the nanotubes and change their electrical property. Secondly, the as-produced nanotubes are in the form of bundles and are mixtures of semiconducting and metallic nanotubes. For many electronics applications, pure individual semiconducting nanotubes are desired. So, size- and structure-specific fabrication methods must be developed in order for the full potential of the materials to be realized.

While little progress has been made toward selective growth of metallic or semiconducting nanotubes, some recent progress has been made toward diameter control of individual SWNTs using monodispersed nanoparticles [25–28] and the separation of semiconducting and metallic nanotubes [29–32]. For example, metallic SWNTs can be separated from semiconducting SWNTs in solution using alternating current dielectrophoresis [32]. Single-stranded DNA can self-assemble into a helical structure around an individual SWNT and was found to be sequence dependent [29]. The electrostatics of the DNA-CNT hybrid depends on the tube diameter and electronic properties, which enables the separation of nanotubes by anion exchange chromatography [29]. Some surfactants such as octadecylamine can be used for sorting the nanotubes [31]. A bulk selective enrichment in metallic or semiconducting nanotubes based on the chemical discrimination in the charge-transfer complex formation between bromine and the nanotubes was also reported [30]. All these post-synthetic separation methods are solution-based approaches.

Thirdly, perhaps more importantly, in order to compete with Si technology for practical applications in the future, the process to organize a large amount nanotube devices and integrate them into logic circuits on the substrate has to be developed. This requires the fabrication of high density SWNT arrays with a specific location and orientation on the substrate. Much effort has been focused in this direction. For example, electric fields guided growth [33, 34] of nanotubes on an Si wafer has been reported. A solution-based large-area assembly of nanotubes on gold surfaces has also been demonstrates recently [35]. However, there are still many unsolved problems for obtaining pre-designed complex structures using solution-based approaches. On the other hand, direct CVD growth of SWNTs on flat surfaces has several advantages over deposition of nanotubes from solutions for device fabrication. For example, the nanotube samples are usually very clean after surface growth, containing only SWNTs and nanoparticles, no sonication or oxidative purification steps need to be involved, which greatly reduces the possibility of defect formation on SWNTs. Recent transport measurements by several research groups have shown that devices made of CVD nanotubes grown directly on substrates tend to exhibit better performance than those produced using SWNTs prepared by other methods (Fig. 4.1) [7, 10, 13, 36, 37]. In this article, we will present recent progress in controlled growth of SWNTs on surfaces using the CVD growth process. In particular, we will discuss the growth of long and well aligned SWNTs using a special "fast heating" CVD method [38, 39] more detail. The growth mechanism of nanotubes in the process and the advantages of nanotubes with controllable orientations for device fabrication and their electrical properties will also be discussed.

4.2 Diameter Control

Theoretical calculations have shown that the electronic structures of SWNTs depend strongly on their diameter and chirality. Therefore the diameter control of SWNT is very important for both research and industrial application

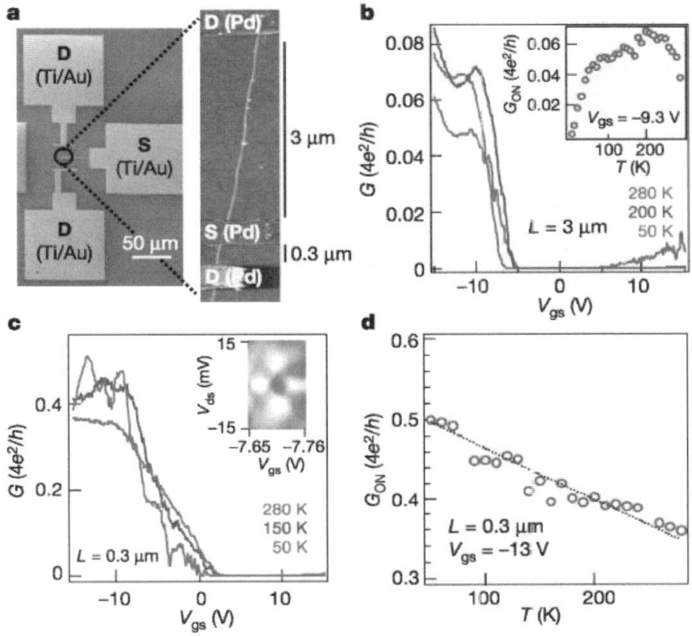

Fig. 4.1. (**a**) A scanning electron microscope (SEM) image (left) and atomic force microscope (AFM) image (right) of representative Pd-contacted long ($L < 3\,\mu m$) and short ($L < 300\,nm$) back-gated SWNT devices formed on the same nanotubes directly grown on SiO2/Si substrates using the CVD approach. Ti/Au metal bonding pads were used to connect to the Pd source (S) and drain (D) electrodes. The devices were annealed in Ar at 225°C for 10 min after fabrication. (**b**) G (at low S–D) bias V_{ds} versus gate voltage V_{gs} for a 3-mm-long SWNT ($d = 3.3\,nm$) device recorded at various T. Inset, G_{ON} versus T for the device. (**c**) G versus V_{gs} for a 300-nm-long tube section on the same tube as for (**b**) at various T. Differential conductance dI_{ds}/dV_{ds} versus V_{ds} and V_{gs} (inset, measured by a lock-in technique) at $T < 1.5\,K$ shows a Fabry-Perot-like interference pattern (bright peak $G < 4e^2/h$, dark region $G < 0.5 * 4e^2/h$). (**d**) G_{ON} versus T for the $L < 300\,nm$ semiconducting tube down to 50 K. Data and caption reproduced from [10] (see Color Plates, p. 340)

purposes. The diameter of SWNTs is believed to be determined by the size of the catalyst nanoparticles. For most catalysts used for SWNT synthesis, the size of the metal/metal oxide nanoparticle cannot be well controlled. As a result, the SWNTs produced usually exhibit a broad distribution of diameters.

One method to solve this problem is to use preformed monodispersed nanoparticles as a catalyst. Dai and coworkers at Stanford loaded iron atoms into the cores of horse spleen apoferritin, an iron-storage protein [40]. Iron oxide nanoparticles were obtained by heating the artificial ferritin to 800°C in air. The size of the resulting nanoparticles can be controlled by the number of

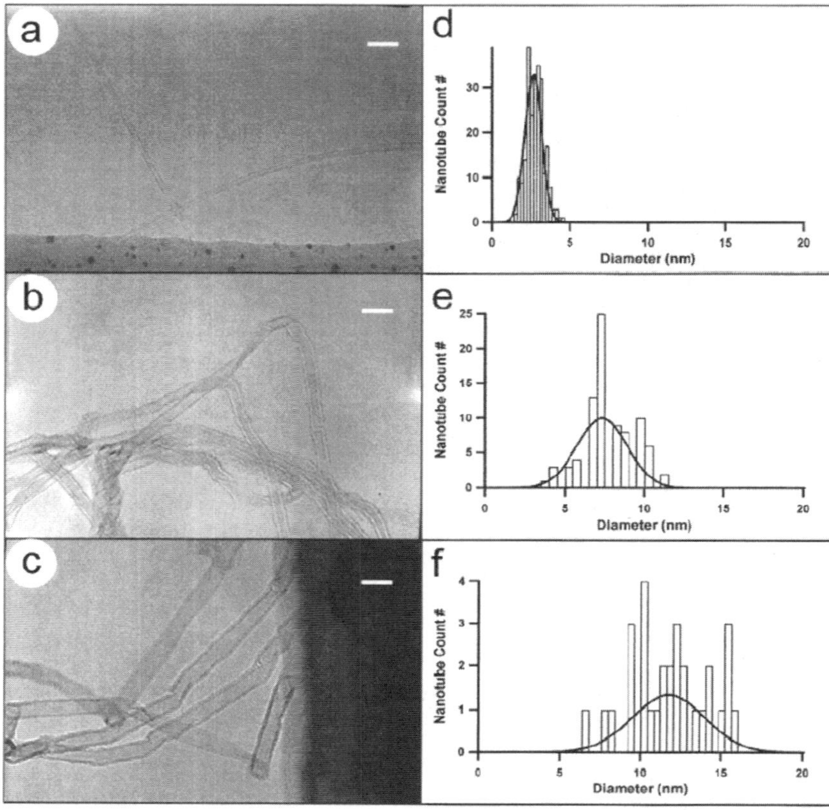

Fig. 4.2. TEM images of carbon nanotubes grown using (**a**) 3, (**b**) 9, and (**c**) 13 nm average diameters iron nanoclusters. Scale bars are 20 nm for (**a**), (**b**), and (**c**). Corresponding histograms of the nanotube diameters are plotted in (**d**), (**e**), and (**f**), respectively. The solid lines correspond to Gaussian fits. Reproduced from [26]

the iron atoms in the protein. Dai's group also developed a method to make uniform catalyst nanoparticles using a dendrimer as a template [25]. Lieber's group at Harvard also demonstrated the dependence of nanotube size on the size of catalyst nanoparticles (Fig. 4.2) [26].

However, the catalyst nanoparticles produced using the described methods still have a distribution range, making the precise control of the nanotube diameter difficult. More recently, a family of metal-containing nanocluster molecules was found to be good catalysts for the CVD growth of SWNTs [27]. These clusters have a well-defined chemical composition and are thus identical to each other. An example of the clusters has the molecular formula: $[H_xPMo_{12}O_{40} \subset H_4Mo_{72}Fe_{30}(CH_3COO)_{15}O_{254}(H_2O)_{98}] \cdot 60H_2O$ which contains 84 Mo atoms and 30 Fe atoms in each molecule. It is a member of a large family of molecular nanoclusters based on molybdenum oxide frameworks [41–46].

To grow nanotubes using the cluster, the samples were first annealed in the air for 5 min at 700°C to decompose the organic ligand of the molecular cluster and yield clusters made of Fe and Mo oxides, and then reduced in H_2 for 5 min at 900°C to form metal nanoclusters containing only Fe and Mo. Subsequently, CVD growth of SWNTs was performed with methane as feeding gas for 15 min at the same temperature. The diameter of the produced SWNTs agrees well with the hypothesis that they are grown from individual nanoclusters. According to the radii of the atom Fe (126 pm) and the atom Mo (140 pm), if all the 84 Mo atoms and 30 Fe atoms in one nanocluster were left and packed tightly together after being reduced in hydrogen at 900°C, the diameter of the residue should be about 1.3 nm. This estimate

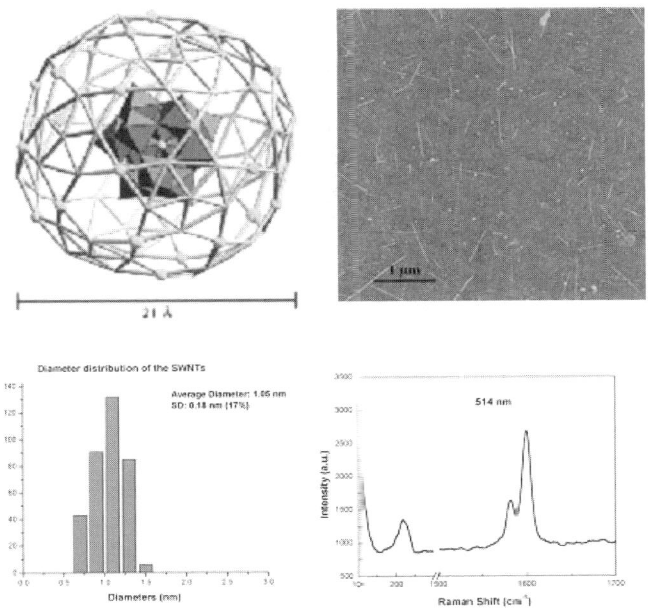

Fig. 4.3. Top left: structure of the molecular cluster used as catalyst for SWNT growth. Top right: AFM image of SWNTs grown using the molecular cluster as catalyst. Bottom left: the diameter distribution of the SWNTs grown on the chemically attached Fe/Mo nanoclusters on the silicon dioxide surfaces. The sample size is more than 350. The average diameter is 1.0 nm and the standard deviation is 0.18 or 17%. Bottom right: the Raman spectrum of the SWNTs grown upon the chemically attached Fe/Mo nanoclusters on the silicon nitride membranes. The excitation wavelength was 514.5 nm by an argon ion laser at a power of 120 mW. Reproduced from [27]

reasonably explains the observation that the SWNTs with a diameter of 1.2 nm are dominant in this distribution.

However, some metal atoms could have been evaporated during the growth process. As a result, the nanoclusters became even smaller than 1.3 nm. That is why there are SWNTs with diameters of less than 1.2 nm. As for the SWNTs with diameters of more than 1.2 nm, a possible explanation is that a small number of nanoclusters aggregated together to form bigger particles. Nevertheless, these results have demonstrated that the diameter of the produced SWNTs can be controlled by choosing the right catalysts and reaction conditions (Fig. 4.3) [27].

4.3 Orientation Control

The capability of controlling position and orientations of nanotubes on a substrate is extremely important for large-scale fabrication of nanotube devices. Current methods make the devices rely on the mapping of nanotubes on a substrate and then design an electrode pattern for each sample to connect nanotubes to the electrodes. Since every sample has nanotubes with different orientations and locations, the whole process has to be repeated for each sample, which is not practical for making a large amount of devices. However, if the location and orientations of nanotubes can be controlled, we can use the same electrode structure for many samples, which significantly reduces the difficulty in scale of the device production.

Since the location of nanotubes can be controlled by the location of the catalysts, which can be deposited on the substrate using e-beam lithography and other lithographic techniques, the remaining challenge is to control the orientations of the nanotubes during growth. In order to achieve orientation control, an external guidance must be applied for the SWNT growth. The first direct growth of suspended SWNTs was demonstrated by Dai's group at Stanford [47]. In their reported process, lithographically patterned silicon pillars were first fabricated on a substrate. Then contact printing was used to transfer catalyst precursor materials onto the tops of pillars. Finally, methane CVD produced suspended SWNTs with their orientation directed by the pattern of the pillars. The same research group also applied a lateral electric field (1 V/µm) during CVD growth and the SWNTs exhibited alignment guided by the field (Fig. 4.4) [48].

This field-alignment effect originates from the high polarizability of SWNTs. Aligning torques and forces on the nanotubes are generated by induced dipole moments. Another promising way of aligning SWNTs was recently developed by Jie Liu's group at Duke University, using the flow of the feeding gas to grow SWNTs along specific directions [38]. Such a process has offered the unique opportunity not only to produce well aligned nanotubes, but also to understand in more detail the growth mechanism of nanotubes. For example, the length of the nanotubes produced by the method can reach

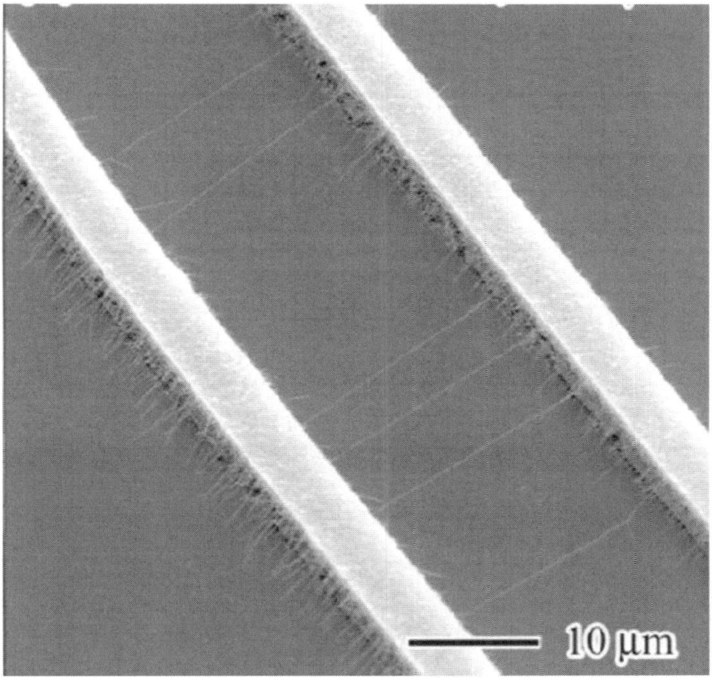

Fig. 4.4. SEM image of SWNTs grown under an electric field. Reproduced from [48]

centimeter length in a short growth time (20 min) rather than several tens of microns in most previous cases. Understanding the mechanism would enable us to significantly increase the yield of nanotubes. The rest of the chapter will discuss this new process in more detail.

4.4 Growth of Superlong and Well-Aligned SWNTs on a Flat Surface by the "Fast-Heating" Process

In this new approach [38, 39], Fe-Mo bimetallic nanoparticles were used as catalysts and a CO/H_2 mixture was used as the feeding gas. An Si wafer containing patterned catalysts was heated from room temperature to reaction temperature (900°C) in a few seconds by a quick transfer of the sample from the outside of a heated furnace to the center (Fig. 4.5). Such a "fast-heating" CVD process produces long and aligned SWNTs as shown in Fig. 4.6a.

The lengths of the nanotubes can reach 1.5 cm for a 20 minutes growth process. More importantly, since no electrodes are needed to guide the growth direction of the nanotubes, multidimensional nanotube structures can be easily prepared by the repeated growth of nanotubes on the wafer along different directions (Fig. 4.6b).

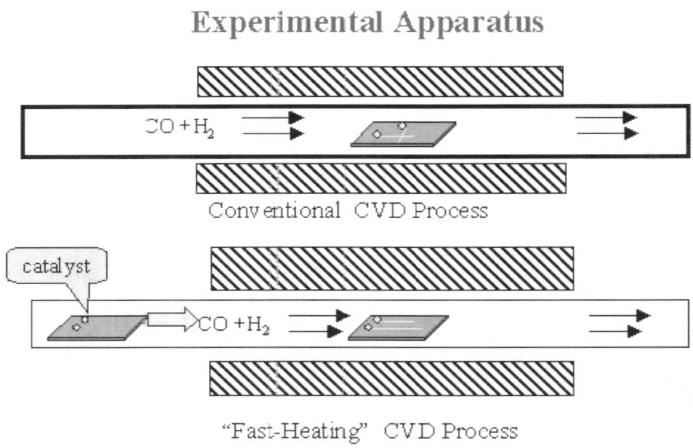

Fig. 4.5. Schematic diagram of the "fast-heating" CVD process

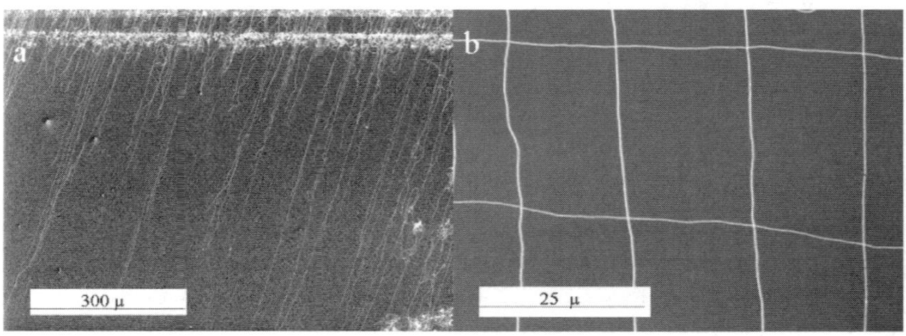

Fig. 4.6. (**a**) SEM image of long and aligned SWNTs grown from "fast heating" CVD process and (**b**) two-dimensional nanotube network fabricated by a multi-step growth process

AFM height measurements show that the diameter of the nanotubes range from 0.8 nm and 2.5 nm (Fig. 4.7a). Raman spectra (Fig. 4.7b) also confirm the SWNT structure. A strong G mode vibration is observed at 1586 and 1563 cm^{-1}, which relates to the in-plane vibration of a highly curved graphene sheet and the characteristic SWNT radial breathing mode is observed at 181 cm^{-1}. The average diameter is calculated to be 1.25 nm. It was found that the nanotubes produced from the "fast-heating" process has a stronger time dependence than the random short nanotubes from conventional CVD process. For the "fast-heating" process, the longer the growth time the longer nanotubes. For example, 2.1 mm long nanotubes can be produced in 10 minutes growth (Fig. 4.8a) while 3.7 mm long nanotubes can be produced in 20 minutes growth (Fig. 4.8b) under the same growth conditions.

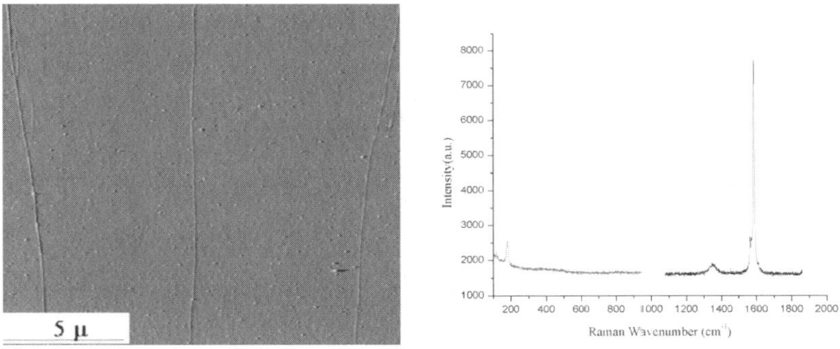

Fig. 4.7. (**a**) AFM image and (**b**) Raman spectra of the nanotubes grown with the "fast heating" process

It has been proven that such a "fast-heating" process is not limited to the specific type of catalysts and feeding gas as described above for growing long and well-oriented SWNTs on a surface. Such SWNTs can be produced by several different catalysts, including Fe/Mo nanoparticles, pure Fe nanoparticles, Fe/Pt nanoparticles, and water soluble molecular Fe/Mo clusters, and several different carbon sources, including CO, hydrocarbons such as CH_4 and alcohols such as CH_3OH (Fig. 4.9) [49].

However, for a different carbon source, it was found that the growth rates of nanotubes were different. For example, we have noted that the growth rate of nanotubes could be as high as 20 μm/second when CH_4 was used as the carbon source. The longest nanotube we have prepared is about 1.5 cm after 20 minutes growth. The location of nanotubes can be controlled by the position of catalysts, which can be patterned by various lithographic methods. Many patterning techniques such as photolithography, soft-lithography and ink-jet printing can be applied to pattern catalysts for both MWNT [50–52] and SWNT growth [47, 53, 54].

Among various technologies photolithography is one of the most important and powerful patterning techniques for microfabrication, and is widely used in the integrated circuits industry. The patterned resist on substrates from photolithography can act as removable mask for the deposition of materials (e.g. the evaporation of a metal or spinning solution). Large scale patterning of the catalyst can be achieved by this technique. A more convenient method to deposit catalysts on a suitable substrate is direct ink printing: a solution of catalysts can be directly drawn on the substrate from an ink pen (Fig. 4.10a) [55], a PDMS stamp (micro-contact printing) (Fig. 4.10b) or possibly by high resolution inkjet printing in the future. The control of the growth direction of a long nanotube by gas flow offers a feasible way to generate a crossed-bar structure with any angles with both multi-junction or suspended crossed-network frames by simply changing the wafer and apply-

Fig. 4.8. (**a**) SEM image of a 2.1 mm long nanotube grown in 10 min and (**b**) SEM image of a 3.7 mm long nanotube grown in 20 min under the same conditions

ing multiple patterning catalysts and multiple growth processes. Figure 4.11 shows the SEM image of the two-crossed parallel nanotube arrays by a two-step growth process using Fe/Mo nanoparticles as catalyst and CO/H_2 as feeding gas.

4.5 Growth Mechanism

Obtaining insights of the growth mechanism of such long nanotubes is both of scientific importance for understanding how the nanotubes grow and technical importance for designing future experiments to achieve a higher yield and better control of the nanotube growth. For nanotube growth in the CVD process, two main mechanisms were proposed previously with the "tip-growth" mechanism well demonstrated for certain multi-walled carbon nanotubes (MWNTs) [56–60] and "base-growth" for both MWNTs and SWNTs [40]. Previous reports have demonstrated that the "base-growth" mechanism is most likely the mechanism for SWNTs grown in the conventional CVD process [40]. However, the base growth mechanism cannot be used to satisfactorily explain the growth of long nanotubes under fast-heating condi-

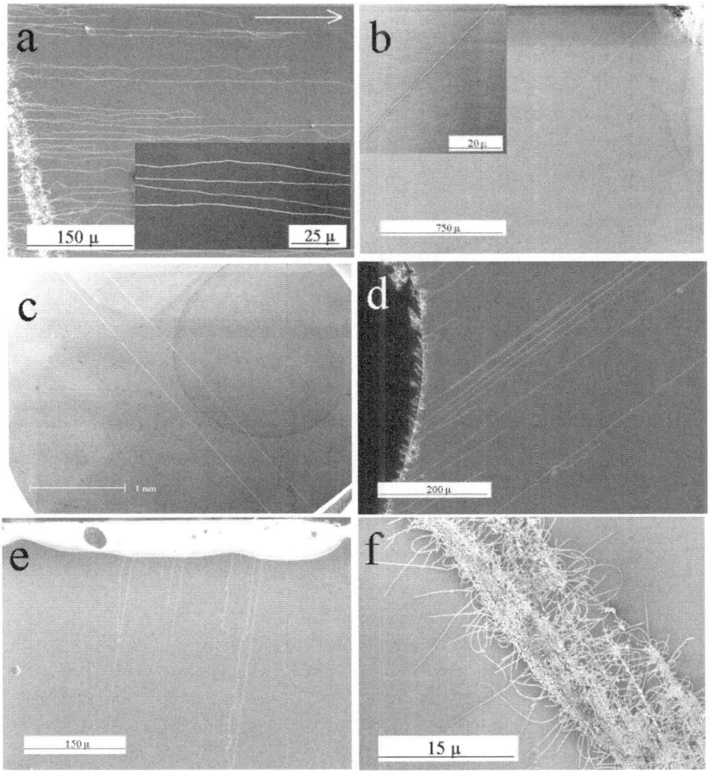

Fig. 4.9. SEM images of SWNTs on the surface (**a**) oriented long SWNT arrays from the "fast-heating" growth process. Catalyst: Fe/Mo nanoparticles; CO/H_2, 900°C, 10 minutes. Inset is the enlarged image; (**b**) a 2.1 mm individual straight SWNT. Catalyst: Fe/Mo nanoparticles; CH_4/H_2, 900°C, 10 minutes; (**c**) individual parallel straight SWNTs having 3.9 mm in length. Catalysts: Fe/Pt nanoparticles; CO/H_2, 900°C, 20 minutes; (**d**) SWNTs arrays. Catalyst: identical Fe/Mo molecular cluster, CO/H_2, 900°C, 10 minutes; (**e**) long SWNTs. Catalysts: Fe/Mo nanoparticles; $H_3OH/H_2/Ar$, 900°C, 20 minutes; (**f**) random short SWNTs from conventional growth process using Fe/Mo nanoparticles as catalysts and CO/H_2 as feeding gas at 900°C for 10 min

tions. Based on experimental results, a modified tip-growth mechanism was proposed to explain the effect of "fast-heating" as shown schematically in Fig. 4.12 [49].

In this mechanism, the nanotubes grow up at the initial stage of the growth, with the catalysts on their tips. The convection flow caused by the temperature difference between the substrate and the feeding gas is a critical factor that helps nanotubes growing away from the substrate. Such convection flow does not exist if the sample is heated slowly, since the whole system

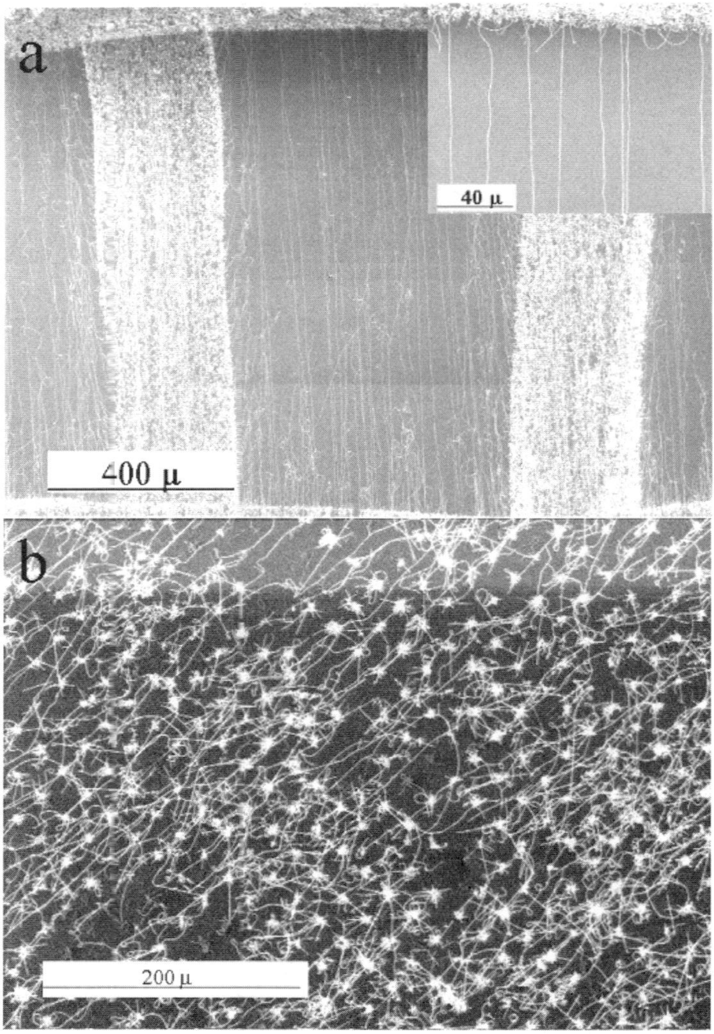

Fig. 4.10. SEM images of long nanotubes grown from catalysts patterned on a substrate using (**a**) a simple ink pen drawing and (**b**) PDMS stamping

would be in thermal equilibrium and the difference in temperature between the substrate and the gas would be small. Once the nanotube grows up into a fast-flowing gas stream above the surface, they are caught by the wind of the flowing gas and keep floating in the gas flow, resulting in the alignment and continuous growth up to centimeter length. The evidence for this mechanism is from experimental observations. Since the orientation of the nanotubes is determined by the direction of the gas flow, it implies that the nanotubes

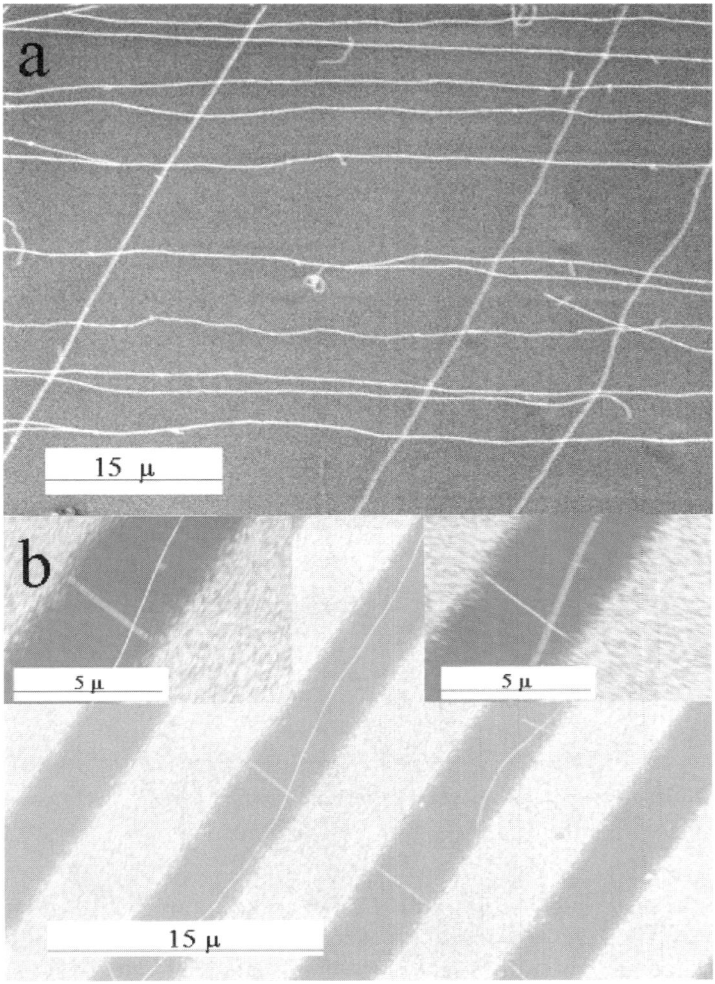

Fig. 4.11. (**a**) SEM image of a two dimensional nanotube network with the nanotubes in contact with each other at the junction and (**b**) a similar 2-D network with the top layer of nanotubes suspended over the bottom layer. The insets in (**b**) are two SEM images focused on either the top or the bottom nanotube

have to be floating in the gas flow during growth rather than growing on the surface of the substrates and sliding along the surface. The initial growth away from the substrate is confirmed by the observation of nanotubes grown over a barrier on the substrate and across trenches (Fig. 4.13).

In this experiment, the catalysts were deposited in the trench fabricated on a Si wafer with a depth of 800 nm. Several nanotubes shown in the figure were observed to grow over the barrier. If nanotubes grew along the surface,

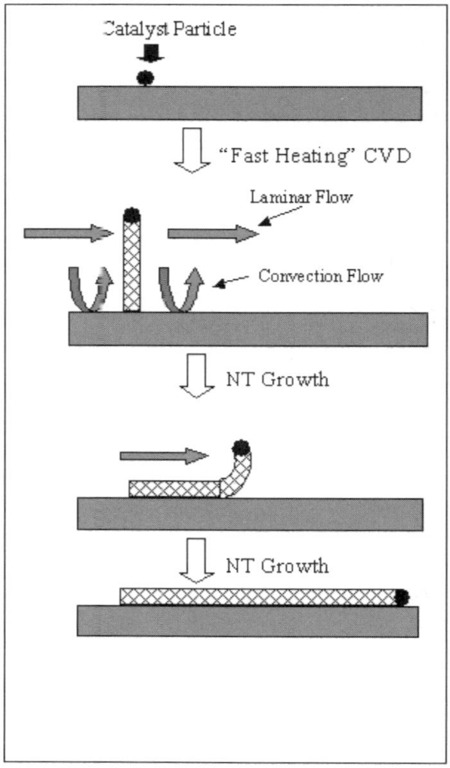

Fig. 4.12. Schematic diagram of the growth mechanism for the "fast heating" CVD process

they would be stopped by the vertical barrier on the surface. Additionally, the nanotubes have to be floating to grow across a trench as shown in Fig. 4.13b. These results demonstrated that at the initial stage of the growth, the nanotubes grew up from the surface. Additionally, the results also suggested that the "base-growth" mechanism may not be appropriate for the growth of these nanotubes, since the whole nanotubes up to centimeter long have to be floating in order to remain relatively straight. Taking Fig. 4.13b as an example, if the nanotubes grew with a "base-growth" mechanism, once part of the nanotube touched the edge of the step, the nanotube section between the catalyst and the edge of the step would not be straight, as shown in the figure since the nanotubes would keep growing longer, making the nanotubes curvy in the region. Based on such a hypothesis, these nanotubes should be grown in a "tip-growth" mechanism with the active catalysts on the tips of the growing nanotubes. Direct evidence for the "tip-growth" mechanism under "fast heating" conditions also came from the AFM observations. It was

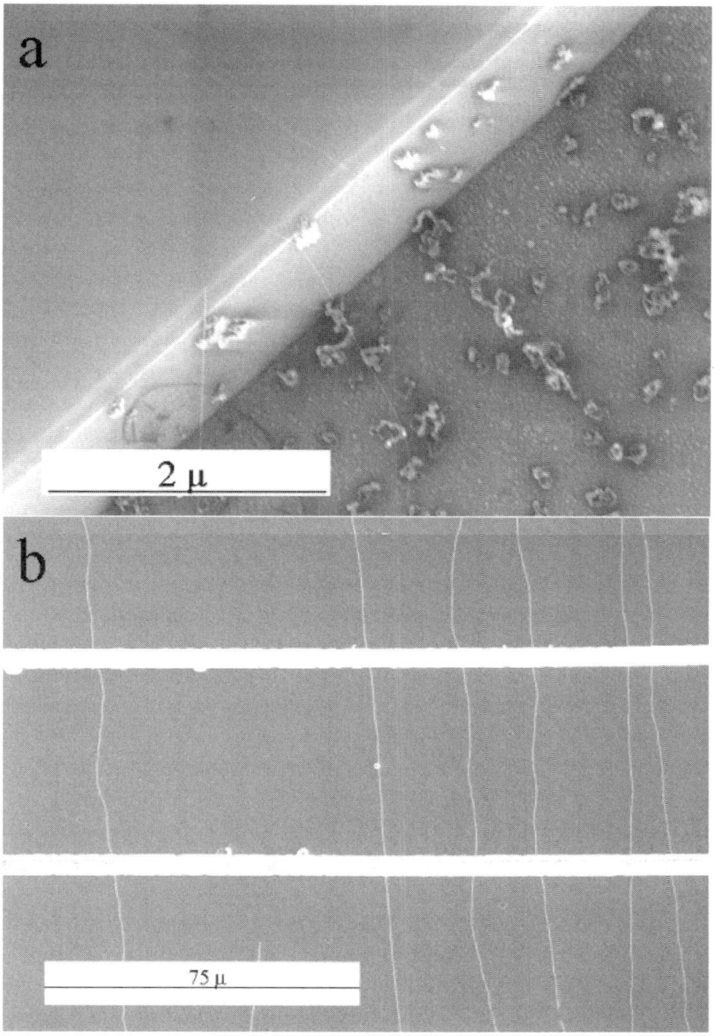

Fig. 4.13. SEM images of long nanotubes (**a**) grown over a 800 nm high barrier and (**b**) grown across 5 micron wide trenches

found that every tip of the long nanotubes had a nanoparticle as indicated in Fig. 4.14.

The size of the particles on the tip of the nanotubes was normally larger than the nanotube diameter. This may be due to the amorphous carbon coating around the catalyst during the cooling process. According to the proposed mechanism, the length of the nanotubes could be unlimited. The difference in lengths between nanotubes grown using different CVD processes can be

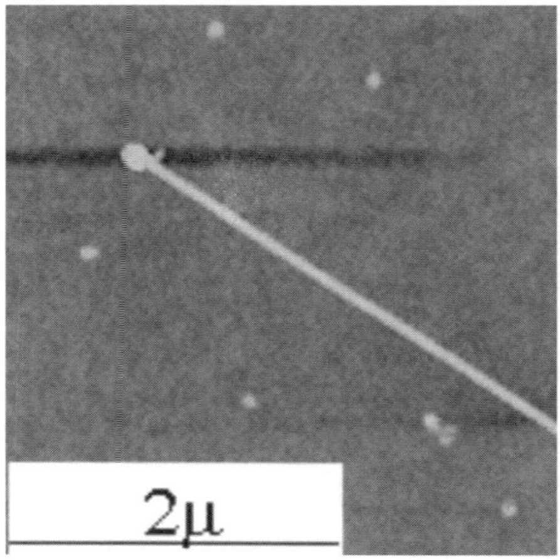

Fig. 4.14. AFM image of oriented long nanotubes with the catalyst particle on the tip

explained by taking into account of the difference between "tip-growth" and "base-growth" mechanisms. In the "base-growth" mechanism, the catalysts stay on the substrate throughout the growth process. Two possible reasons limited the growth of the nanotubes. One is the termination of nanotube growth because of the strong Van der Waals interaction between the nanotubes and the substrate surface when the nanotubes reach a certain length. For the base growth mechanism, since the whole nanotubes need to slide on the surface, once they rest on the surface, the nanotube/substrate interaction would increase as a function of the length. The growth would eventually stop when the force needed to move the whole nanotube became energetically unfavorable.

For the tip-growth mechanism, this would not present a problem since the catalysts were on the tip of the nanotubes. The other reason for the length difference between the two growth methods may be the diffusion of the feeding gas to the surface of the catalysts. The flow rate of the feeding gas on the substrate surface is much lower than above the surface. There are several possible reasons for this low efficiency. First, not all nanotubes grown in the process were grown by the tip-mechanism. Only a small fraction of nanoparticles having weak interaction with surface initiated nanotubes under the tip-growth mechanism, while the majority of the catalysts still nucleate nanotubes under the "base-growth" mechanism. This hypothesis was confirmed by SEM and AFM observations that there were many short and randomly oriented nanotubes close to the patterned catalyst area. These short nanotubes are presumably grown from the base-growth mechanism as demonstrated by the AFM observation that these nanotubes do not have nanoparticles on their tips. Second, some of nanotubes stopped growing because the grow-

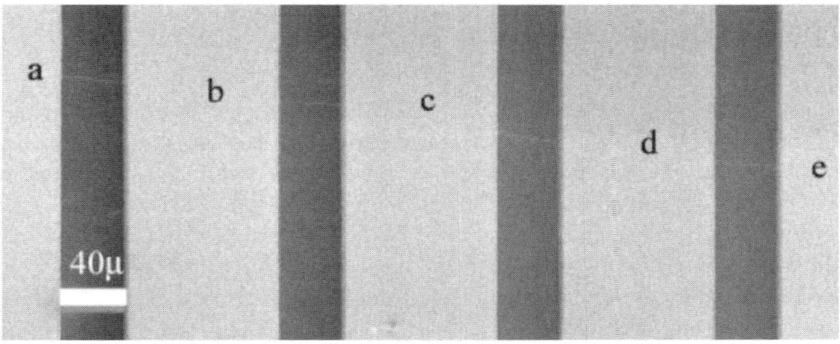

Fig. 4.15. SEM image of a long nanotube with multiple electrodes evaporated on

ing ends landed on the substrate rather than floating in the gas flow. This could explain why long nanotubes have different lengths ranging from several hundreds microns to millimeters.

4.6 Advantages of Long and Oriented Nanotubes for Device Applications

Long and well-oriented SWNTs on the surface have great advantages for nanotube-based device fabrication. Since the locations and orientations of the nanotubes on the substrate can be controlled, it is much easier to make contact with these nanotubes using a standard lithographic method with a predesigned electrode pattern. It is also possible to use a shadow mask to fabricate the devices without involving any photoresist during the device fabrication process. The commonly used photoresist for photolithography is very hard to remove from nanotubes, causing undesired high resistance contacts between nanotubes and metal electrodes. Additionally, it is possible to fabricate multiple devices on the same long nanotube as shown in Fig. 4.15.

Experimental results show that multiple devices fabricated on the same long nanotube have reproducible device performances.

4.7 Summary

In this chapter, we have discussed the direct growth of SWNTs on suitable substrates for device fabrication and, in particular, discussed a new "fast-heating" CVD process, which produces long and well-oriented SWNTs on the surface. Different catalysts, including Fe, Fe/Mo, Fe/Mo nanoparticles and Fe/Mo molecular nanoclusters, and different carbon sources, including carbon monoxide, hydrocarbons such as methane and alcohols such as methanol, can be used in the process. The capability of controlling the orientations and locations of nanotubes has been demonstrated by experimental data. Two dimensional SWNT crossed structures can easily be fabricated using the process. The length and alignment of nanotubes has been found to be due to a different growth mechanism of nanotubes under "fast-heating" conditions. These long and well-oriented SWNT arrays with controllable directions and locations offer a great advantage for nanoscale device fabrication, representing a significant step towards the control of SWNT structures for nanoelectronics applications. Further investigation on improving the growth efficiency and more precise control of the structure and helicity should be the focus of future research in the field.

References

1. S.J. Tans, M.H. Devoret, H.J. Dai, A. Thess, R.E. Smalley, L.J. Geerligs, C. Dekker, Nature **386**, 474 (1997)
2. A. Bachtold P. Hadley, T. Nakanishi, C. Dekker, Science **294**, 1317 (2001)
3. A. Bachtold P. Hadley, T. Nakanishi, C. Dekker, Physica E-Low-Dimensional Systems Nanostructures **16**, 42 (2003)
4. P. Avouris, R. Martel, V. Derycke, J. Appenzeller, Physica B-Condensed Matter **323**, 6 (2002)
5. V. Derycke, R. Martel, J. Appenzeller, P. Avouris, Nano Letters **1**, 453 (2001)
6. A. Javey, Q Wang, A. Ural, Y.M. Li, H.J. Dai, Nano Letters **2**, 929 (2002)
7. A. Javey, H. Kim, M. Brink, Q. Wang, A. Ural, J. Guo, P. McIntyre, P. McEuen, M. Lundstrom, H.J. Dai, Nature Materials **1**, 241 (2002)
8. J.M. Bonard, H. Kind, T. Stockli, L.A. Nilsson, Solid-State Electronics **45**, 893 (2001)
9. O. Zhou, H. Shimoda, B. Gao, S.J. Oh, L. Fleming, G.Z. Yue, Accounts of Chemical Research **35**, 1045 (2002)
10. A. Javey, J. Guo, Q. Wang, M. Lundstrom, H.J. Dai, Nature **424**, 654 (2003)
11. M. Radosavljevic, S. Heinze, J. Tersoff, P. Avouris, Applied Physics Letters **83**, 2435 (2003)
12. W. Kim, A Javey, O. Vermesh, O. Wang, Y.M. Li, H.J. Dai, Nano Letters **3**, 193 (2003)
13. S. Rosenblatt, Y. Yaish, J. Park, J. Gore, V. Sazonova, P.L. McEuen, Nano Letters **2**, 869 (2002)
14. M. Shim, A. Javey, N.W.S. Kam, H.J. Dai, Journal of the American Chemical Society **123**, 11512 (2001)
15. T. Someya J. Small, P. Kim, C. Nuckolls, J.T. Yardley, Nano Letters **3**, 877 (2003)
16. S.J. Wind, J. Appenzeller, P. Avouris, Physical Review Letters **91**, Art. No. 058301 (2003)
17. M.S. Fuhrer, J. Nygard, L. Shih, M. Forero, Y.G. Yoon, M.S.C. Mazzoni, H.J. Choi, J. Ihm, S.G. Louie, A. Zettl, P.L. McEuen, Science **288**, 494 (2000)
18. K. Besteman, J.O. Lee, F.G.M. Wiertz, H.A. Heering, C. Dekker, Nano Letters **3**, 727 (2003)

19. R.J. Chen, S. Bangsaruntip, K.A. Drouvalakis, N.W.S. Kam, M. Shim, Y.M. Li, W. Kim, P.J. Utz, H.J. Dai, Proceedings of the National Academy of Sciences of the United States of America **100**, 4984 (2003)
20. J. Kong, N.R. Franklin, C.W. Zhou, M.G. Chapline, S. Peng, K.J. Cho, H.J. Dai, Science **287**, 622 (2000)
21. J. Kong, M.G. Chapline, H.J. Dai, Advanced Materials **13**, 1384 (2001)
22. J. Li, Y.J. Lu, Q. Ye, M. Cinke, J. Han, M. Meyyappan, Nano Letters **3**, 929 (2003)
23. A. Modi, N. Koratkar, E. Lass, B.Q. Wei, P.M. Ajayan, Nature **424**, 171 (2003)
24. Q.F. Pengfei, O. Vermesh, M. Grecu, A. Javey, O. Wang, H.J. Dai, S. Peng, K.J. Cho, Nano Letters **3**, 347 (2003)
25. H.C. Choi, W. Kim, D.W. Wang, H.J. Dai, Journal of Physical Chemistry B **106**, 12361 (2002)
26. C.L. Cheung, A. Kurtz, H. Park, C.M. Lieber, Journal of Physical Chemistry B **106**, 2429 (2002)
27. L. An, J.M. Owens, L.E. McNeil, J. Liu, Journal of the American Chemical Society **124**, 13688 (2002)
28. B.C. Liu, S.H. Tang, Z.L. Yu, B.L. Zhang, T. Chen, S.Y. Zhang, Chemical Physics Letters **357**, 297 (2002)
29. M. Zheng, A. Jagota, E.D. Semke, B.A. Diner, R.S. McLean, S.R. Lustig, R.E. Richardson, N.G. Tassi, Nature Materials **2**, 338 (2003)
30. Z.H. Chen, X. Du, M.H. Du, C.D. Rancken, H.P. Cheng, A.G. Rinzler, Nano Letters **3**, 1245 (2003)
31. D. Chattopadhyay, L. Galeska, F. Papadimitrakopoulos, Journal of the American Chemical Society **125**, 3370 (2003)
32. R. Krupke, F. Hennrich, H. von Lohneysen, M.M. Kappes, Science **301**, 344 (2003)
33. E. Joselevich, C.M. Lieber, Nano Letters **2**, 1137 (2002)
34. A. Ural, Y.M. Li, H.J. Dai, Applied Physics Letters **81**, 3464 (2002)
35. S.G. Rao, L. Huang, W. Setyawan, S.H. Hong, Nature **425**, 36 (2003)
36. D. Mann, A. Javey, J. Kong, Q. Wang, H.J. Dai, Nano Letters **3**, 1541 (2003)
37. P.L. McEuen, M.S. Fuhrer, H.K. Park, Ieee Transactions on Nanotechnology **1**, 78 (2002)
38. S.M. Huang, X.Y. Cai, J. Liu, Journal of the American Chemical Society **125**, 5636 (2003)
39. S. Huang, B. Maynor, X. Cai, J. Liu, Advanced Materials **15**, 1651 (2003)
40. Y.M. Li, W. Kim, Y.G. Zhang, M. Rolandi, D.W. Wang, H.J. Dai, Journal of Physical Chemistry B **105**, 11424 (2001)
41. E.K. Achim Muller, Hartmut Bogge, Marc Schmidtmann, Christian Beugholt, Paul Kogerler, Canzhong Lu, Angewandte Chemie International Edition **37**, 1220 (1998)
42. E.K. Achim Muller, Hartmut Bogge, Marc Schmidtmann, Frank Peters, Angewandte Chemie International Edition **37**, 3360 (1998)
43. S.S. Achim Muller, Syed Qaiser Nazir Shah, Hartmut Bogge, Marc Schmidtmann, Shatarupa Sarkar, Paul Kogerler, Bjorn Hauptfleisch, Alfred X. Trautwein, Volker Schunemann, Angewandte Chemie International Edition **38**, 3238 (1999)
44. S.P. Achim Muller, Samar K. Das, Erich Krickemeyer, Hartmut Bogge, Marc Schmidtmann, Bjorn Hauptfleisch, Angewandte Chemie International Edition **38**, 3241 (1999)

45. S.K.D. Achim Muller, Paul Kogerler, Hartmut Bogge, Marc Schmidtmann, Alfred X. Trautwein, Volker Schunemann, Erich Krickemeyer, Wilhelm Preetz, Angewandte Chemie International Edition **39**, 3413 (2000)
46. S.K.D. Achim Muller, Marina O. Talismanova, Hartmut Bogge, Paul Kogerler, Marc Schmidtmann, Serge S. Talismanov, Marshall Luban, Erich Krickemeyer, Angewandte Chemie International Edition **41**, 579 (2002)
47. A.M. Cassell, N.R. Franklin, T.W. Tombler, E.M. Chan, J. Han, H. Dai, Journal of the American Chemical Society **121**, 7975 (1999)
48. Y.G. Zhang, A.L. Chang, J. Cao, Q. Wang, W. Kim, Y.M. Li, N. Morris, E. Yenilmez, J. Kong, H.J. Dai, Applied Physics Letters **79**, 3155 (2001)
49. S. Huang, X. Cai, J. Liu, Nano Letters **4(6)**, 1025 (2004)
50. B.Q. Wei, R. Vajtai, Y. Jung, J. Ward, R. Zhang, G. Ramanath, P.M. Ajayan, Nature **416**, 495 (2002)
51. S.M. Huang, L.M. Dai, A.W.H. Mau, Advanced Materials **14**, 1140 (2002)
52. S.M. Huang, A.H.W. Mau, Applied Physics Letters **82**, 796 (2003)
53. S.M. Huang, A.W.H. Mau, Journal of Physical Chemistry B **107**, 3455 (2003)
54. G. Gu, G. Philipp, X.C. Wu, M. Burghard, A.M. Bittner, S. Roth, Advanced Functional Materials **11**, 295 (2001)
55. S. Huang, Q. Fu, L. An, J. Liu, Physical Chemistry Chemical Physics **6**, 1077 (2004)
56. J. Kong, H.T. Soh, A.M. Cassell, C.F. Quate, H.J. Dai, Nature **395**, 878 (1998)
57. R.T.K. Baker, P.S. Harris, *Formation of Filamentous Carbon*, Vol. 14. New York: Marcel Dekker, 1978
58. R.T.K. Baker, Carbon **27**, 315 (1989)
59. S.B. Sinnott, R. Andrews, D. Qian, A.M. Rao, Z. Mao, E.C. Dickey, F. Derbyshire, Chemical Physics Letters **315**, 25 (1999)
60. J. Han, J.-B. Yoo, C.Y. Park, H.-J. Kim, G.S. Park, M. Yang, I.T. Han, N. Lee, W. Yi, S.G. Yu, J.M. Kim, Journal of Applied Physics **91**, 483. (2002)

5 Nano-Peapods Encapsulating Fullerenes

Toshiya Okazaki and Hisanori Shinohara

This chapter reviews the recent progress on structures and properties of nano-peapods (single-wall carbon nanotubes (SWNTs) encapsulating fullerenes or metallofullerenes) by the high-resolution transmission electron microscope (HRTEM), electron energy-loss spectroscopy (EELS), low-temperature scanning tunneling microscope (STM) and spectroscopy (STS), and field effect transistor (FET) transport measurements together with the details of their synthesis methods. The potential abilities of this novel hybrid material for electronic devices, sample cells for TEM and reaction chambers at a molecular scale are discussed.

5.1 Introduction

Carbon is a remarkable element showing a variety of stable forms. Under the ambient condition and in bulk form, sp^2 bonded graphite is the most stable phase of carbon. At higher temperatures and pressures, however, sp^3 bonded cubic diamond is stable. In addition to the bulk phases, much attention has recently focused on small and nanometer-scale sized carbon materials. In particular, carbon nanotubes (CNTs), a new form of carbon, have been attracted great interest because they are considered as prototypes of one-dimensional nanostructures in nanoscience and nanotechnology. One of the most distinguishing features of these materials is that they can encapsulate metal atom(s) or molecule(s) into their central hollow space, which may lead to a new class of hybrid materials with novel electronic properties.

The filling of the interior space of CNTs was first achieved with multi-wall carbon nanotubes (MWNTs) in 1993. When lead particles were deposited on the walls of MWNTs in order to investigate their surface activity, Ajayan and Iijima accidentally found that many tubes were trapped them inside after annealing [1]. Although the encapsulation occurs on nanometer scale, this phenomenon can be explained by macroscopic capillarity [2]. It was found that only relatively low-surface-tension materials (less than 100–200 mN/m) can be drawn inside MWNTs.

The incorporation of molecules into the hollow space of single-wall nanotubes (SWNTs) was also accidentally discovered [3]. In 1998, Luzzi and co-workers found that C_{60} molecules nested inside SWNT, like many peas

Table 5.1. Smallest limit of tube diameters (d) for C_{60}, C_{70}, C_{78} and C_{84} [9]. Here the relation of $\omega = 246/d$ (ω; RBM frequency / cm^{-1}) was used for calculations

Fullerene	d/nm
C_{60}	1.37
C_{70}	1.45
C_{78}	1.45
C_{84}	1.54

in a pod, when they were studying purified SWNTs by using a transmission electron microscope (TEM). This hybrid nano-carbon material has, therefore, been called "nano-peapods", "nanotube peapods", "fullerene peapods" or merely "peapods". Since then, nano-peapods have attracted the attention of many researchers due to their unique structural properties, which may open up a brand new "nano-world" in both scientific and practical arenas. For instance, the confining geometry on quasi one-dimensional crystallization of fullerenes may provide an ideal system to study the physical and chemical properties of low-dimensional materials. Furthermore, they are also expected to act as building blocks in future nano-electronics because the mechanical and electronic properties of SWNTs frequently undergo considerable modification upon encapsulation, which allows us to finely tune these parameters by altering the encapsulated fullerenes.

A few years after the discovery, several groups independently succeeded in synthesizing nano-peapods with a high efficiency [4–7]. Their synthesis methods are very similar to each other except for small details, in which fullerenes are encapsulated into the hollow space of SWNTs by thermal kinetic energy. It is an astonishingly simple method, which does not require any special apparatus. An researchers can synthesize nano-peapods by this method in the ordinary physical and chemical laboratory.

5.2 High-Yield Synthesis of Nano-Peapods

A bulk production method for nano-peapods has been achieved through the gas phase reaction by several groups [4–7]. The method allows us to produce various fullerene peapods in high yield ($> 80\,\%$). To obtain high-purity and high-yield nano-peapods, one should prepare isolated fullerene molecules and high-purity open-ended SWNTs. Fullerenes are soluble molecules so that one can easily purify them more than 99 % with a well-established separation method, such as a high performance liquid chromatography (HPLC) [8]. On the other hand, SWNTs can be purified by a combination of several methods.

For example, we remove the amorphous carbon materials and metal catalyst particles by refluxing the as-produced SWNTs in acid solutions. The purification method of SWNTs strongly depends on the purity of the pristine SWNT material. (This is also true for MWNTs.) In particular, it is almost

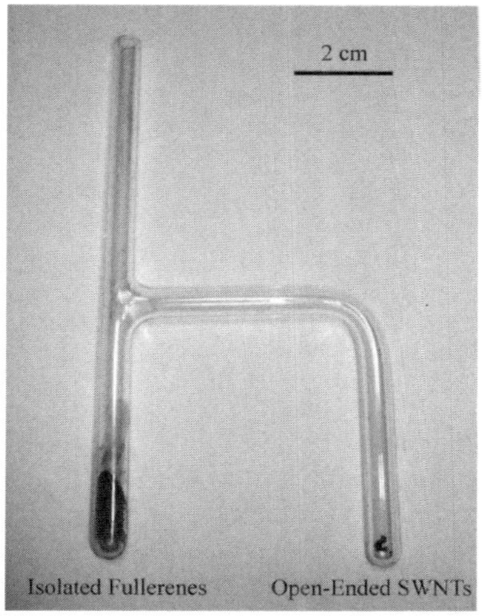

Fig. 5.1. Fullerenes and open-ended SWNTs in a glass ampoule. Prior to the doping, the SWNTs and the fullerenes were degassed at $\sim$ 200°C in a vacuum. High-yield nano-peapods are produced after heating for several days at 400–500°C

impossible to remove graphitic particles or sheets because they are much more stable than SWNTs. Therefore it is important to synthesize SWNTs without such graphitic impurities from the beginning. The acid treatment also promotes opening the cap of SWNTs. Following this treatment, the SWNTs were heated in dry air at 450°C for 30 min to fully open the tube ends because some resultant amorphous carbons clog them.

The distribution of the diameter of SWNTs is another important factor. If the diameter of the SWNT is smaller than that of the fullerenes, encapsulation of the fullerenes will not be possible. The smallest limit of tube diameter was determined for several fullerene molecules based on the shift of the radial breathing mode (RBM) in the Raman spectra [9]. Table 5.1 shows the smallest SWNT diameter obtained that can encapsulate fullerenes from C_{60} to C_{84}. For instance, SWNTs with diameters larger than 1.37 nm are required for making C_{60} nano-peapods, which is almost equal to the sum of the diameter of C_{60} (= 0.71 nm) and twice the interlayer distance of the graphite ($= 0.34 \times 2 = 0.68$ nm).

The doping of fullerenes into the inner hollow space of the so-prepared SWNTs is an easy job. Both the fullerenes and the open-ended SWNTs are placed in a vacuum-sealed glass ampoule (Fig. 5.1). Prior to the doping, the SWNTs and the fullerenes was degassed at $\sim$ 200°C in a vacuum to remove residual gases. One can obtain high-yield nano-peapods after heating the ampoule for two days at $\sim$ 400°C for C_{60}, and $\sim$ 500°C for higher fullerenes and metallofullerenes. A typical high-resolution TEM (HRTEM) image of a rope of $Gd@C_{82}$ nano-peapods is shown in Fig. 5.2 [10]. It shows a crystalline

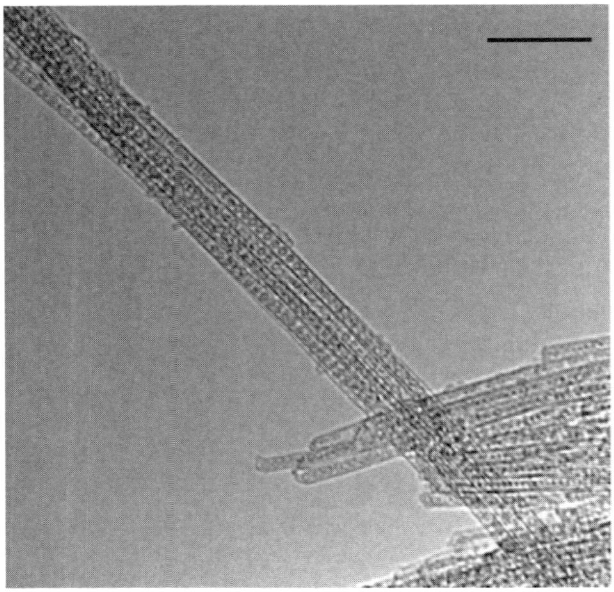

Fig. 5.2. HRTEM images of the bundles of the $(Gd@C_{82})_n@SWNTs$ [10]. It is clear that $Gd@C_{82}$ are close-packed inside the SWNTs. Scale bar = 10 nm

bundle of close-packed tubes, almost all of which contain $Gd@C_{82}$ molecules in a one dimensional array.

Even though the HRTEM image clearly shows that fullerenes are closely packed inside the SWNTs, it is difficult to quantify the filling-yield. Several evaluation criteria have been proposed so far based on Raman scattering, [7] electron energy-loss spectroscopy (EELS) [4, 11]. X-ray diffraction [12] and difference in weight before and after the doping [13]. To date, one of the most reliable evaluation methods is that of utilizing the X-ray diffraction spectrum [12]. The (1 0) peak of SWNTs in the X-ray diffraction spectrum is very sensitive to the molecular encapsulation because the structural factor of SWNTs is drastically changed by the additional electron density from the encapsulated molecules. Thus, a substantial reduction in the intensity of the (1 0) peak is observed upon encapsulation of molecules, which allows us to quantitatively estimate the filling yield.

Here the question arises as to how SWNTs catch fullerenes. Do fullerenes enter the interior space from the open end of the tube or from defects that might be present on the side wall? While the mechanism of fullerene encapsulation is still not fully elucidated, several molecular dynamics simulations have provided important clues to it [14, 15]. In general, the number and the thickness of bundles of SWNTs increase, as the yield and purity of SWNTs increase. In most cases, researchers have used such high-purity SWNTs for synthesizing nano-peapods, most of which formed bundles rather than being isolated. The end of the bundles may not be flat, that is, some SWNTs stick out of the others and some stand back from their neighbors. When sublimating fullerenes approach the end of such bundles, they may first adsorb in the sticking neighborhoods and then enter the inner hollow space of the standing back SWNT through the open end by "head-on" collisions [15]. The probability of encapsulation via "head-on" collision was found to be approximately 100 times larger than those through the defects on the side wall and an isolated open end.

Once nano-peapods are formed, their structure is very stable. Theoretical studies predict that the encapsulation process of C_{60} molecules is exothermic for SWNTs with proper diameter [16, 17]. A decrease of the total entropy is compensated for by the gain in potential energy. For example, each C_{60} molecule is stabilized by 0.51 eV upon encapsulation into (10, 10) SWNT, indicating that it is very difficult to extract C_{60} from the SWNTs by high temperature annealing, for example. Indeed, it has been experimentally observed that C_{60} molecules react with each other within SWNTs before extraction [9]. However, it is possible to take C_{60} molecules out from SWNTs by strong sonication [18]. It is likely that such sonication treatment breaks the side walls of SWNTs [19] and C_{60} molecules can come out through the produced holes.

Besides the gas phase doping, nano-peapods can also be made through liquid phase reactions [20, 21]. The method involves immersing open-ended SWNTs in a saturated solution of fullerenes. Unfortunately, the doping efficiency was as high as $\sim 70\,\%$ at the present moment. The solution method is especially advantageous in the bio-medical use of nano-peapods because it can be applied for thermally unstable molecules such as large organic and biological molecules.

5.3 Packing Alignment of the Fullerenes Inside SWNTs

As the interior space of SWNTs is similar in size to the diameter of fullerenes, the encapsulated fullerenes form quasi-one-dimensional crystals inside SWNTs. Intermolecular distances between adjacent fullerenes have been obtained by electron diffraction patterns (Table 5.2) [4,12,22,23]. In the case of C_{60} nano-peapods, the intermolecular distance of the encapsulated C_{60} molecules is shorter than that of the three-dimensional bulk crystal of C_{60}. This means that the interaction between the adjacent C_{60} molecules in nano-peapods is stronger than that in the bulk crystal. However, the C_{60} molecules do not form covalent bonds because the obtained distance of 0.97 nm is longer than that of the C_{60} polymers. The CNT walls should affect the electronic structure of the encapsulated C_{60}, which results in a subtle structural change.

The elongated C_{70} and C_{80} (D_{5d}) molecules exhibit different intermolecular distances depending on their orientations inside the SWNTs (Ta-

Fig. 5.3. HRTEM images of the isolated $(C_{80}~(D_{5d}))_n$@SWNT with (**a**) 1.29 and (**b**) 1.43 in diameters [22]

Table 5.2. Intermolecular distances of various fullerenes in their nano-peapod structures, together with those in the bulk crystal and C_{60} polymer for [4, 22, 23]. All measurements were carried out at room temperature

Fullerene (Symmetry)	In nano-peapods /nm	In 3D bulk crystal /nm	In polymerized crystal /nm
C_{60}	0.97 ± 0.02	1.002	0.91
C_{70}	1.02 ± 0.04	1.044	
$C_{78}~(C_{2v}(3))$	1.00 ± 0.02		
$C_{80}~(D_{5d})$	1.08 ± 0.04		
$C_{82}~(C_2)$	1.10 ± 0.03	1.14	
Sm@C_{82}(I)	1.10 ± 0.02		
Gd@C_{82}	1.10 ± 0.03		
La@C_{82}(I)	1.11 ± 0.03		
$C_{84}~(D_2/D_{2d})$	1.10 ± 0.03	1.12	
Sc_2@C_{84}	1.10 ± 0.03	1.12	

ble 5.2) [22, 24]. Figure 5.3 shows HRTEM images of $(C_{80}(D_{5d}))_n$@SWNTs with different diameters. It is clearly seen that C_{80} have to lie inside the SWNTs when the diameter of the SWNTs is smaller than the longer axis of the $C_{80}(D_{5d})$ molecules (Fig. 5.3a). In contrast, $C_{80}(D_{5d})$ fullerenes are "standing" to the tube axis when they are encapsulated in SWNTs with a larger diameter (Fig. 5.3b). Such variations are likely to be reflected in the intermolecular distance.

Generally endohedral metallofullerenes undergo charge transfer in the ground states. Two or three valence electrons transfer from the encaged atom to fullerenes. For instance, the electronic structure of Sm@C_{82} is well represented as Sm^{2+}@C_{82}^{2-} as a result of the transfer of the two electrons [25, 26]. On the other hand, three electrons transfer in Gd@C_{82} and La@C_{82} leading to Gd^{3+}@C_{82}^{3-} and La^{3+}@C_{82}^{3-} configurations [27, 28]. The obtained intermolecular distance of Sm@C_{82} inside the SWNTs is 1.10 nm, which is identical to that of C_{82} in $(C_{82})_n$@SWNTs and Gd@C_{82} in (Gd@$C_{82})_n$@SWNTs (Table 5.2) [4]. It is likely that the intermolecular distance inside the SWNT is not sensitive to the electronic structure and is rather determined by the size of the fullerene cage.

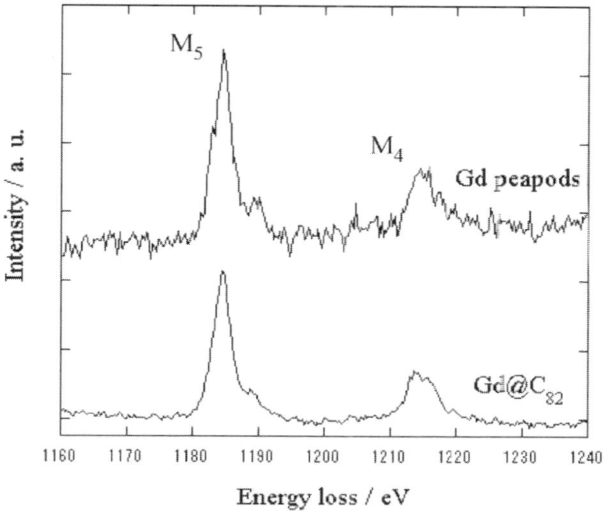

Fig. 5.4. EELS spectrum obtained from the bundle of $(Gd@C_{82})_n$@SWNT together with those of Gd^{3+} in the intact $Gd@C_{82}$ [10]

5.4 Electronic Structures of Nano-Peapods

The electronic structures of doubly (i.e., first by a fullerene and then by a CNT) encapsulated atoms in nano-peapods have been investigated by EELS. Especially, the peak positions of the M-edge of the lanthanoid metal atoms have served as guides for determining the valence state and consequently the amount of charge transfer [25–27]. An EELS spectrum observed from the bundles of $(Gd@C_{82})_n$@SWNTs is shown in Fig. 5.4 together with the reference spectra of Gd^{3+} in the intact (bulk) $Gd@C_{82}$. A somewhat poor experimental signal to noise ratio for the EELS spectra of $(Gd@C_{82})_n$@SWNTs is due to the extremely small number of Gd atoms present in this structure and also to the remarkably weak electron beam employed to prevent the irradiation damage. The highest peak positions for the M_5 and M_4 edges for $(Gd@C_{82})_n$@SWNTs are experimentally measured as 1184 eV and 1214 eV, respectively. The peak positions of Gd M_{45} edges for $(Gd@C_{82})_n$@SWNTs are very similar to those for the intact $Gd@C_{82}$, indicating that the valence state for the encaged Gd atoms of $Gd@C_{82}$ remains unchanged upon encapsulation by SWNTs.

The SWNTs show novel electronic properties depending on diameters and chirality. All armchair SWNTs are metallic, but 2/3 of the possible zigzag and chiral SWNTs are semi-conductive [29]. This has been confirmed experimentally by atomically resolved STM and STS [30, 31]. When the Gd-metallofullerenes are encapsulated inside the SWNT, the electric structure of both the Gd-metallofullerene and the SWNT can be perturbed substantially.

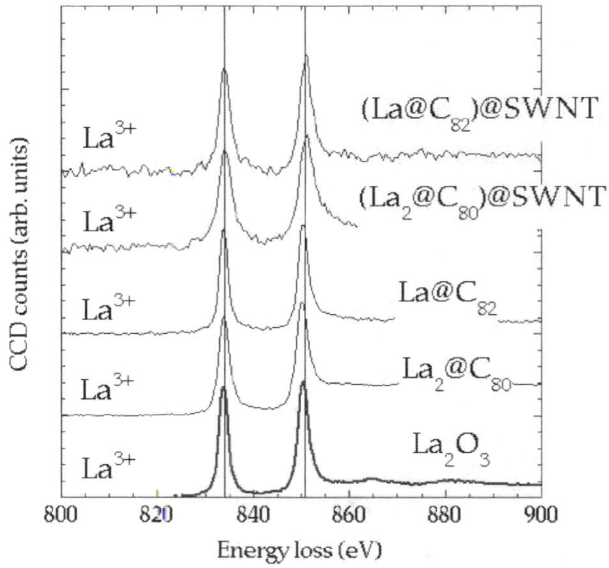

Fig. 5.5. EELS spectra obtained from the bundle of $(La@C_{82})_n$@SWNTs and $(La_2@C_{80})_n$@SWNTs together with those of La^{3+} in the intact $La@C_{82}$, $La_2@C_{80}$ and La_2O_3 [33]

However, the observed peak positions and shapes of the EELS spectra are almost identical to each other and do not depend on the choice of the bundles. (A few tens of the bundles were examined.) This proves that the Gd atom in $(Gd@C_{82})_n$@SWNTs takes +3 state irrespective of the physical property of SWNT.

Similar experiments have been carried out for $(Sc@C_{82}(I))_n$@SWNTs [32], $(Ti_2C_2@C_{78})_n$@SWNTs [33], $(La@C_{82})_n$@SWNTs [33], $(La_2@C_{80})_n$@SWNTs [33], $(Ce_2@C_{80})_n$@SWNTs [33], $(Sm@C_{82})_n$@SWNTs [6,26] and $(Gd_2@C_{92})_n$@SWNTs [34]. Figure 5.5 shows the EELS spectra of $La@C_{82}$ nano-peapods and $La_2@C_{80}$ nano-peapods together with those of the intact $La@C_{82}$, $La_2@C_{80}$ and La_2O_3 for references of La^{3+}. La-metallofullerenes are one of the first metallofullerenes produced and a typical example for an encaged metal atom having the electronic structure of M^{3+} (M = metal atom) [8]. In Fig. 5.5, the peak positions of La M_{45} edges in all spectra are almost identical to each other. Even in the other nano-peapods, the EELS spectra of the encapsulated atoms in the nano-peapods were same as those in the corresponding intact metallofullerenes and insensitive to the chirality of SWNTs. The encapsulated atoms of Sc, Ti, La, Sm and Gd still remain Sc^{2+}, Ti^{2+}, La^{3+}, Sm^{2+} and Gd^{3+} states, respectively, in their nano-peapods structures.

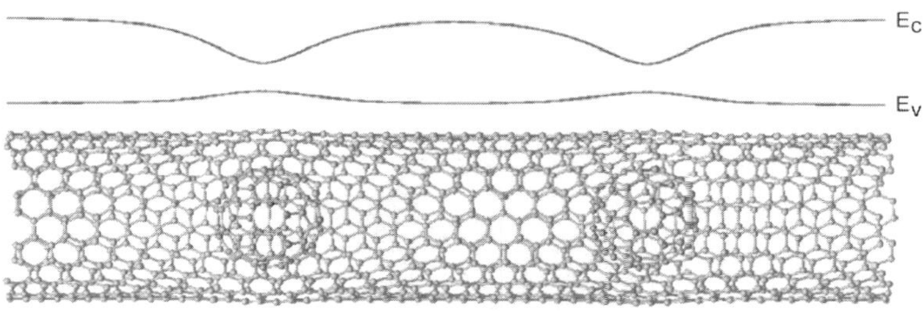

Fig. 5.6. Illustration of two Gd@C_{82} endohedral metallofullerenes in a (11,9) SWNT. Minute elastic strain is expected around the metallofullerenes. The schematic representation of the modulation of the conduction and valence bands is also shown

The electronic structures of the outer SWNTs are, in contrast, severely modified upon fullerene insertion. Low-temperature STM and STS studies performed on (Gd@C_{82})$_n$@SWNTs and (C_{60})$_n$@SWNTs revealed significant band-gap modulations [18,35]. As is schematically shown in Fig. 5.6, the band gap of a semiconducting SWNT has been altered by the insertion of Gd@C_{82}. Previous TEM images and EELS spectra suggested that the metallofullerenes can be spaced regularly as close as 1.1 nm in a high-density nano-peapod structure (Fig. 5.2, Table 5.2) [4, 5], while 1.1–3 nm spacing was often observed in a low-density nano-peapod. The low density of the encapsulated Gd@C_{82} was achieved by sonicating the high-density nano-peapods [18]. In the Gd@C_{82} peapod samples, about 10 % of over 200 SWNT images showed locally modified semiconducting band gaps [18].

STM and STS ($\mathrm{d}I/\mathrm{d}V$) give bias-dependent topographic images and the local density of the states (LDS) of carbon nanotubes near the fermi level, respectively. Figure 5.7 shows a typical STM image (upper) and the corresponding spatial variation of $\mathrm{d}I/\mathrm{d}V$ along the tube axis at a temperature of $\sim$ 5 K. The thermal drift is less than 0.1 nm/hour at this temperature. The chirality of the nanotube is determined to be (11,9) from a measured diameter of 1.4 nm (nominally 1.375 nm) and a chiral angle of 27 (nominally 26.7°). In Fig. 5.7, two strong Van Hove singularity peaks corresponding to conduction and valence band edges are clearly seen with two smaller ones at higher bias voltages. The x-axis is along the tube axis, the y-axis the energy, and the z-axis $\mathrm{d}I/\mathrm{d}V$ as shown at the bottom part. The original band gap of 0.43 eV is narrowed down to 0.17 eV where the fullerene is expected to be located.

There are several theoretical studies which explain the modulation of the band gap [16, 17, 36–39]. SWNTs have the maximum electron density inside the wall, which is called "nearly free-electron (NFE)" states. Upon insertion of fullerenes into a proper semi-conducting SWNT, the NFE state couples with the π state of the fullerenes leading to novel electronic states between the valence and the conduction bands of the original SWNT. The degree

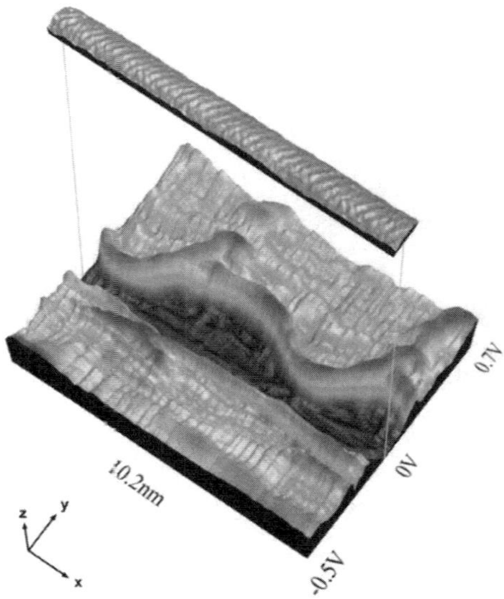

Fig. 5.7. 3D representation of a (11,9) SWNT topograph. The corresponding 512 dI/dV spectra at the center of the tube along the tube. The x-axis indicates the position along the tube, the y-axis the energy and the z-axis dI/dV [10] (see Color Plates, p. 341)

of the hybridization and the energy level of the derived state depend on the encapsulated fullerenes and the chirality of SWNT, which means that we can perform 'local band gap engineering' at the site where a fullerene is endothermally inserted.

5.5 Transport Properties of Nano-Peapods

Such band-gap modulation leads to a substantial change in the transport properties of SWNTs. The electric transport properties of several nano-peapods are investigated by using these as channels in the field effect transistor (FET), which has a back gate structure [40, 41]. The current versus gate voltage (I_D–V_{GS}) characteristics of several FETs are shown in Fig. 5.8. In this case, an individual SWNT or an individual nano-peapod was used as a the FET channel. The diameter of the SWNTs ranges from 1.3 to 1.6 nm. Figure 5.8a shows the $I_D - V_G$ curve of unfilled-SWNT-FET. Starting from a negative gate voltage V_{GS}, the current I_D steeply decreases by three orders of magnitude towards to 0V as the gate voltage increases. This implies that the SWNT-FET was p-type dominated by positive carriers (holes).

As is the case for C_{60}-peapod-FET (Fig. 5.8b), the I_D–V_{GS} curves show a hole transport region for $V_{GS} < 10$ V, undergoing an insulating region and switching to the electron transport region for $V_{GS} > 35$ V. This indicates that

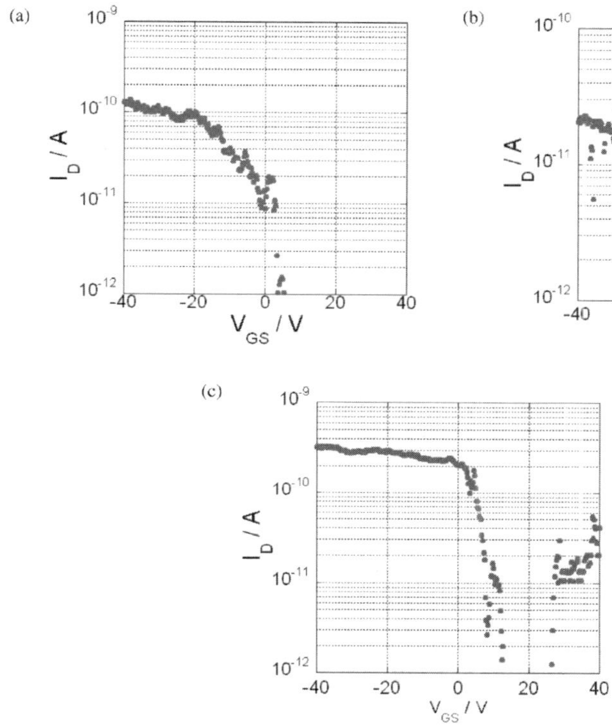

Fig. 5.8. I_D–V_{GS} curves of (**a**) unfilled SWNT-FET, (**b**) C_{60}-peapod-FET and (**c**) Gd@C_{82}-peapods-FET at source-drain bias of 1 mV. All transport measurements were carried out at 23 K

C_{60} peapods exhibit ambipolar FET behavior with both n- and p-channels easily accessible by simple electrostatic gates. Similar results were obtained from more than 10 independent devices composed of individual C_{60} nano-peapods.

Such ambipolar behavior is very salient in Gd@C_{82} peapods (Fig. 5.8c). Starting from the positive gate voltage V_{GS}, the current I_D first decreases, then becomes insulating at $V_{GS} \sim 10$ V, and increases again at $V_{GS} \sim 25$ V. The widths of the insulating regions are different between C_{60}-peapod-FET ($\Delta V_{GS} \sim 25$ V) and Gd@C_{82}-peapod-FET ($\Delta V_{GS} \sim 15$ V).

The difference in the FET characteristics observed here can be explained as follows. A typical semi-conducting SWNT with a diameter of ~ 1.4 nm has a band gap of around 0.6 eV [30]. To show the n-type characteristics, sufficiently high gate voltages are needed to electrostatically shift down the conduction band of the SWNT by 0.6 eV. In contrast, C_{60} peapods and Gd@C_{82} peapods have smaller band gaps [18, 35]. For instance, STM/STS studies revealed that electronic structures of the SWNT are strongly modulated by encapsulating $Gd@C_{82}$ (Fig. 5.7). A band gap of ~ 0.5 eV is narrowed down to ~ 0.1 eV at sites where $Gd@C_{82}$ are inserted. The small band gap can lead to significant carrier transport through both the conduction and valence bands. In fact, ambipolar FET behavior has already been reported previously with large-diameter (3–5 nm) SWNTs that have small band gaps (~ 0.2 eV). Compared to $Gd@C_{82}$ peapods, the doping effect of C_{60} peapods may be much smaller [35]. The C_{60} peapods have, therefore, a larger band gap resulting in the wider insulating region (ΔV_{GS}).

5.6 Nano-Peapod as a Sample Cell at Nanometer Scale

Since SWNTs are extremely thin and almost transparent to high-energy electron beams, TEM can allow us to observe molecular structures encapsulated in SWNTs. The HRTEM images of the isolated $(Sm@C_{82})_n@SWNT$ and $(Gd@C_{82})_n@SWNT$ are shown in Fig. 5.9 [4, 23].

The parallel lines denote the walls of the SWNTs. The inner circles are interpreted as the image of C_{82} fullerenes. The dark spot inside the circle can be ascribed to the encapsulated metal ion. This spot usually locates near the fullerene cage rather than the center of the cage, which is consistent with the known structure of the reported mono-metallofullerenes [8]. As compared with the $(Gd@C_{82})_n@SWNT$ case (Fig. 5.9b), the dark spots in the HRTEM image of $(Sm@C_{82})_n@SWNT$ (Fig. 5.9a) are oftentimes missing. The results suggest that the interaction between $Sm@C_{82}$ and SWNT is weaker than that between $Gd@C_{82}$ and SWNT. The electronic structures of $Sm@C_{82}$ and $Gd@C_{82}$ are well represented as $Sm^{2+}@C_{82}^{2-}$ and $Gd^{3+}@C_{82}^{3-}$, respectively. The ability of metallofullerenes to rotate in the SWNT may depend on this difference in the electronic structure.

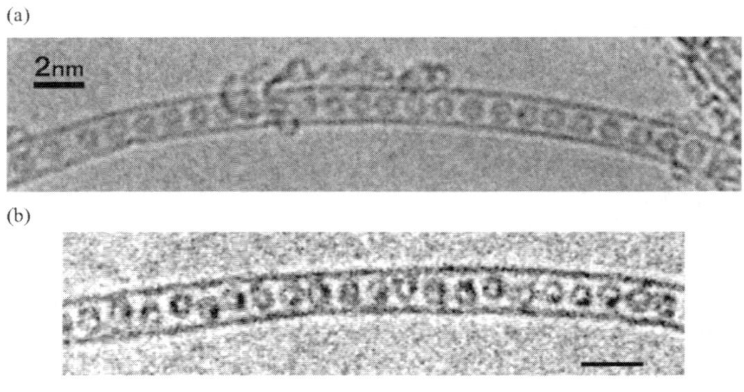

Fig. 5.9. HRTEM images of (**a**) the isolated $(Sm@C_{82})_n@SWNT$ and (**b**) the isolated $(Gd@C_{82})_n@SWNT$. Scale bars = 2 nm

(a)

(b)

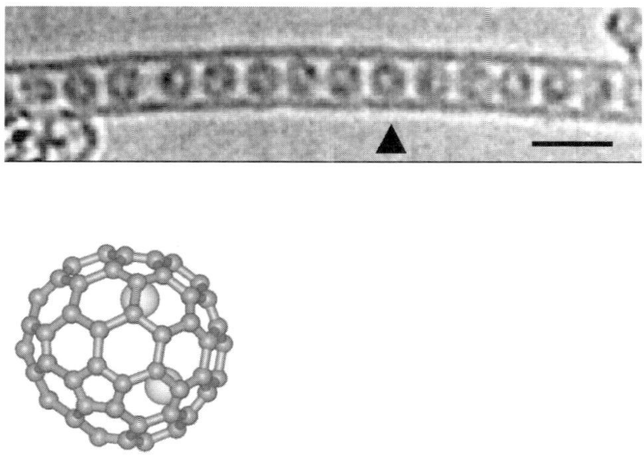

Fig. 5.10. (**a**) HRTEM images of the isolated $(Sc_2@C_{84}(I))_n@SWNT$ (scale bar = 2 nm) and (**b**) the expected molecular structure of $Sc_2@C_{84}(I)$. Two Sc atoms locate on the C_s plane of C_{84}

By utilizing SWNT as a nanometer-scale sample cell, the locations of the Sc atoms in $Sc_2@C_{84}$ (I) was determined [32]. Figure 5.10a shows a HRTEM image for an individual $Sc_2@C_{84}$ (I) peapod. Previous ^{13}C-NMR and ^{45}Sc-NMR studies suggested that $Sc_2@C_{84}$ (I) has C_s symmetry and the location of two Sc atoms are not equivalent [42,43]. However, the precise positions of Sc atoms inside the cage have remained unclear. In Fig. 5.10a, it is clearly seen that two Sc atoms locate near the edge rather than the center of the cage. Simulations for HRTEM images revealed that the two Sc atoms are located at 0.20 ± 0.02 nm from the molecular center with a Sc-Sc distance of 0.35 ± 0.02 nm (Fig. 5.10b). The determined Sc atom positions are in good agreement with those theoretically predicted (0.3486 nm for the Sc-Sc distance) [44].

Intramolecular motion of encapsulated Gd atoms inside C_{92} fullerenes was also observed by using a nano-peapod strucutre [34]. At room temperature, the elongated dark spots were obtained due to a rapid motion of Gd atoms along the fullerene cage. This motion was significantly depressed at the lower temperature (~ 100 K), indicating that the Gd atoms are thermally activated to librate around the equilibrium position inside the cage.

5.7 Peapod as a "Nano-Reactor"

The interior space of SWNTs may offer a unique field for chemical reactions. Because the size of the hollow space is almost same as that of organic molecules, chemical reactions should be restricted in one direction parallel to the tube axis, which may lead to unexpected products. Double-wall carbon nanotubes (DWNTs) have been synthesized from C_{60} nano-peapods by utilizing the interior space of SWNTs as a template [45]. When C_{60} nano-peapods are annealed at $\sim 1200°C$ in vacuum, fusion reactions of C_{60} occur. The reaction is confined to one direction by the surrounding wall, which results in the formation of an inner tube. The radial breathing modes of the produced inner tubes can be observed as well as those of the outer tubes, suggesting that perfect DWNTs are synthesized as a result of the reaction.

C_{60} polymers were produced inside SWNTs by doping potassium to C_{60} nano-peapods [46]. The polymerization was initiated by the electron transfer from the doped potassium to the encapsulated C_{60} molecules. The electronic configuration of the C_{60} in the polymer chain was presumed to be C_{60}^{6-} from the Raman spectra and theoretical calculations.

As described in the previous section, SWNT is an ideal nanometer-scale sample cell for direct imaging of molecular structures with TEM. By using this fascinating feature, the direct observation of chemical reactions occurring within SWNTs can be allowed by modern analytical techniques based on TEM [6, 47]. Figure 5.11a shows HRTEM images of an individual $(Sm@C_{82})_n@SWNT$ at the beginning of the observation (before the reaction) [6]. The $Sm@C_{82}$ molecules are perfectly aligned in a one-dimensional way with a constant intermolecular spacing.

Following the irradiation for a period of 4 min, a molecular rearrangement in the nanotubes starts to occur (Fig. 5.11b). Some $Sm@C_{82}$ molecules are already so close to each other as to form dimers or clusters as seen in the HRTEM image. The appearance of dark spots by Sm atoms in Fig. 5.11b suggests that the rotational motion of Sm-metallofullerenes was frozen due to the formation of intermolecular covalent bondings. Further exposure for ~ 10 min resulted in coalescence or fusion, which generates the dimers and clusters of the Sm-metallofullerenes (Fig. 5.11c). These were finally converted to much longer nano-encapsulates after ~ 20 min irradiation (Fig. 5.11 d). The dark spots observed near the wall of the newly produced capsules correspond to the Sm atoms.

Further information on such reaction dynamics can be obtained by simultaneous in-situ EELS measurements (Fig. 5.12) [6]. The EELS spectrum observed from the bundles of the intact $(Sm@C_{82})_n@SWNTs$ is shown Fig. 5.12a together with the reference spectra of Sm^{2+} in $Sm@C_{82}$ and Sm^{3+} in Sm_2O_3. The peak positions of Sm M_{45} edges for $(Sm@C_{82})_n@SWNTs$ before the reaction are the same as those for the intact $Sm@C_{82}$. After electron beam irradiation for a period of ~ 10 min, the peak widths of M_{45} edges became broader, showing additional intense peaks overlapping with the main peaks (Fig. 5.12b). At the same time, the intensity of the divalent Sm^{2+} peaks decreased with the irradiation, and finally new M_{45} peaks appeared (Fig. 5.12c). In particular, the M_5 edge at ~ 1080 eV can be clearly seen in the spectrum; the peak position for this M_{45} edge corresponds to that of the trivalent Sm^{3+}.

At this point the valency of the Sm atoms has transformed from +2 to +3 with a time scale of ~ 10 min and the conversion was completed within

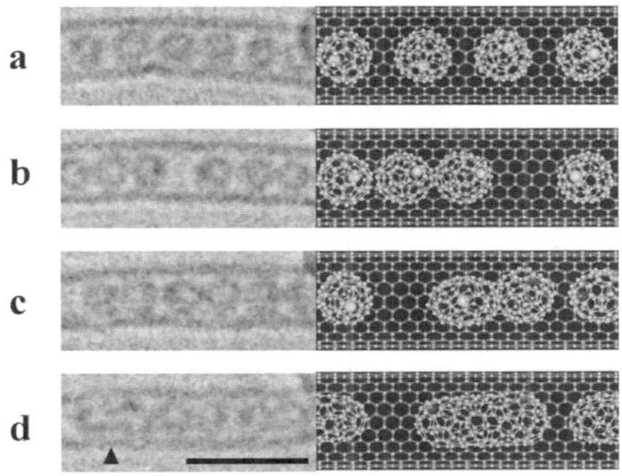

Fig. 5.11. Time evolution of the HRTEM images of ($Sm@C_{82}$)@SWNTs (left) and the schematic illustrations (right). The HRTEM images were observed after (**a**) ~ 0 min, (**b**) ~ 4 min, (**c**) ~ 10 min and (**d**) ~ 20 min irradiations, respectively. The produced nanocapsules can be clearly seen inside the SWNT (*arrows*). Scale bar = 3 nm (see Color Plates, p. 341)

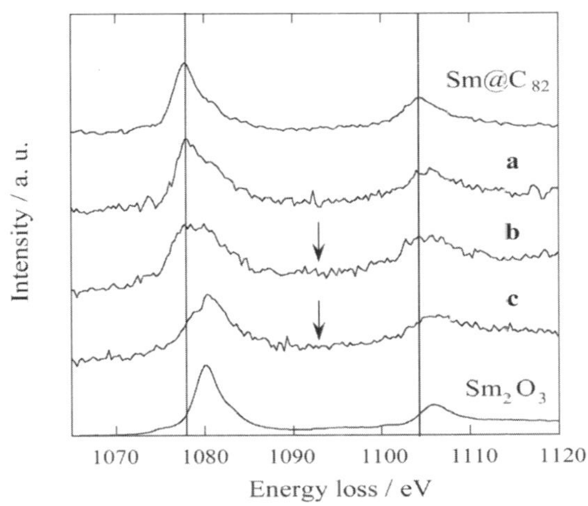

Fig. 5.12. Time evolution of the EELS spectrum for $(Sm@C_{82})_n$@SWNTs, together with reference spectra for the divalent Sm^{2+} in the intact $Sm@C_{82}$ molecule (top) and the trivalent Sm^{3+} in Sm_2O_3 (bottom) [6]. The EELS spectra were taken at ~ 10 min intervals

~ 20 min irradiation. The time-dependent HRTEM and EELS measurements indicate that the oxidation state of Sm has transformed from +2 to +3 as a result of a new chemical bond formation due to Sm carbides upon coalescence of the fullerene cages (Fig. 5.11c). Sm atoms may not be encaged in the nanocapsule but should at least be incorporated into the cage network, since the oxidation state of the Sm carbides (SmC_2) is +3 and those of various Sm-metallofullerenes are known to be +2, irrespective the fullerene size. The Sm atoms tend to be incorporated into the carbon networks to form very stable carbide bonds and do not form Sm clusters.

5.8 Conclusions

Nano-peapods are novel hybrid materials of carbon nanostructures, which were obtained in bulk in the last year of the 20th century, 2000. Since then, as seen in this chapter, many investigations have revealed various aspects of nano-peapods. Even so, interest in nano-peapods is still growing. There remain many potentially important and intriguing topics. For example, encapsulated atoms of metallofullerenes in nano-peapods can be viewed as qubits in quantum computation [21].

Although we have focused on fullerene nano-peapods, non-fullerene nano-peapods have been also studied [48–51]. Sloan and co-workers have produced novel low-dimensional inorganic materials that form an unprecedented structure in the bulk phase [48, 49]. A novel ice phase was also found inside SWNTs [50]. At room temperature, the confined water molecules behave like liquid water. As the temperature decreases, liquid-like water is transformed into an "ice nanotube" phase at 235 K.

The ability to control materials at a molecular level can be realized by nano-peapods. Nano-peapods will play an important role in the era of molecular electronics.

Acknowledgements

We thank Prof. S. Iijima (Meijo University, NEC, AIST), Prof. S. Bandow (Meijo Univ.), Dr. K. Suenaga (AIST), Dr. K. Hirahara (Meijo Univ.), Dr. Y. Lian (Peking Univ.), Prof. Z. Gu (Peking Univ.), Prof. Y. Kuk (Seoul National Univ.), Dr. Y. Ohno (Nagoya Univ.) and Prof. T. Mizutani (Nagoya Univ.) for their help with the studies contained in this chapter. This work was supported by the JST CREST Program on Novel Carbon Nanotube Materials and the Special Coordination Funds of MEXT.

References

1. P.M. Ajayan and S. Iijima, *Nature*, **361**, 333 (1993)
2. E. Dujardin, T.W. Ebbesen, H. Hiura and K. Tanigaki, *Science*, **265**, 1850 (1994)
3. B.W. Smith, M. Monthioux and D.E. Luzzi, *Nature* **396**, 323 (1998)
4. K. Hirahara, K. Suenaga, S. Bandow, H. Kato, T. Okazaki, H. Shinohara and S. Iijima, *Phys. Rev. Lett.*, **85**, 5384 (2000)

5. K. Suenaga, M. Tencé, C. Mory, C. Colliex, H. Kato, T. Okazaki, H. Shinohara, K. Hirahara, S. Bandow and S. Iijima, *Science*, **290**, 2280 (2000)
6. T. Okazaki, K. Suenaga, K. Hirahara, S. Bandow, S. Iijima and H. Shinohara, *J. Am. Chem. Soc.*, **123**, 9673 (2001)
7. H. Kataura, Y. Maniwa, T. Kodama, K. Kikuchi, K. Hirahara, K. Suenaga, S. Iijima, S. Suzuki, Y. Achiba and W. Krätschmer, *Synthetic Metals*, **121**, 1195 (2001)
8. H. Shinohara, *Rep. Prog. Phys.*, **63**, 843 (2000)
9. S. Bandow, M. Takizawa, H. Kato, T. Okazaki, H. Shinohara S. Iijima, *Chem. Phys. Lett.*, **347**, 23 (2001)
10. T. Okazaki, T. Shimada, K. Suenaga, Y. Ohno, T. Mizutani, J. Lee, Y. Kuk and H. Shinohara, *Appl. Phys. A*, **76**, 475 (2003)
11. X. Liu, T. Pichler, M. Knupfer, M.S. Golden, J. Fink, H. Kataura, Y. Achiba, K. Hirahara and S. Iijima, *Phys. Rev. B*, **65**, 045419 (2002)
12. H. Kataura, Y. Maniwa, M. Abe, A. Fujiwara, T. Kodama, K. Kikuchi, H. Imahori, Y. Misaki, S. Suzuki and Y. Achiba, *Appl. Phys. A*, **74**, 349 (2002)
13. B.W. Smith, R.M. Russo, S.B. Chikkannanavar and D.E. Luzzi, *J. Appl. Phys.*, **91**, 9333 (2002)
14. S. Berber, Y.-K. Kwon and D. Tománek, *Phys. Rev. Lett.*, **88**, 185502 (2002)
15. H. Ulbricht, G. Moos and T. Hertel, *Phys. Rev. Lett.*, **90**, 095501 (2003)
16. S. Okada, S. Saito and A. Oshiyama, *Phys. Rev. Lett.*, **86**, 3835 (2001)
17. S. Okada, S. Saito and A. Oshiyama, *Phys. Rev. B*, **67**, 205411 (2003)
18. J. Lee, H. Kim, S.-J. Kahng, G. Kim, Y.-W. Son, J. Ihm, H. Kato, Z.W. Wang, T. Okazaki, H. Shinohara and Y. Kuk, *Nature*, **415**, 1005 (2002)
19. A. Koshio, M. Yudasaka, M. Zhang and S. Iijima, *Nano Lett.*, **1**, 361 (2001)
20. M. Yudasaka, K. Ajima, K. Suenaga, T. Ichihashi, A. Hashimoto and S. Iijima, *Chem. Phys. Lett.*, **380**, 42 (2003)
21. F. Simon, H. Kuzmany, H. Rauf, T. Pichler, J. Bernardi, H. Peterlik, L. Korecz, F. Fülöp and A. Jánossy, *Chem. Phys. Lett.*, **383**, 362 (2004)
22. K. Hirahara, K. Suenaga, S. Bandow, H. Kato, T. Okazaki, H. Shinohara and S. Iijima, *Phys. Rev. B.*, **64**, 115420 (2001)
23. T. Okazaki, K. Suenaga, K. Hirahara, S. Bandow, S. Iijima and H. Shinohara, *Physica B*, **323**, 97 (2002)
24. Y. Maniwa, H. Kataura, M. Abe, A. Fujiwara, R. Fujiwara, H. Kira, H. Tou, S. Suzuki, Y. Achiba, E. Nishibori, M. Takata, M. Sakata and H. Suematsu, *J. Phys. Soc. Jpn.*, **72**, 45 (2003)
25. T. Okazaki, Y. Lian, Z. Gu, K. Suenaga and H. Shinohara, *Chem. Phys. Lett.*, **320**, 435 (2000)
26. T. Okazaki, K. Suenaga, Y. Lian, Z. Gu and H. Shinohara, *J. Chem. Phys.*, **113**, 9593 (2000)
27. K. Suenaga, H. Kato, H. Shinohara and S. Iijima, *Phys. Rev. B*, **62**, 1627 (2000)
28. E. Nishibori, M. Takata, M. Sakata, H. Tanaka, M. Hasegawa and H. Shinohara, *Chem. Phys. Lett.*, **330**, 497 (2000)
29. R. Saito, M. Fujita, G. Dresselhaus and M.S. Dresselhaus, *Appl. Phys. Lett.*, **60**, 2204 (1992)
30. J.W.G. Wildoer, L.C. Venema, A.G. Rinzler, R.E. Smalley and C. Dekker, *Nature*, **391**, 59 (1998)
31. T.W. Odom, J. Haung, P. Kim and C.M. Lieber, *Nature*, **391**, 62 (1998)

32. K. Suenaga, T. Okazaki, C.-R. Wang, S. Bandow, H. Shinohara and S. Iijima, *Phys. Rev. Lett.*, **90**, 055506 (2003)
33. K. Suenaga, K. Hirahara, S. Bandow, S. Iijima, T. Okazaki, H. Kato and H. Shinohara, *Core-Level Spectroscopy on the Valence State of Encaged Metal in Metallofullerene-Peapods*. Proceedings of the 15th International Winterschool, AIP Conference Proceedings Vol. 591. H. Kuzmany, J. Fink, M. Mehring, S. Roth (Eds.), American Institute of Physics, New York, pp. 256–260, 2001
34. K. Suenaga, R. Taniguchi, T. Shimada, T. Okazaki, H. Shinohara and S. Iijima, *Nano Lett.*, **3**, 1395 (2003)
35. D.J. Hornbaker, S.-J. Kahng, S. Misra, B.W. Smith, A.T. Johnson, E.J. Mele, D.E. Luzzi and A. Yazdani, *Science*, **295**, 828 (2002)
36. C.L. Kane, E.J. Mele, A.T. Johnson, D.E. Luzzi, B.W. Smith, D.J. Hornbaker and A. Yazdani, *Phys. Rev. B*, **66**, 235423 (2002)
37. T. Miyake and S. Saito, *Solid State Comm.*, **125**, 201 (2003)
38. Y. Cho, S. Han, G. Kim, H. Lee and J. Ihm, *Phys. Rev. Lett.*, **90**, 106402 (2003)
39. M. Otani, S. Okada and A. Oshiyama, *Phys. Rev. B*, **68**, 125424 (2003)
40. T. Shimada, T. Okazaki, R. Taniguchi, T. Sugai, H. Shinohara, K. Suenaga, Y. Ohno, S. Mizuno, S. Kishimoto and T. Mizutani, *Appl. Phys. Lett.*, **81**, 4067 (2002)
41. T. Shimada, Y. Ohno, T. Okazaki, T. Sugai, K. Suenaga, S. Kishimoto, T. Mizutani, T. Inoue, R. Taniguchi, N. Fukui, H. Okubo and H. Shinohara, *Physica E*, **21**, 1089 (2004)
42. M. Inakuma, E. Yamamoto, T. Kai, C.-R. Wang, T. Tomiyama and H. Shinohara, *J. Phys. Chem. B*, **104**, 5072 (2000)
43. Y. Miyake, S. Suzuki, Y. Kojima, K. Kikuchi, K. Kobayashi, S. Nagase, M. Kainosho, Y. Achiba, Y. Maniwa and K. Fisher, *J. Phys. Chem.*, **100**, 9579 (1996)
44. S. Nagase and K. Kobayashi, *Chem. Phys. Lett.* **276,** 55 (1997)
45. S. Bandow, M. Takizawa K. Hirahara, M. Yudasaka and S. Iijima, *Chem. Phys. Lett.*, **337**, 48 (2001)
46. T. Pichler, H. Kuzmany, H. Kataura and Y. Achiba, *Phys. Rev. Lett.*, **87**, 267401 (2001)
47. D.E. Luzzi and B.W. Smith, *Carbon*, **38**, 1751 (2000)
48. J. Sloan, A.I. Kirkland, J.L. Hutchison and M.L.H. Green, *Chem. Comm.*, 1319 (2002)
49. E. Philp, J. Sloan, A.I. Kirkland, R.R. Meyer, S. Friedrichs, J.L. Hutchison and M.L.H. Green, *Nature Mater.*, **2**, 788 (2003)
50. Y. Maniwa, H. Kataura, M. Abe, S. Suzuki, Y. Achiba, H. Kira and K. Matsuda, *J. Phys. Soc. Jpn.*, **71**, 2863 (2002)
51. T. Takenobu, T. Takano, M. Shiraishi, Y. Murakami, M. Ata, H. Kataura, Y. Achiba and Y. Iwasa, *Nature Mater.*, **2**, 683 (2003)

6 The Selective Chemistry of Single Walled Carbon Nanotubes

M.S. Strano, M.L. Usrey, P.W. Barone, D.A. Heller, and S. Baik

Functionalization of solution-phase carbon nanotubes has led to promising developments in nanotube science. Covalent interactions place organic functional groups on the nanotube sidewall, and certain moieties, such as diazonium salts exhibit preferential reactivity based on electronic structure. Concurrent developments in spectroscopic tools facilitate the monitoring of these reactions as well as the nanotubes' local environment. Non-covalent interactions on the nanotube surface are diverse and selective, including solvatochromatic shifts and charge transfer. This review describes covalent and non-covalent nanotube chemistries as well as methods of spectroscopic analysis.

6.1 Introduction: Advances in Carbon Nanotube Characterization

Carbon nanotubes are expected to comprise an important class of materials for novel applications in electronics, optics, and high-strength materials [1–3]. For the development of these applications, researchers invariably must understand the surface chemistry of nanotubes in solution for subsequent processing. Solution-phase chemical processing enables manipulation, separation, purification and provides direct routes to more complex materials. Understanding the physical and chemical interactions of such 1-D electronic materials under these conditions has proven difficult, however. The problem is not straightforward: carbon nanotubes invariably exist after synthesis as aligned aggregates or bundles that are tightly bound by an estimated 500 eV/micron of tube length [4, 5]. The problem of poor dispersion at the single nanotube level has greatly inhibited their study in solution. Recent breakthroughs allowing single tube dispersion have changed this situation, and from these techniques a wealth of knowledge regarding the optical and chemical properties has been obtained. In particular, advances in spectroscopic characterization including band-gap fluorescence and the assignments of metallic [6] and semi-conducting [7] spectral features for single walled carbon nanotubes allow for detailed, chirality based characterization of samples with great efficiency. Raman spectroscopy has been used extensively to characterize solid [8,9] and solution phase [10] carbon nanotube systems because it has the ability to probe distinct populations that have inter-band transitions in resonance with the excitation laser. In this way, Raman excitation profiles can be used to probe the unique geometric dependence of these transitions.

Early work using Raman spectroscopy to probe inter-band transitions has established the technique as a valuable tool for characterizing carbon nanotubes [9]. Kukovecz and co-workers [8] examined HiPco [11] produced carbon nanotubes using a series of excitation wavelengths. The authors quantitatively modeled peak shapes of tangential and disorder modes with success and described the radial breathing mode diameter dependence using a semi-empirical relation. Canonico et al. [12] measured Raman excitation profiles on laser-oven prepared nanotubes and used the line-shape to characterize armchair species. These earlier studies involved aggregated nanotube ropes, however, and in this state nanotubes have been shown to experience strong perturbations to their electronic structure that complicate the interpretation of such profiles. Dresselhaus and co-workers [13–19] have pioneered experimental techniques enabling single nanotube spectroscopy to circumvent this limitation while Yu and Brus [20, 21] profiled bundle ends using micro-Raman spectroscopy to assign features using the results of the tight binding description of graphene [22–24]. This work uses a solution phase dispersion of carbon nanotube recently shown [25] to yield individually isolated species that fluoresce in the near-infrared. This fluorescence is a key indicator of dispersion and isolation. Hence, this method provides an opportunity to extend the results of past researchers to examine the resonant phonon spectra of a large number of isolated nanotubes in response to changes to their chemical environment.

Carbon nanotubes have a unique electronic structure that follows from the quantization of the electronic wave vector through a conceptual rolling of a graphene plane into a cylinder forming the carbon nanotube. The vector in units of the hexagonal elements connecting a continuous path tracing the nanotube circumference defines the nanotube chirality in terms of two integers: n and m [2, 3]. When |n-m| = 3q where q is an integer, the nanotube is metallic or semi-metallic while remaining species are semi-conducting with a diameter and chirality dependent band-gap. In the Raman spectrum, phonon modes at low Raman shift [8, 26] identified as radial breathing modes (RBMs) correspond to a uniaxial expansion and contraction and have shifts strongly dependent on nanotube diameter [2, 18]. In this way, nanotubes of a distinct chiral vector can be identified in the Raman spectrum readily and tracked as the excitation wavelength is varied through the absorption maxima of the nanotubes.

In this work, we review the covalent and non-covalent interactions of individual nanotubes in solution. We limit the discussion to samples where the nanotubes have been dispersed and purified as to yield mostly individually dispersed species. Such samples demonstrate well-resolved absorption spectra and band-gap fluorescence. The subject is divided into covalent interactions and non-covalent. The latter is then divided into what we consider charge transfer chemistries, where the interaction involves a shifting of the Fermi level of the nanotubes, and solvatochromic interactions, where the presence of the interaction is detected as a characteristic shift in either the fluorescence or the absorption transition.

6.2 Selective Covalent Chemistry of Single-Walled Carbon Nanotubes

6.2.1 Motivation and Background

The realistic application of single-walled carbon nanotubes is limited by problems in the manipulation and processing of these promising materials. Covalent chemistry can provide the means to increase solubility, allow better dispersion, and facilitate the addition to surfaces or other molecules, including polymers and proteins. By controlling the moiety covalently bonded to the nanotube sidewall, different applications can be realized, including electronic devices, polymer composites, and nanotube sensors. For the former, a covalent attachment could allow the nanotubes to assemble onto the substrate. High-strength polymer composites require that the stress of the polymer matrix be transferred to the imbedded nanotubes. This can only be successfully accomplished if the nanotubes are covalently linked to the polymer matrix, which can be accomplished by reacting functional groups on the nanotube sidewall with the matrix material. Sensors might utilize covalently attached moieties in the molecular recognition process.

Obviously, there is great promise for single-walled carbon nanotube covalent chemistry. Although this field is not well-developed at this point, there are a number of successful covalent reactions in the literature, which are reviewed in the next section. The remainder of the next section is devoted to the selective chemistries currently under investigation.

6.2.2 Review of Carbon Nanotube Covalent Chemistry

The chemistries described here, which focus on the exohedral (outside wall) functionalization of single-walled carbon nanotubes, can be divided into two main classifications. First, oxidation or ozonation is used to generate defect sites in the nanotube sidewall or at the open tube ends, and then functionalize those sites [27–32]. These oxygenated functional groups can be converted into other derivatives, including amides. The next classification focuses on the direct chemical modification of the nanotube sidewall using addition reactions. This classification includes fluorination [33] and aryl diazonium salt addition [34, 35].

The first successful covalent chemistry was conducted under oxidizing conditions, involving sonication in either sulfuric acid/nitric acid or sulfuric acid/hydrogen peroxide mixtures [27, 28]. These procedures not only purified the nanotube material of carbon impurities, but also reduced their lengths. Following this process, a variety of oxygenated functional groups are found on the tube ends or at highly reactive defect sites along the nanotube sidewall [36]. These groups include carboxylic acids, esters, quinones, and anhydrides. Most often, carboxylic acid groups are "activated" by treatment with thionyl chloride and then converted to derivative functionalities, such as amides. Carboxylic acid functional groups also provide access to peptide chemistry.

Although this chemistry is useful for some applications, including chemical force microscopy [37], the restriction of the moieties to the tube ends and a limited number of defect sites does not change the bulk properties of the nanotube material. For example, the spectroscopic signatures of material functionalized in this manner do not differ significantly from un-reacted nanotube material [29, 30].

The first extensive sidewall reaction was the direct fluorination of a single-walled carbon nanotube 'bucky paper', as reported by Margrave et al. in 1998. In this case, a variety of spectroscopic techniques were utilized to confirm widespread covalent functionalization [33]. Infrared spectroscopy confirmed the presence of covalently bound fluorine. More importantly, the signature Raman spectrum of un-reacted nanotubes was found to be drastically altered following functionalization. Section 6.2.4 describes the analysis of covalent functionalization using Raman spectroscopy in more detail.

Furthermore, it was found that the fluorine moieties could be removed using treatment in hydrazine and the spectroscopic properties of the nanotubes almost completely restored. Conversely, the fluorine moieties can be displaced by strong nucleophiles, such as metal alkoxides and grignard reagents [33].

Other chemistries providing functionalization of single-walled carbon nanotubes include reaction with dichlorocarbene [38], or carbene derivatives [39, 40], and reaction with nitrenes [40, 41]. These chemistries report low degrees of functionalization, and therefore do not affect the spectroscopic signatures similar to oxidation functionalization above.

In conclusion, the covalent chemistry of single-walled carbon nanotubes is an incredibly promising field that exhibits great potential for the applications described. However, this field has been limited to the functionalization of carbon nanotube ropes and bundles. For many of the desired applications, the covalent modification of individually dispersed carbon nanotubes in solution is important. In the following section, such reaction chemistry is further investigated. Future applications might also require covalent reactions to be conducted on only nanotube species with a particular diameter or length.

6.2.3 The Pyramidalization Angle Formalism for Carbon Nanotube Reactivity

Fullerene chemistry has shown that the reactivity of these molecules with respect to addition chemistries is strongly dependent on the curvature of the carbon framework [42]. In comparison single-walled carbon nanotubes consist only of graphite-like sidewalls and are only moderately curved [43]. Consequently, nanotubes are expected to be less reactive than most fullerene molecules due to larger curvature but more reactive than a graphene sheet due to pyramidalization and misalignment of the π-orbitals [43–47].

In any case, the strain of the carbon framework is reflected in the pyramidalization angle, θ_P, of the carbon constituents. This angle equals the difference between the π-orbital axis vector of a conjugated carbon atom and a normal right angle (90 °). This vector is defined as one which makes equal angles ($\theta_{\sigma\pi}$) to the three σ-bonds of a conjugated carbon [48]. Trigonal carbon atoms (sp^2-hybridized) prefer a planar orientation with $\theta_P = 0$ °. Tetragonal carbon atoms (sp^3-hybridized) possess a pyramidalization angle of 19.47 ° due to a larger relative π-orbital axis vector angle. By comparing θ_P of a particular carbon atom with these values, it can be estimated which geometry is most appropriate for that atom. All carbons in the fullerene C_{60} have $\theta_P = 11.6$ ° which indicates a sp^3-hybridized state is most appropriate [44].

The (5,5) single-walled carbon nanotube has a $\theta_P \approx 6$ ° for the sidewall and $\theta_P \approx 11.6$ ° for the endcap [44]. Values for other distinct (n, n) nanotube species show a trend of increasing θ_P (sidewall) with decreasing n. Therefore, it can be inferred that chemical reactivity in covalent additions is directly related to nanotube diameter. Furthermore, following the reaction of a single carbon atom, the reactivity of the remaining atoms is increased [44]. This occurs because as the strain is relaxed at one point on the nanotube sidewall, the total strain is divided among the remaining sp^2-hybridized carbons. Therefore, this so-called strain relief can increase the rate of addition reactions in single-walled carbon nanotubes [47, 49, 50].

6.2.4 The Selective Covalent Chemistry of Single-Walled Carbon Nanotubes

The main barrier to the extensive application of single-walled carbon nanotubes is the inability to manipulate according to electronic structure [1]. All standard preparative techniques produce polydisperse mixtures of metallic, semimetallic, and semiconducting fractions [51–53]. In this section, novel nanotube chemistry is presented which provides electronic structure selectivity in the covalent functionalization of single-walled carbon nanotubes.

As discussed above, most developed mechanisms for nanotube selectivity in covalent reactions are based upon the carbon pyramidalization angle [43, 44]. This indicates that nanotube diameter is the reaction index. There is no direct evidence of the existence of covalent reaction pathways utilizing nanotube band gap as a reaction index. However, previous experimentation displayed that protonation of single-walled carbon nanotubes dispersed in solution is highly selective depending upon the nanotube band gap [10]. This non-covalent, reversible interaction allows the possibility for covalent, irreversible reactions to also demonstrate selectivity based upon electronic structure.

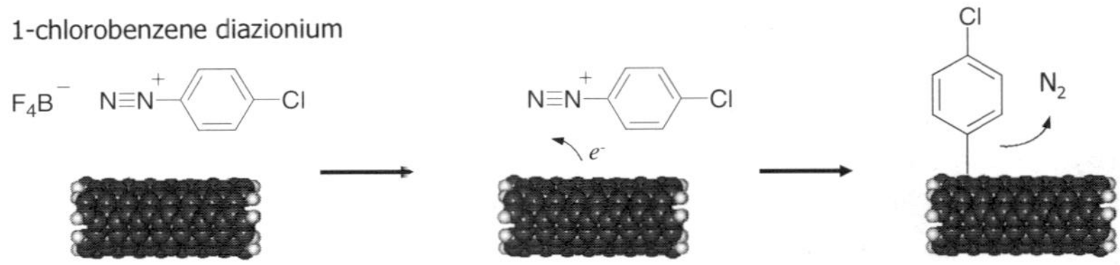

Fig. 6.1. Diazonium reagents extract electrons from the nanotube, releasing N_2 gas, and forming a C-C covalent bond

The selective pathways described here offer a unique method for the manipulation of nanotubes of distinct electronic structures by selective reaction of metallic nanotubes. This method of control should allow the separation of metallic nanotubes with high selectivity and scalability. Separation will allow the manufacture of electronic devices with specific electronic properties.

Water-soluble diazonium salts [54] have been shown to react with carbon nanotubes [34, 55, 56] via charge transfer, extracting electrons and forming a covalent aryl bond. Figure 6.1 illustrates this reaction. During reaction, the diazonium reagent becomes part of a charge-transfer complex at the surface with electron donation from the nanotube stabilizing the transition state.

Recently, it was found that under certain controlled conditions, a reaction pathway exists in which the diazonium salt reacts selectively with the metallic nanotubes to the near exclusion of the semi-conductors [57]. The covalent bond forms with high affinity for electrons with energies near the Fermi level (E_F). Figure 6.2 shows the density of states for semiconducting and metallic nanotubes as well as the region of interest at E_F. Since metallic nanotubes have a greater electron density near E_F, this translates into higher reactivity. This reactivity is amplified once the bond symmetry of the nanotube is

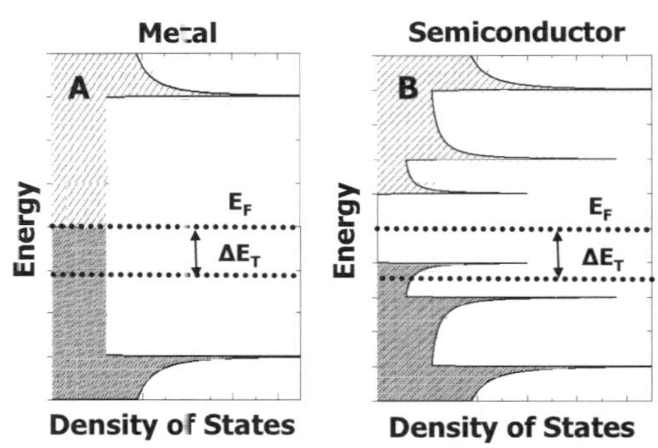

Fig. 6.2. The extent of electron transfer depends on the density of allowed electronic states within ΔE_T of the Fermi level (E_F). This leads to higher reactivity of metallic and semimetallic species versus semiconductor species

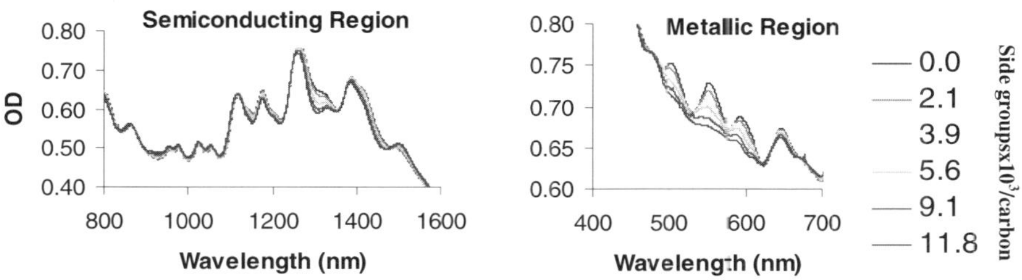

Fig. 6.3. UV-vis-nIR absorption spectra of SDS-suspended single-walled carbon nanotubes at various steady state points in the 4-chlorobenzene-diazonium selective reaction. Comparison indicates high selectivity for metallic species over semiconducting species under controlled conditions

initially disrupted by the formation of the defect causing adjacent carbons to increase in reactivity. This trend continues as the whole nanotube is reacted.

Spectroscopic evidence of selective functionalization can be found from ultraviolet-visible-near infrared (UV-vis-NIR) absorption spectra. Figure 6.3 shows steady state absorption spectra of individually dispersed nanotubes in aqueous solution following successive, equal additions of 4-chlorobenzene diazonium tetrafluoroborate. These spectra monitor the valence (v) to conduction (c) electronic transitions, denoted E_{nn} where n is the band index. The E_{11} transitions for the metallic and semimetallic nanotubes occur from approximately 440 to 645 nm. The E_{11} and E_{22} transitions for the semiconducting nanotubes are found from 830 to 1600 nm and 600 to 800 nm, respectively. Since individual peaks in an absorption spectra represent a distinct, identifiable collection of nanotubes, this allows the direct monitoring of the valence electrons for each. As covalent bonds are formed, electrons along the nanotube are localized, and the absorption maxima decay. Therefore, a relative decrease in metallic absorption features over semiconducting features represents preferential functionalization of the metallic nanotubes. Selectivity is also verified by the preservation of band gap fluorescence of the semiconducting nanotubes, which is known to be highly sensitive to chemical defects in the nanotube sidewall. Also worthy of note is that this selective decay is distinct from the similar, reversible behavior observed in band-gap-selective protonation [10].

Raman spectroscopy offers another method of monitoring the selective reaction. Functionalization increases the intensity of the 'D' peak, a phonon mode at 1330 cm^{-1}, also known as the disorder mode. This mode corresponds to the conversion of a sp^2 hybridized carbon to a sp^3 hybridized carbon on the surface during the formation of the aryl-nanotube bond [51–53]. Essentially, the integrated intensity of the 'D' peak measures covalent bonds made with the nanotube surface. Figure 6.4 illustrates the Raman spectrum at 785 nm excitation, showing the growth of the D peak with increasing reaction. This feature does not increase as a result of surfactant [25] or hydronium

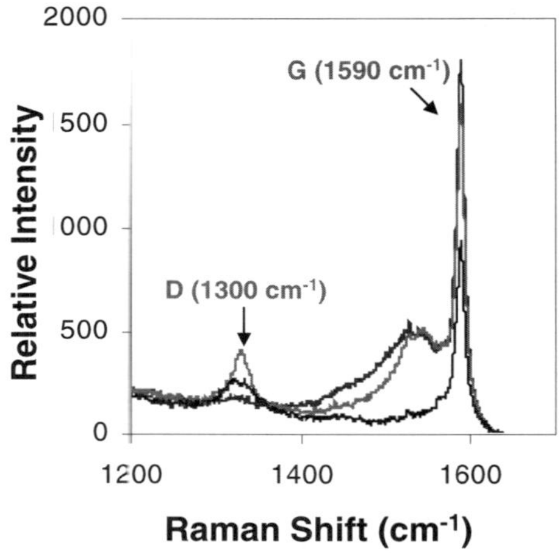

Fig. 6.4. Raman spectrum at 785 nm excitation showing the increase in the D peak (or disorder mode) with increasing functionalization

ion [10] adsorption on the nanotube surface. As the disorder mode increases due to functionalization, the C-C tangential mode decreases due to the loss of electronic resonance. Thus, analysis of steady state Raman spectra during selective reaction allows correlation between the number of sidewall functionalization events and the intensity of the disorder mode.

Covalent functionalization of the nanotube sidewall disrupts the symmetry of the Raman radial breathing mode (RBM), and this (n, m)-specific feature decays with increasing reaction [34, 55, 56]. Again, analogous behavior is not observed with surface adsorption, which decreases the Raman resonance enhancement without destroying the RBM [58]. Figure 6.5 shows the low-wavenumber solution-phase Raman spectra of the reaction mixture at 532 nm at three distinct steady state points during reaction. The relative rates of decay for the features in Fig. 6.5 show unprecedented reactivity differences between semimetallic species. The complete decay of all metallic modes, as seen in Fig. 6.5c, leaves only one low-wavenumber RBM, which has been previously assigned to the (9,2) semiconductor RBM [10]. This is in agreement with Fig. 6.3, as complete decay of all metallic and semimetallic absorption features leaves one semiconducting feature also belonging to the (9,2).

The reacted material was thermally treated at 300 °C in an atmosphere of inert gas. This cleaves the aryl chloride moieties from the nanotube sidewall and restores the spectroscopic features of pristine nanotubes [55]. Raman spectra taken before reaction (solid line) and following thermal treatment (dotted line) at 633 nm are shown in Fig. 6.6. This excitation wavelength was chosen because it probes both semiconducting (gray notation) and metallic (black notation) features [6]. The radial breathing modes are nearly com-

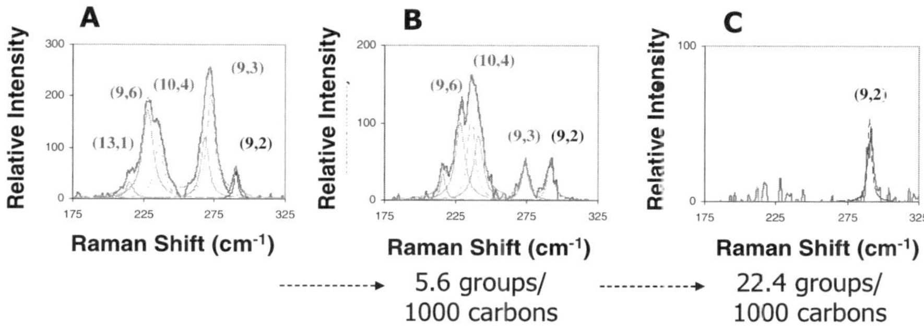

Fig. 6.5. Low wavenumber Raman spectra at 532 nm excitation at three steady state points during selective 4-chlorobenzenediazonium reaction. (**a**) Initial solution, shows four metallic nanotube species (gray) and one semiconducting nanotube species (black) probed at this wavelength. (**b**) Following attachment of 5.6 groups per 1000 carbon atoms, decay of the metallic species is evident, while the (9,2) semiconductor mode is unchanged. (**c**) Following attachment of 22.4 groups per 1000 carbons, all metallic modes have completely decayed, leaving only the (9,2) semiconductor. This supports the UV-vis-nIR data displayed in Fig. 6.3

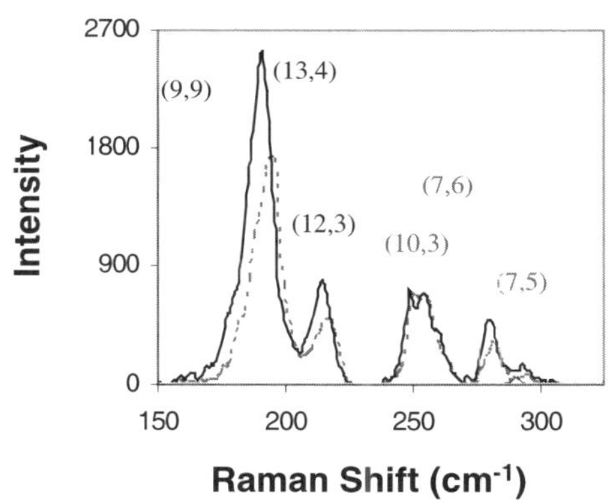

Fig. 6.6. Low wavenumber Raman spectra at 633 nm excitation, which probes both metallic (black) and semiconducting (gray) species, before functionalization (*solid line*) and after recovery and thermal pyrolysis (*dotted line*). Restoration of RBM features indicates the chemistry is reversible and optical features can be recovered

pletely restored, indicating the chemistry is reversible. In addition, the corresponding absorption features are recovered, signifying the restoration of the intrinsic electronic structure of the pristine nanotube. This result shows that this covalent, selective chemistry can be utilized as a reversible route to manipulate nanotubes of a particular electronic structure and then allow the recovery of the initial optical and electronic properties.

6.2.5 Spectroscopic Tools for Understanding Selective Covalent Chemistry

The development of solution phase dispersion methods [6, 36], spectroscopic identification and monitoring [25], and spectral assignments of single-walled carbon nanotubes [6, 7], allows carbon nanotube surface chemistries to be studied in great detail. The next step is to link these sidewall reactions and pathways to the unique electronic structure of the nanotubes. Using Raman spectroscopy, nanotubes possessing a distinct chiral vector can be monitored during reactions by probing the radial breathing modes of the nanotubes that have electronic transitions in resonance with the excitation laser wavelength. These modes are distinct for nanotubes of specific diameter and electronic structure [26].

This analysis requires an accurate mapping of carbon nanotube electronic transitions as a function of diameter is required [59]. The first organization of this kind is attributed to H. Kataura, using single-electron theory. Using the Tight Binding Approximation, the electronic structure was described as a graphene sheet with electronic wavevectors quantized into unique integers [60]. This predicts the assignment of metallic and semiconducting electronic transitions based upon the chirality vector. However, the actual nanotube structure is more complicated than this simple formalism due to trigonal warping and the warping of the Brillion zone due to curvature [22]. Orbital overlap introduces effects from multiple neighbors [24] and breaks the Tight Binding Approximation. Finally, single-electron models cannot include multiple electron effects, such as exciton interactions. This complicates the direct comparison of experimental results and theoretical predictions [61,62]. Not surprisingly, assignments based upon single-electron theory have not been successful [6, 7, 25].

In response to this problem, an experimentally based Kataura plot has been generated [63]. All experiments were conducted using surfactant-suspended single-walled carbon nanotubes to avoid the perturbation of electronic transitions caused by nanotube ropes. Spectrofluorimerty has been utilized to map the electronic transitions of the semiconductors, denoted $vn \rightarrow cn$, as defined above in Sect. 6.2.4. This mapping illustrated considerable deviations from the conventional Kataura plot [7]. Raman excitation profiles, using metallic absorbing excitations, have been used to map the $v1 \rightarrow c1$ transitions for the metallic nanotubes [6]. All experimental transitions were correlated using semiempirical models derived from asymptotic diameter expansions in the graphene dispersion relation.

Figure 6.7 is a compilation of these results for carbon nanotubes up to 2 nm in diameter for the first and second semiconductor transitions and the first metallic and semimetallic transitions. The metallic transitions split into low- and high-energy transitions for nonarmchair ($n{\neq}m$) metallic nanotubes [22,24,60]. These nonarmchair species are actually semimetallic due to

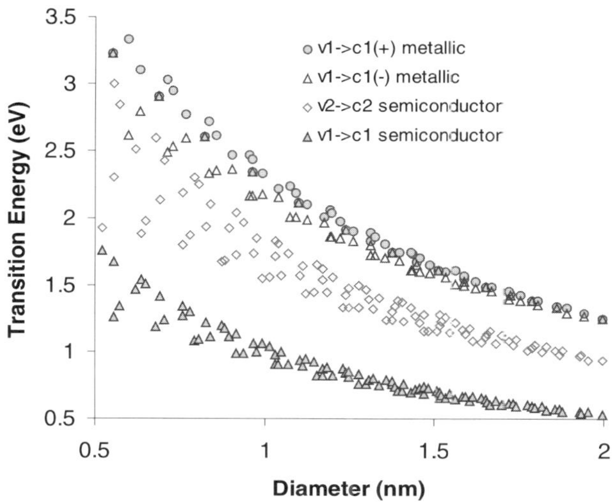

Fig. 6.7. Experimental mapping of transition energy versus diameter for metallic and semiconducting single-walled carbon nanotubes. Metallic transitions split into low- and high- energy transitions

curvature-induced electronic gap. This experimental mapping allows the interpretation of optical experiments involving single-walled carbon nanotubes.

When compared to the theoretically based Kataura plot, this experimental mapping shows an increased deviation from the "armchair curve". This curve is a limiting inverse diameter curve providing the limit of the transition energy as diameter approaches infinity. Specifically, small diameter transitions show notably "intermixing" in strong contrast to single-electron theory. It is this "intermixing" which allows the probing of selective reaction chemistries using Raman spectroscopy.

The experimental Kataura plot (Fig. 6.7) can be used to determine which nanotubes are in resonance with a particular excitation wavelength. In turn, the coordinating Raman radial breathing modes [19, 53] can be calculated from this derived relationship [8]

$$\omega_{RBM} = A + \frac{B}{d_t} \tag{6.1}$$

where $\omega_{RBM}$is the location of the RBM in Raman shift (cm^{-1}), d_t is the nanotube diameter, and A and B are 223.5 nm cm^{-1} and 12.5 cm^{-1}, respectively. It is important to note that the values for constants A and B in this expression are valid for HiPco samples. Figures 6.8, 6.9, 6.10, and 6.11. represent determinations of Raman RBM features from the experimental mapping for four excitation wavelengths; 830 nm, 785 nm, 633 nm, and 532 nm, respectively.

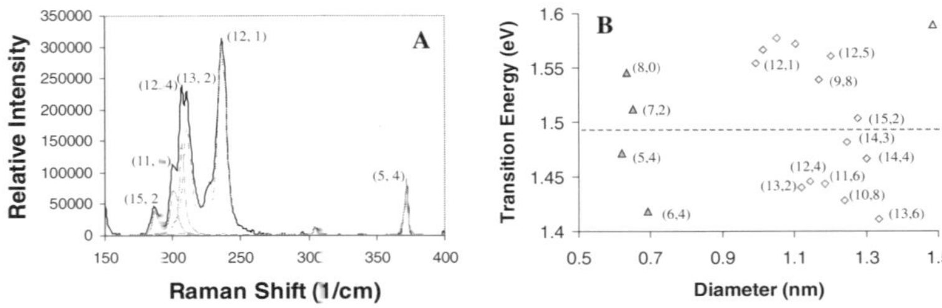

Fig. 6.8. (**a**) Low wavenumber Raman spectrum at 830 nm excitation. (**b**) Expanded view of Fig. 6.7 comparing laser excitation energy ($\sim$ 1.50 eV) to experimentally predicted transition energies

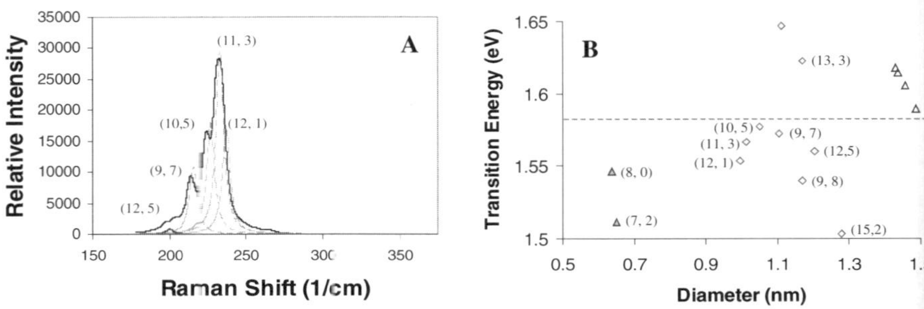

Fig. 6.9. (**a**) Low wavenumber Raman spectrum at 785 nm excitation. (**b**) Expanded view of Fig. 6.7 comparing laser excitation energy (1.58 eV) to experimentally predicted transition energies

Figure 6.8a shows the low wavenumber Raman shift spectrum for an aqueous solution of single-walled carbon nanotubes. At 830 nm, the Kataura plot predicts resonance with the $v1 \rightarrow c1$ transition of the (5,4) species as well as with the $v2 \rightarrow c2$ transitions of a cluster of semiconducting nanotubes. Figure 6.8b shows an expanded version of Fig. 6.7 surrounding the laser excitation energy. This shows that the mapping is able to predict the Raman RBM shift associated with each nanotube species with a transition within the resonance window (40 meV) of the laser excitation energy [6, 7].

Figure 6.9, parts a and b, show a similar comparison for a Raman excitation wavelength of 785 nm. At this wavelength, $v2 \rightarrow c2$ semiconducting transitions are exclusively visible. This comparison shows an excellent agreement between theoretical prediction and experimental mapping. Figure 6.10, parts a and b, illustrates the experimental predictions for a Raman spectrum at 633 nm. At this excitation, both large diameter metallic nanotubes (shown in bold) and small diameter semiconducting nanotubes are resonant. Consequently, this excitation wavelength provides a unique view for researchers interested in metallic/semiconductor separation [64–66] or electronic structure selective chemistries. The experimental mapping is able to describe this tran-

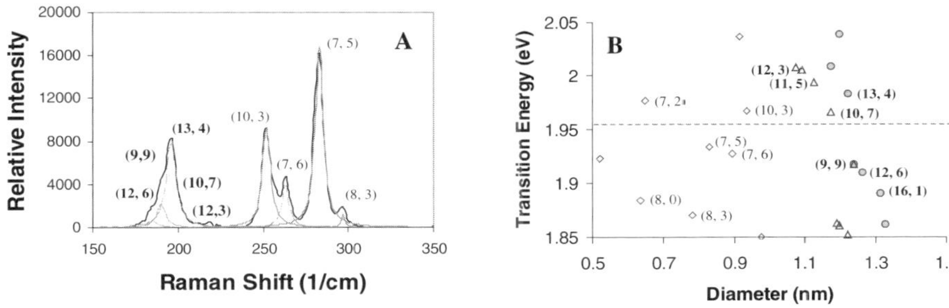

Fig. 6.10. (**a**) Low wavenumber Raman spectrum at 633 nm excitation. (**b**) Expanded view of Fig. 6.7 comparing laser excitation energy (1.96 eV) to experimentally predicted transition energies

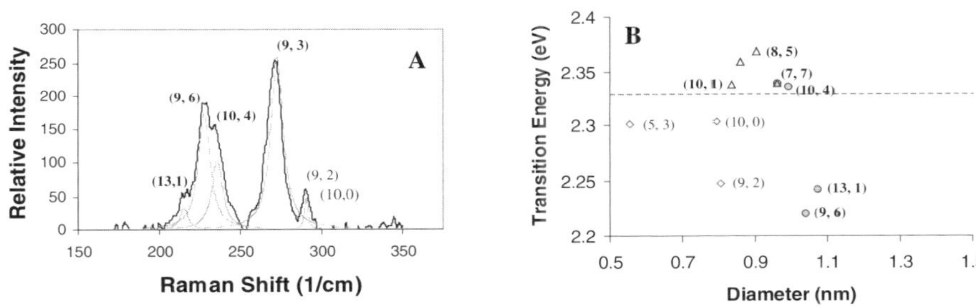

Fig. 6.11. (**a**) Low wavenumber Raman spectrum at 532 nm excitation. (**b**) Expanded view of Fig. 6.7 comparing laser excitation energy (2.33 eV) to experimentally predicted transition energies

sition region with excellent accuracy. Figure 6.11, parts a and b, at 532 nm probes mainly the $v1 \rightarrow c1$ transitions for metallic nanotubes with only one or two small diameter semiconductors present; the (9,2) and (10,10). The latter two species have been shown, in previous selective protonation work [10], to be large band gap semiconductors. In this case, the comparison between theoretical and experimental predictions is acceptable, although several metallic species are only partially resonant.

As described in the previous section, this experimental mapping of nanotube transition energies can be used to monitor selective nanotube surface chemistries. Specifically, Raman spectroscopy at 633 nm excitation is ideal for probing electronic structure selective pathways because of its resonance with both a collection of metallic and semiconducting nanotubes (HiPco material) [52].

In conclusion, Raman spectroscopy is a valid tool for correlating the single-walled carbon nanotube transition energies with the radial breathing modes of unique, chiral nanotubes. These results are compiled in an experimental mapping (Fig. 6.7) which replaces theoretically based plots widely used in the interpretation of nanotube experiments. This revised Kataura plot allows researchers to probe electronic structure selective chemistries of particular nanotube species for the first time. This has been further demonstrated in Sect. 6.2.4, with the reaction of 4-chlorobenzenediazonium with an aqueous nanotube solution. Other reaction chemistries that benefit from this more effective gauge are surface oxidations that favor smaller diameters [67] or nitric acid treatments [58].

6.3 Selective Non-covalent Chemistry: Charge Transfer

6.3.1 Single-Walled Nanotubes and Charge Transfer

Charge transfer has played an important role in understanding carbon nanotube electronic structure. For example, controlling the potential of single-walled carbon nanotubes in an electrolyte solution allows the electronic structure to be charged electrochemically. This charging causes reversible bleaching of the nanotube optical transitions [68]. Kavan et al. monitored these transitions using UV-vis-nIR absorption in an acitonitrile electrolyte solution and in an ionic liquid with an applied voltage and found that electronic states of larger diameter tubes are depleted/filled using lower potentials than the corresponding states of small diameter tubes [69,70]. This diameter selective charging can also be seen via Raman spectroscopy. The radial breathing modes of metallic nanotubes are more sensitive to electrochemical doping than those of semi-conducting tubes; this sensitivity is manifest as a greater decrease in the Raman scatter of the metals than the semi-conductors at higher potentials.

Adsorbates particularly oxygen, also play an important role in mediating charge transfer in nanotubes. Nanotube conductance is highly sensitive to oxygen adsorption/desorption [71, 72] and is believed to be the reason for nanotube p-type character. Shim and Siddons showed this importance in examining photoinduced conductivity changes in nanotube transistors. Irradiating a carbon nanotube transistor with UV light, they observed a large decrease in conductance; however, the conductance fully recovered upon exposure to oxygen [73]. It is important to note that the photodesorption of oxygen does not cause a change in dopant levels, thus ruling out oxygen as a p-dopant.

6.3.2 Selective Protonation of Single-Walled Carbon Nanotubes in Solution

Recent advancements in nanotube dispersion have led to the ability to monitor the chemical behavior of carbon nanotubes in solution. Because these individually suspended nanotubes exhibit band gap fluorescence [25] and enhanced resonant Raman scattering, both of which are good indicators of the

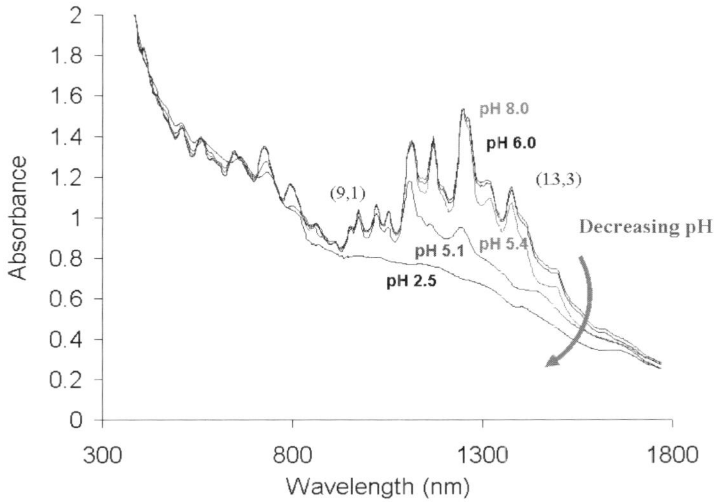

Fig. 6.12. Absorbance of individually dispersed carbon nanotubes in 1 wt.% SDS during a titration from pH 8.0 to pH 2.5. The E11 transitions can be seen to broaden and decrease more than the E22 transitions

environment surrounding the nanotube, it is possible to monitor changes in the nanotube microenvironment by observing changes in nanotube fluorescence or Raman scattering. Using these techniques, Strano et al. discovered that nanotubes suspended in sodium dodecyl sulfate (SDS) and water exhibit reversible protonation when in the presence of ambient O_2 [10]. Additionally, the protonation is selective with decreasing pH, favoring metallic nanotubes followed by semi-conducting nanotubes with large band gap nanotubes protonating last.

In the case of semi-conducting nanotubes, the reaction can be followed by monitoring changes in nanotube absorption as a function of pH. Protonation of the nanotube causes valence electrons to become localized and no longer able to participate in photoabsorption; therefore, the absorption intensity due to the highest lying valence electrons is decreased. This can be seen in Fig. 6.12, where the pH is cycled from 8 to 2.5. The absorption of the first Van Hove, E_{11}, transitions from 800–1600 nm can be seen to broaden and decrease in intensity with decreasing pH, while the second Van Hove, E_{22}, transitions from 500 – 900 nm are less affected. The reversibility of the reaction suggests a nanotube–proton equilibrium interaction leading to a protonated nanotube complex with a decreased absorption cross section.

$$n\left[H^{+}\right] + SWNT \underset{K_P}{\longleftrightarrow} [SWNT - H] \tag{6.2}$$

Here n[H^+] is the number of free protons in solution, [SWNT – H] is the nanotube, proton complex and K_p is the equilibrium constant for the

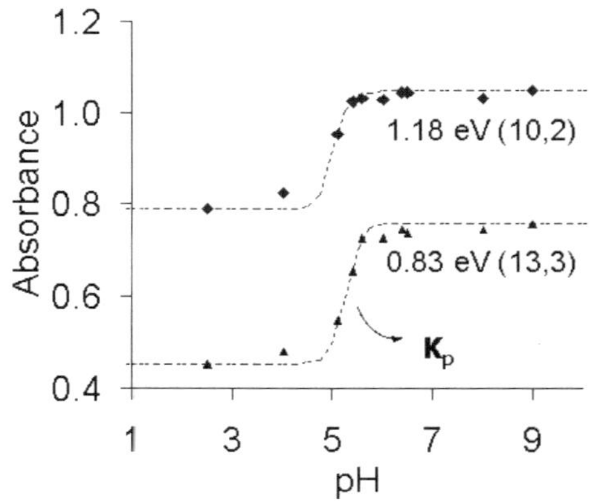

Fig. 6.13. Absorbance versus pH for the (10,2) and (13,3) nanotubes used to calculate the equilibrium constant, Kp. Values for ln(Kp) range from −36.39 to −33.97 for the (12,5) and the (8,3) nanotubes respectively

reaction. By scaling the absorption, the fraction of reacted nanotubes can be determined and then substituted into the equilibrium relation to yield:

$$\frac{A(pH) - A_p}{A_d - A_p} = \frac{K_p}{[H^+]^n + K_p} = \frac{K_p}{10^{-n(pH)} + K_p} \tag{6.3}$$

where A_p and A_d are the protonated and deprotonated absorption intensities, respectively. Plotting absorption versus pH, Fig. 6.13, and fitting the data with the above equation, n was determined to be 3. Interestingly, the rates of reaction were also selective with $\ln(K_p)$ ranging from −36.39 to −33.97 for the (12,5), 0.83 eV band gap, and the (8,3), 1.3 eV band gap, nanotubes, respectively.

The metallic nanotubes are much more difficult to monitor via spectroscopy due to the broad continuum of electronic states near the Fermi energy. For all nanotubes, the tangential mode Raman scattering splits into high- and low-frequency phonons; however, for metallic nanotubes, the low-frequency phonons couple to the electronic states near the Fermi level leading to a broadening and a low-frequency shit in the Breit – Wigner – Fano (BWF) line shape adjacent to the Lorentzian line shape of the remaining tangential modes. In Fig. 6.14c as solution pH decreases and electrons are withdrawn from the nanotube, the coupling of the low-frequency phonons to the continuum electronic states is disrupted causing the BWF line shape to shift to higher frequencies but maintaining the same intensity. However, the Lorentzian line shape does decrease with increasing acidity because Raman scattering is coupled to the absorption strength for some of the semiconducting nanotubes. The frequency shift of the BWF line shape, seen in Fig. 6.14a, can be used to calculate an equilibrium constant for metallic nanotubes using:

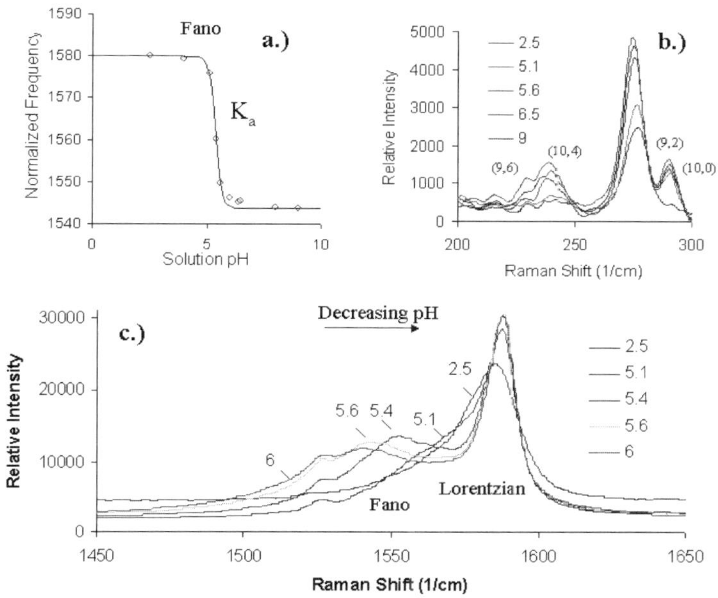

Fig. 6.14. Normalized frequency of the Breit - Wigner - Fano (Fano) line shape versus pH, used to calculate the equilibrium constant of the metals (**a**). Radial breathing mode (**b**) and tangential mode (**c**) of SWNT excited at 532 nm. The metals, (9,6) and (10,0), respond at higher pH than the semi-conductors, (9,2) and (10,0). The Fano line shape reflects phonon coupling to electronic continuum of metallic nanotubes

$$\frac{\omega(pH)-\omega_p}{\omega_d-\omega_p}=1-\frac{K_p}{[H^+]^n+K_p} \tag{6.4}$$

which is analogous to the previous equation with ω_p and ω_d being the peak normalized frequencies for the protonated and deprotonated state.

Further protonation selectivity can be seen in the radial breathing modes (RBMs) of a solution excited at 532 nm, of which the (9,2) and (10,0) semiconducting nanotubes are nearly resonant. In Fig. 6.14b the RBM peaks due to the (9,2) and (10,0) nanotubes decay selectively as proton concentration is increased. Metallic peaks due to the (10,4), (10,1) and (9,6) are also resonant at this excitation and respond to higher pH than the semi-conducting nanotubes.

Two other factors that are important to this reaction are the presence of oxygen and the type of surfactant used to suspend the nanotubes. For an acidic sample the absorption intensity can be restored to initial intensity by purging the solution of oxygen with a UV lamp under a shelter gas. This irradiation appears to decompose the nanotube/ proton complex and to remove adsorbed oxygen, effectively preventing the reformation of the nanotube/proton complex. The importance of oxygen in the reaction should not

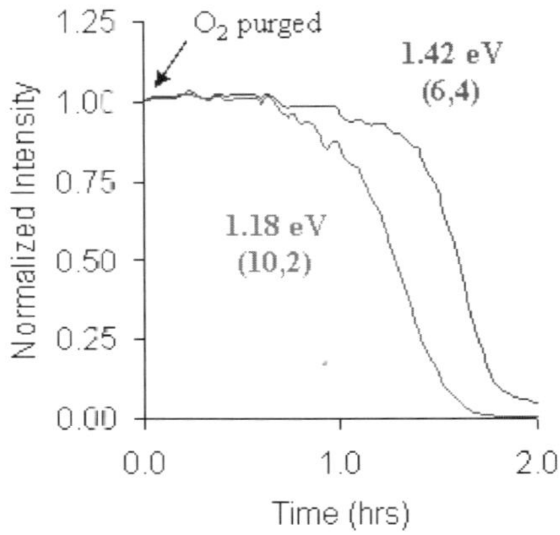

Fig. 6.15. Transient response of the normalized fluorescence intensity, initially purged of O_2, after O_2 is reintroduced. The solution pH is 2

come as a surprise when also considering the results of Shim and Siddons. It is likely that oxygen acts as a catalyst to the reaction by lowering the reaction barrier or by participating in the complex itself. Nanotube fluorescence of an oxygen purged sample at pH 2 can be seen to restore to the pH selective level in Fig. 6.15, further solidifying the importance of oxygen to the reaction mechanism. In this case the reaction rate is limited by oxygen adsorption kinetics and band gap selectivity can be seen. It should be noted that the fluorescence diminishes earlier than Raman scattering during oxygen readsorption, reflecting the greater sensitivity of fluorescence to the highest lying electrons becoming localized.

The other important factor affecting the protonation of the nanotube sidewall is the surfactant system used to suspend the nanotube. This is due to the fact that the electronic double layer formed at the surfactant/water interface influences local hydronium ion concentration. Generally for ionic surfactants, the ionic moieties of the surfactant layer are able to act as ion exchange sites. Molecules that are able to compete favorably for these sites cause the local hydronium ion concentration to decrease. As an example, PVP adsorbs strongly to SDS at the surfactant/water interface via charge transfer from sulfate groups. This adsorption causes the interfacial pH to be lowered due to a competitive exchange of sulfate groups. This can be further seen in the partial restoration of fluorescence and absorption in SDS suspended nanotubes at pH 2 after the addition of 0.1% PVP. This restoration can be generalized to occur for all species that complex to the surfactant. In the presence of the complexing agent, the surfactant shifts the spectroscopic response by changing the interfacial pH by ΔpH toward more neutral conditions.

The shift in localized pH for different surfactant systems can be seen in Table 6.1 with ΔpH values compared to findings for SDS and SDS/PVP mixtures. With nanotubes as the chromophore, PVP and PEO produce sim-

Table 6.1. Comparison of local pH changes in different surfactant systems

Surfactant System	wt %	ΔpH	ΔpH from SDS
sodium dodecyl sulfate (SDS)	1	2	0
poly (vinylpyrrolidone) 40 kDa/1% SDS	1	0.2	1.8
poly (ethylene oxide) 70 kDa/1% SDS	0.1	−0.1	2.1
poly (vinylpyrrolidone) 40 kDa/1% SDS	0.1	−0.15	2.15
1-vinyl-2-pyrrolidone/1% SDS	0.67	−0.4	2.4

ilar ΔpH values. They saturate the surface of the micelle and do not cause further pH shifting for molecular weights above 40 kDa or concentrations above 0.1 wt %. However, 1-vinyl-2-pyrrolidone, the monomer of PVP, does demonstrate increasing pH shift with increasing concentration and saturates only with full restoration. Long-chain alcohols display similar behavior, but they do not occupy exchange sites, instead they displace a sulfate group upon adsorption to the micelle layer. This displacement is equivalent to a loss of exchange sites as the SDS is displaced. The conclusion drawn from the observation of different surfactant systems is that SDS shifts the reaction behavior to a more neutral pH thereby obscuring the true equilibrium

6.3.3 Selective Protonation of Single-Walled Carbon Nanotubes Suspended in DNA

While various SDS/polymer mixtures exhibit different nanotube protonation behavior, the use of a totally different surfactant also causes nanotube protonation behavior to change. Nanotubes were recently suspended with DNA by Ming Zheng and co-workers [74] creating a novel new nanotube suspension. We observed similar selective protonation of DNA suspended nanotubes as that of SDS suspended nanotubes. In Fig. 6.16 we see absorption due to the first Van Hove transition, 800 – 1600 nm, exhibits both a broadening and an intensity decrease as the pH is dropped from 10.5 to 2.5, while the change in the absorption due to the second Van Hove transitions, 500 – 900 nm, is much less severe. Again, this reaction is reversible and the absorption recovers upon pH increase.

Monitoring the nanotube fluorescence gives further evidence for the selective protonation of DNA suspended nanotubes as shown in Fig. 6.17. The fluorescence intensity slightly increases as pH is decreased from 9 to 6.4 followed by a marked decrease in fluorescence intensity from pH 6.4 to 2.5. However, of note is that the fluorescence of the DNA suspended nanotubes did not quench completely, as was the case for SDS suspended nanotubes, but was still present at pH 2.5. The fluorescence due to the (7,5) nanotube also diminishes at a higher rate than does the fluorescence from the (6,5) nanotube. Therefore, selective protonation of nanotubes still occurs in different surfactant suspensions, but the nature of the surfactant also plays a role in the reaction.

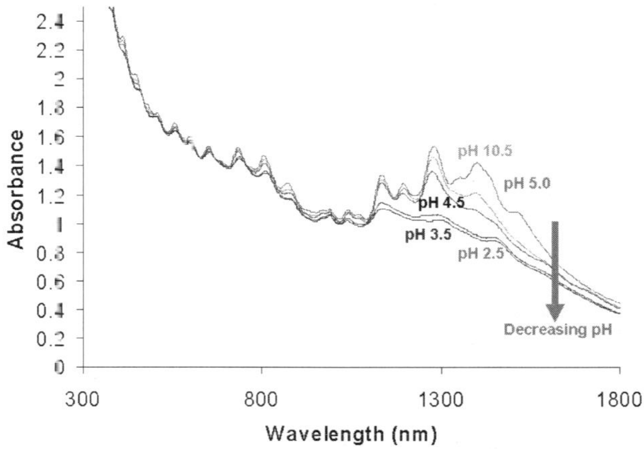

Fig. 6.16. Absorbance of carbon nanotubes dispersed with DNA, 30 mer GC repeating sequence, during titration from pH 10.5 to pH 2.5. The first Van Hove transitions respond at a higher pH than the second Van Hove transitions

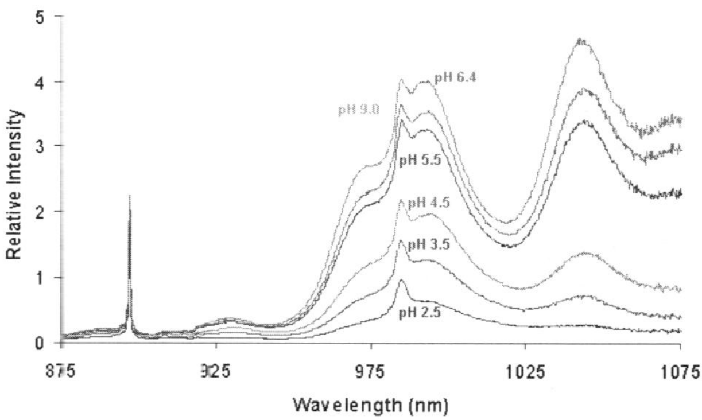

Fig. 6.17. Fluorescence intensity of DNA suspended nanotubes as pH is decreased from 9.0 to 2.5. Fluorescence intensity shows a slight increase from pH 9.0 to pH 6.4, and a sharp decrease from pH 6.4 to pH 2.5

6.4 Selective Non-covalent Chemistry: Solvatochromism

6.4.1 Introduction and Motivation

Carbon nanotubes exhibit significant solvatochromic changes which complicate spectral analysis but provide useful information regarding nanotubes' chemical and physical environment. For instance, spectral changes may denote concentration of the surfactant used to suspend the tubes [25] and relate information about dispersion/aggregation state [25]. For bulk processing techniques and reaction chemistries on the surfaces of carbon nanotubes, it

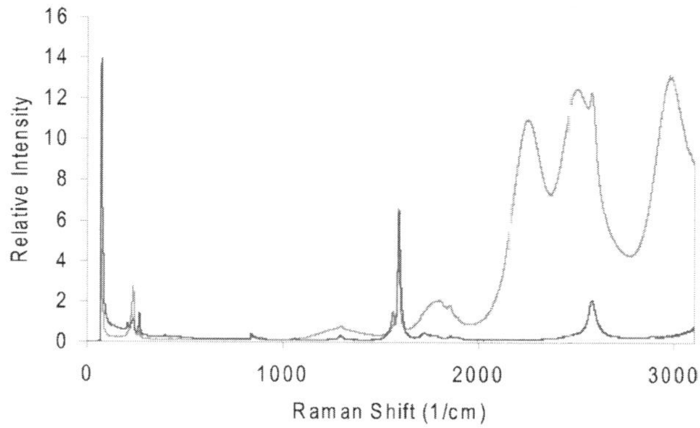

Fig. 6.18. Raman/fluorescence spectra of carbon nanotubes in (**a**) 0 mM surfactant before sonication and (**b**) 34 mM sodium dodecyl sulfate after sonication

is necessary to study the dispersion of nanotubes in solution. Therefore, examination of solvochromatic changes in aqueous surfactants will aid this line of inquiry.

6.4.2 Fluorescence Intensity Changes

Upon synthesis, aggregated nanotubes are bound by an estimated 500 eV per μm of tube length [4, 5]. This "bundling" behavior limits access to the tube's surface and also quenches nanotube fluorescence because of energy transfer to neighboring bundled tubes [25]. The latter fact, however, provides a method of discriminating individual tubes dispersed in solution from nanotube bundles. Ultrasonication in aqueous surfactants is the primary mode of dispersing nanotubes bound by these large energies [5, 25, 75]. Figure 6.18 shows a Raman/fluorescence spectrum at 785 nm excitation of a nanotube solution in 0mM sodium dodecyl sulfate (SDS) concentration, and after 5 hrs of sonication in 34 mM SDS. During the process of ultrasonication, the fluorescent features grow as a function of time. By adding surfactant and sonicating, the fluorescence intensity increases and reaches a steady state in 20 min to 1 hr. Figure 6.19 shows the growth of the (8,3) peak.

6.4.3 Wavelength Shifts

In addition to intensity changes, nanotube dispersion induces wavelength shifts in fluorescence. Upon adding an aliquot of surfactant, the fluorescence peaks immediately shift to higher energy. These changes occur systematically; as each aliquot is added, all peaks shift by approximately 8 meV.

Figure 6.20 shows SWNT solutions at varying concentrations of surfactant centrifuged at 30,000 RPM to remove bundles and diluted with deionized

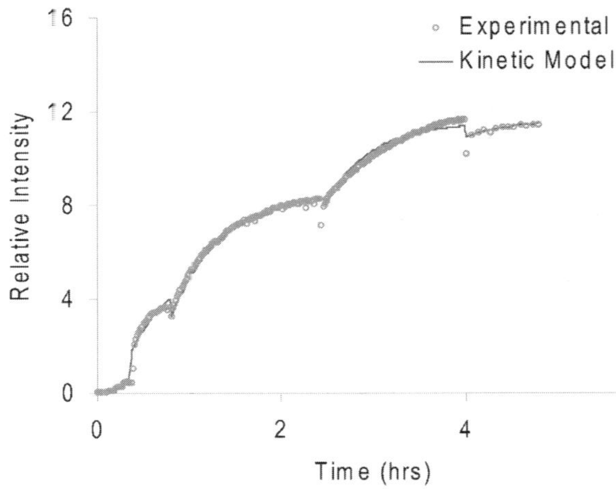

Fig. 6.19. The (8,3) nanotube's fluorescence as a function of sonication time and SDS concentration. Each curve represents a distinct surfactant concentration. Following each surfactant addition, the intensity approaches a steady-state value. Surfactant additions were made at the five discontinuities shown with 0.5, 2.7, 4, 6.2, 6.3 mM as the total concentration after each addition

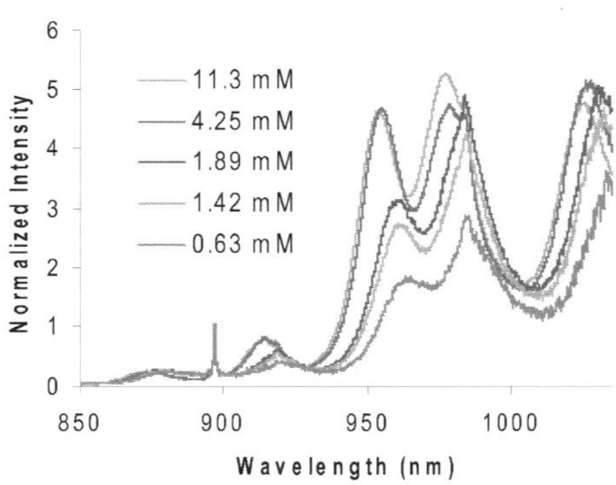

Fig. 6.20. Fluorescence spectra of nanotubes in SDS shows a solvatochromatic shift in to shorter wavelength with increasing concentration of surfactant

water. Fluorescence was then measured with respect to SDS concentration. Above SDS's critical micelle concentration (CMC), no solvatochromatic shift occurred. However, upon dilution to below the CMC (8.1 mM in water at 298 K), fluorescence peaks shifted reversibly to lower energy. Addition of SDS to the sample resulted in shifting of the peaks back to their previous positions. These changes are metastable, however, and equilibration overnight results in aggregation of the nanotubes. Suspending nanotubes in other surfactants: anionic, cationic, and nonionic, results in similar fluorescence spectra to the

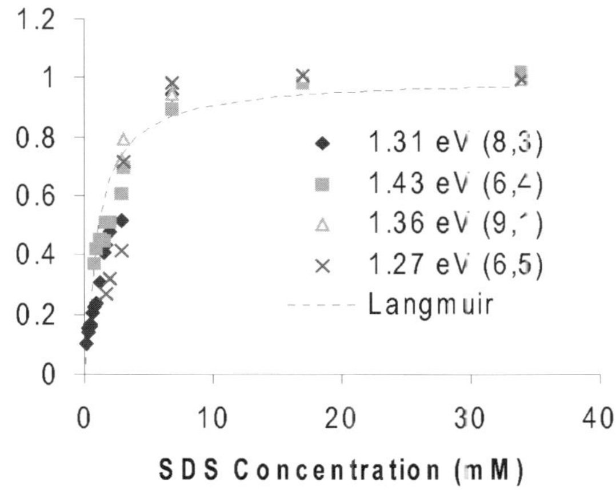

Fig. 6.21. A Langmuir isotherm is traced out by measuring the relative fluorescence peak shift scaled by maximum observable shift for four fluorescent features. We assume that peak shift is directly proportional to surfactant surface coverage

one generated in SDS, though fluorescence peak positions are red-shifted or blue-shifted in comparison [25].

Using solvatochromatic shifts in fluorescence, a Langmuir adsorption isotherm can be drawn by normalizing with respect to the maximum spectral shift in a solution of SWNT in 1% SDS/H_2O (Fig. 6.21).

$$\frac{\Delta E}{\Delta E_{\max}} = \frac{K_{SDS}C_{SDS}}{K_{SDS}C_{SDS}+1} \tag{6.5}$$

The ΔE is the peak's shift relative to the shift in a 1% SDS/H_2O solution and ΔE_{max} is the maximum shift. The equilibrium constant for SDS adsorption, K_{ads}, is 0.98 L/mol [76]. The fluorescence intensity increases and shifts to higher energy when the nanotube surface decreases its association with water, lessening the possibility of nonradiative energy transfer processes.

In order to model the process of bundled nanotubes versus those in solution, we can write

$$\frac{dC_1}{dt} = k_z C_b C_{SDS} - \frac{k_z}{K_{eq}} C_1 \tag{6.6}$$

where C_I is the concentration of individual tubes in solution and C_b is the concentration of bundled tubes. The constant k_z is the rate at which nanotubes will "unzip" from the bundle, and K_{eq} is the equilibrium constant of bundling. This equation provides the kinetic model

$$C_T = C_b(t) + C_I(t) \tag{6.7}$$

By integrating these equations, we obtain

$$C_1(t) = \frac{K_{eq}C_T C_{SDS}(\gamma(t) - 1) + C_1(0)C_{SDS}(1 + K_{eq}C_{SDS})}{\gamma(t)(1 + K_{eq}C_{SDS})} \tag{6.8}$$

where

$$\gamma(t) = e^{k_z(C_{SDS}+1/K_{eq})t} \tag{6.9}$$

This equation models the process of producing fluorescent nanotubes during sonication and predicts the fluorescence saturation observed at an SDS concentration given by

$$C_{1,max} = C_T \frac{K_{eq}C_{SDS}}{K_{eq}C_{SDS} + 1} \tag{6.10}$$

6.4.4 Changes to the Raman Spectrum

Raman spectroscopy is used to measure the compositions of SWNT samples.

The tangential mode (G-peak) near 1593 cm^{-1} is used to identify the presence of semiconducting or metallic nanotubes, as metallic tubes cause the mode to split, giving a Breit-Wigner-Fano (BWF) line-shape in the Raman spectrum. This line-shape is caused when the tangential phonon couples to the continuum of electronic states at the Fermi level [19, 26, 77]. The radial breathing modes (RBM) which appear between 200 cm^{-1} and 400 cm^{-1} are diameter dependent and can be used to measure particular nanotube species that are resonant at a specific excitation wavelength [3, 7, 19, 26].

Despite the valuable information concerning SWNT sample composition which Raman spectra provide, environmental conditions in the sample may alter the spectra, complicating use of the technique. The strong resonant enhancement and concomitant coupling between strength and energy of nanotube inter-band transitions cause spectral changes as physical [25] or chemical [10] conditions perturb the sample [3, 6]. Raman scattering can be resonantly enhanced or shift out of resonance by nanotube bundling. For instance, the tangential mode at 1593 cm^{-1} increases in intensity during sonication of a sample of nanotubes [25]. In addition, the aforementioned BWF feature may shift to become indistinguishable from the Lorentzian component of the tangential mode with decreasing pH [10].

The effects of aggregation are illustrated on the RBMs of three different HiPco SWNT samples, excited at 785 nm, are shown in Fig. 6.22. Figure 6.22a shows nanotubes in a 1% SDS/$H_2$0 solution at pH 10. Figure 6.22b shows solid HiPco SWNT used as received and containg 28.5% Fe catalyst by weight. Figure 6.22c contains material from Fig. 6.22a which was deposited on a glass slide and heated until the water evaporated. Illustrations beside the spectra depict the believed aggregation state of the material which increases from Fig. 6.22a to Fig. 6.22c. The spectra reveal a substantial decrease

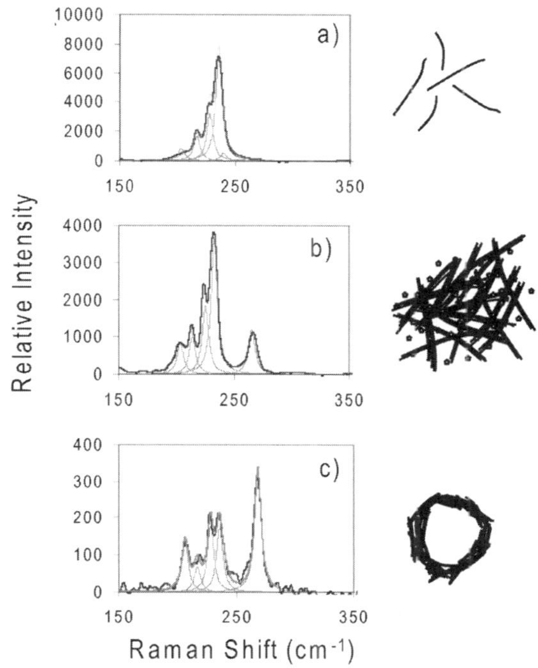

Fig. 6.22. Radial breathing modes of HiPco carbon nanotubes taken at three different states of aggregation: (**a**) Individual tubes suspended in SDS. (**b**) Raw HiPco material as received. (**c**) Dry SDS-suspended tubes heated on a hot plate

in the 234 cm^{-1} mode and simultaneous increase in the 267 cm^{-1} mode. The 267 cm^{-1} feature corresponds to the (10,2) nanotube and undergoes a v2→c2 transition at 734 nm [7]. When bundled, this peak increases as the feature shifts to lower energy and comes into resonance with the 785 nm excitation laser. The 234 cm^{-1} feature corresponds to the (11,3) nanotube with a v2→c2 transition at 792 nm and shows significant enhancement when isolated.

The aggregation process can be modeled by using the spectral assignments of the six nanotubes present [63], and a simple Gaussian distribution in nanotube diameters d_t (σ = 0.2 nm, d_{mean} = 0.93 nm) (Fig. 6.23). We assume that inter-band transitions can be simplified as a delta function at energy $E_{(n,m)}$ and that each spectral feature is Lorentzian with shift given by $223.5/d_t + 12.5$.

$$I(E_{Laser}) \propto \frac{\frac{1}{\sigma\sqrt{4\pi}} \exp\left[-\frac{(d_t - d_{mean})^2}{4\sigma^2}\right]}{\left(\left(E_{Laser} - E_{(n,m)} + \Delta E\right)^2 + \frac{\Gamma^2}{4}\right)\left(\left(E_{Laser} - E_{Phonon} - E_{(n,m)} + \Delta E\right)^2 + \frac{\Gamma^2}{4}\right)} \quad (6.11)$$

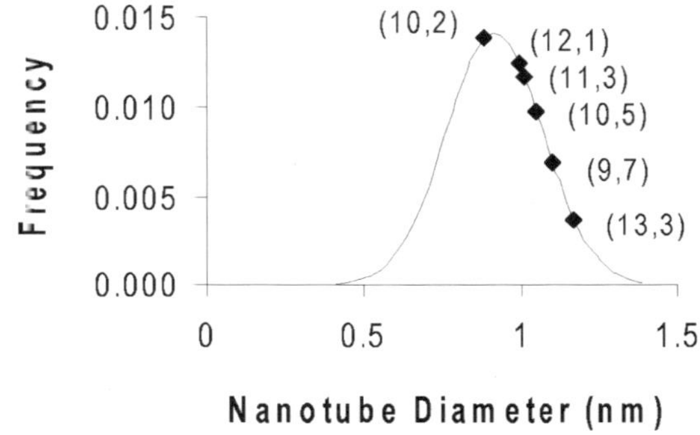

Fig. 6.23. Gaussian distribution of nanotube diameters highlighting the relative frequencies of six species

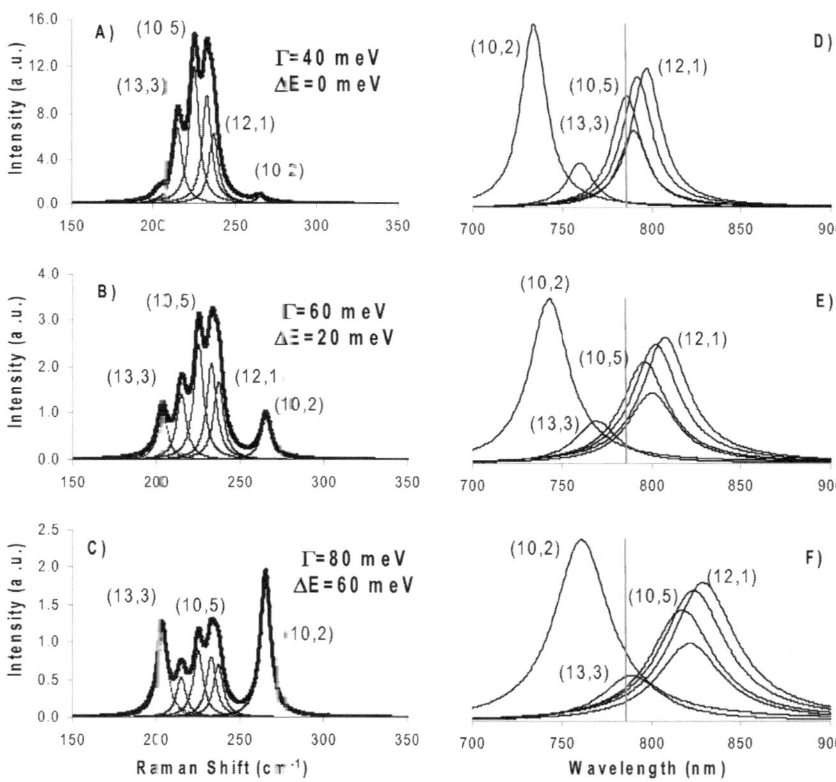

Fig. 6.24. (**a–c**) Simulated radial breathing mode spectra of aggregated nanotubes using (6.11) (**d–f**) Spectral shift and broadening of the same nanotube species observed in absorption spectra

In this expression, E_{laser} is the excitation energy while ΔE and Γ are the energy shift upon roping and peak broadening due to environmental conditions surrounding the nanotube. We assume that ΔE is the same for all transitions. Using this expression, we can simulate the Raman spectra observed in Fig. 6.24. These simulated changes reflect those observed experimentally (Fig. 6.24a through c). For example, the 267 cm^{-1} mode, which is the (10,2) tube, is initially off resonance with the 785 nm excitation laser. As seen in Fig. 6.24 a through c, the transitions shift to lower energy with increasing aggregation state. These shifts in transition energy cause the (10,2) nanotube to come into resonance with the excitation laser. As described by ab-initio calculations [78], upon alignment and aggregation into flush contact, the nanotubes' k_z band structure is changed as an orthogonal electronic dispersion is created. The nanotube absorption spectrum is broadened and red-shifted by aggregation, which can affect the Raman spectrum by bringing a nanotube into or out of resonance with the excitation laser. The model above qualitatively predicts these aggregation effects.

References

1. Avouris, P.: Acc. Chem. Res. **35**, 1026, (2002)
2. Dresselhaus, M.S., Dresselhaus, G., and Eklund, P.C. Science of fullerenes and carbon nanotubes (San Diego: Academic Press, 1996)
3. Saito, R., Dresselhaus, G., and Dresselhaus, M.S. Physical Properties of Carbon Nanotubes (London: Imperial College Press, 1998)
4. Thess, A., Lee, R., Nikolaev, P., Dai, H.J., Petit, P., Robert, J., Xu, C.H., Lee, Y.H., Kim, S.G., Rinzler, A.G., Colbert, D.T., Scuseria, G.E., Tomanek, D., Fischer, J.E., and Smalley, R.E.: Science **273**, 483,(1996)
5. O'Connell, M.J., Boul, P., Ericson, L.M., Huffman, C., Wang, Y.H., Haroz, E., Kuper, C., Tour, J., Ausman, K.D., and Smalley, R.E.: Chemical Physics Letters **342**, 265,(2001)
6. Strano, M.S., Doorn, S.K., Haroz, E.H., Kittrell, C., Hauge, R.H., and Smalley, R.E.: Nano Letters **3**, 1091,(2003)
7. Bachilo, S.M., Strano, M.S., Kittrell, C., Hauge, R.H., Smalley, R.E., and Weisman, R.B.: Science **298**, 2361,(2002)
8. Kukovecz, A., Kramberger, C., Georgakilas, V., Prato, M., and Kuzmany, H.: Eur. Phys. J. B **28**, 223, (2002)
9. Rao, A.M., Richter, E., Bandow, S., Chase, B., Eklund, P.C., Williams, K.A., Fang, S., Subbaswamy, K.R., Menon, M., Thess, A., Smalley, R.E., Dresselhaus, G., and Dresselhaus, M.S.: Science **275**, 187, (1997)
10. Strano, M.S., Huffman, C.B., Moore, V.C., O'Connell, M.J., Haroz, E.H., Hubbard, J., Miller, M., Rialon, K., Kittrell, C., Ramesh, S., Hauge, R.H., and Smalley, R.E.: Journal of Physical Chemistry B **107**, 6979, (2003)
11. Bronikowski, M.J., Willis, P.A., Colbert, D.T., Smith, K.A., and Smalley, R., E. (2001): Journal of Vacuum Science & Technology,(2001)
12. Canonico, M., Adams, G.B., Poweleit, C., Menendez, J., Page, J.B., Harris, G., van der Meulen, H.P., Calleja, J.M., and Rubio, J.: Physical Review B **65**, art. no.201402.(2002)

13. Saito, R., Grueneis, A., Cancado, L.G., Pimenta, M.A., Jorio, A., Dresselhaus, G., Dresselhaus, M.S., and Souza, A.G.: Molecular Crystals and Liquid Crystals **387**, 287, (2002)
14. Saito, R., Jorio, A., Hafner, J.H., Lieber, C.M., Hunter, M., McClure, T., Dresselhaus, G., and Dresselhaus, M.S. : Physical Review B **6408**, art. no.085312, (2001)
15. Jorio, A., Fantini, C., Dantas, M.S.S., Pimenta, M.A., Souza, A.G., Samsonidze, G.G., Brar, V.W., Dresselhaus, G., Dresselhaus, M.S., Swan, A.K., Unlu, M.S., Goldberg, B.B., and Saito, R.: Physical Review B **66**, art. no.-115411, (2002)
16. Jorio, A., Matinaga, F.M., Righi, A., Dantas, M.S.S., Pimenta, M.A., Souza, A.G., Mendes, J., Hafner, J.H., Lieber, C.M., Saito, R., Dresselhaus, G., and Dresselhaus, M.S.: Brazilian Journal of Physics **32**, 921, (2002)
17. Jorio, A., Souza, A.G., Dresselhaus, G., Dresselhaus, M.S., Swan, A.K., Unlu, M.S., Goldberg, B.B., Pimenta, M.A., Hafner, J.H., Lieber, C.M., and Saito, R.: Physical Review B **65**, art. no.155412,(2002)
18. Dresselhaus, M.S., Dresselhaus, G., Jorio, A., Souza, A.G., Pimenta, M.A., and Saito, R. (2002): Accounts of Chemical Research **35**, 1070,(2002)
19. Dresselhaus, M.S., Dresselhaus, G., Jorio, A., Souza, A.G., and Saito, R.: Carbon **40**, 2043,(2002)
20. Yu, Z.H., and Brus, L.E.: Journal of Physical Chemistry B **105**, 6831,(2001)
21. Yu, Z.H., and Brus, L.E.: J. Phys. Chem. A. **104**, 10995, (2000)
22. Saito, R., Dresselhaus, G., and Dresselhaus, M.S.: Physical Review B **61**, 2981,(2000)
23. Reich, S., Maultzsch, J., Thomsen, C., and Ordejon, P.: Physical Review B **66**, 035412,(2002)
24. Reich, S., and Thomsen, C.: Physical Review B **62**, 4273,(2000)
25. O'Connell, M.J., Bachilo, S.M., Huffman, C.B., Moore, V.C., Strano, M.S., Haroz, E.H., Rialon, K.L., Boul, P.J., Noon, W.H., Kittrell, C., Ma, J.P., Hauge, R.H., Weisman, R.B., and Smalley, R.E.: Science **297**, 593, (2002)
26. Sauvajol, J.L., Anglaret E., Rols, S., and Alvarez, L. : Carbon **40**, 1697,(2002)
27. Liu, J., Rinzler, A.G., Dai, H.J., Hafner, J.H., Bradley, R.K., Boul, P.J., Lu, A., Iverson, T., Shelimov, K., Huffman, C.B., Rodriguez-Macias, F., Shon, Y.S., Lee, T.R., Colbert, D.T., and Smalley, R.E.: Science **280**, 1253,(1998)
28. A.G. Rinzler, J.L., H. Dai, P. Nikolaev, C.B. Huffman, F.J. Rodriguez-Macias, P.J. Boul, A.H. Lu, D. Heymann, D.T. Colbert, R.S. Lee, J.E. Fischer, A.M. Rao, P.C. Eklund and R.E. Smalley: Applied Physics A **67**, 29,(1998)
29. Mawhinney, D.B., Naumenko, V., Kuznetsova, A., Yates, J.T., Liu, J., and Smalley, R.E.: Chemical Physics Letters **324**, 213, (2000)
30. Mawhinney, D.B., Naumenko, V., Kuznetsova, A., Yates, J.T., Liu, J., and Smalley, R.E.: Journal of the American Chemical Society **122**, 2383,(2000)
31. Kuznetsova, A., Popova, I., Yates, J.T., Bronikowski, M.J., Huffman, C.B., Liu, J., Smalley, R.E., Hwu, H.H., and Chen, J.G.G. : Journal of the American Chemical Society **123**, 10699,(2001)
32. Deng, J.P., Mou, C.Y., and Han, C.C.: Fullerene Science and Technology **5**, 1033,(1997)
33. Mickelson, E.T., Huffman, C.B., Rinzler, A.G., Smalley, R.E., Hauge, R.H., and Margrave, J.L.: Chemical Physics Letters **296**, 188,(1998)
34. Bahr, J.L., Yang, J.P., Kosynkin, D.V., Bronikowski, M.J., Smalley, R.E., and Tour, J.M.: Journal of the American Chemical Society **123**, 6536, (2001)

35. Bahr, J.L., and Tour, J.M.: Chemistry of Materials **13**, 3823, (2001)
36. Chen, Y., Haddon, R.C., Fang, S., Rao, A.M., Lee, W.H., Dickey, E.C., Grulke, E.A., Pendergrass, J.C., Chavan, A., Haley, B.E., and Smalley, R.E.: Journal of Materials Research **13**, 2423, (1998)
37. Wong, S.S., Woolley, A.T., Joselevich, E., Cheung, C.L., and Lieber, C.M.: Journal of the American Chemical Society **120**, 8557,(1998)
38. Hu, H., Zhao, B., Hamon, M.A., Kamaras, K., Itkis, M.E., and Haddon, R.C.: Journal of the American Chemical Society **125**, 14893,(2003)
39. Chen, J., Hamon, M.A., Hu, H., Chen, Y.S., Rao, A.M., Eklund, P.C., and Haddon, R.C.: Science **282**, 95, (1998)
40. Holzinger, M., Vostrowsky, O., Hirsch, A., Hennrich, F., Kappes, M., Weiss, R., and Jellen, F.: Angewandte Chemie-International Edition **40**, 4002,(2001)
41. Holzinger, M., Abraha, J., Whelan, P., Graupner, R., Ley, L., Hennrich, F., Kappes, M., and Hirsch, A.: Journal of the American Chemical Society **125**, 8566,(2003)
42. Hirsch, A.: Angewandte Chemie-International Edition **41**, 1853,(2002)
43. Chen, Z.F., Thiel, W., and Hirsch, A.: Chemphyschem **4**, 93,(2003)
44. S. Niyogi, M.A.H., H. Hu, B. Zhao, P. Bhowmik, R. Sen, M.E. Itkis, and R.C. Haddon: Accounts of Chemical Research **35**, 1105,(2002)
45. Hamon, M.A., Itkis, M.E., Niyogi, S., Alvaraez, T., Kuper, C., Menon, M., and Haddon, R.C.: Journal of the American Chemical Society **123**, 11292,(2001)
46. Haddon, R.C.: Journal of the American Chemical Society **112**, 3385,(1990)
47. Haddon, R.C.: Science **261**, 1545,(1993)
48. Haddon, R.C.: Journal of Physical Chemistry A **105**, 4164,(2001)
49. Taylor, R., and Walton, D.R.M.: Nature **363**, 685,(1993)
50. Hirsch, A.: The Chemistry of the Fullerenes (Stuttgart: Thieme,1994)
51. M.S. Dresselhaus, G.D., P.C. Eklund: Science of Fullerenes and Carbon Nanotubes (San Diego, CA: Academic Press,1996)
52. Bronikowski, M.J., Willis, P.A., Colbert, D.T., Smith, K.A., and Smalley, R.E.: Journal of Vacuum Science & Technology a-Vacuum Surfaces and Films **19**, 1800,(2001)
53. R. Saito, G.D., M.S. Dresselhaus: Physical Properties of Carbon Nanotubes (London: Imperial College Press,1998)
54. Bravo-Diaz, C., Soengas-Fernandez, M., Rodriguez-Sarabia, M.J., and Gonzalez-Romero, E.: Langmuir **14**, 5098,(1998)
55. Bahr, J.L., and Tour, J.M.: Journal of Materials Chemistry **12**, 1952,(2002)
56. Dyke, C.A., and Tour, J.M.: Journal of the American Chemical Society **125**, 1156,(2003)
57. Strano, M.S., Dyke, C.A., Usrey, M.L., Barone, P.W., Allen, M.J., Shan, H.W., Kittrell, C., Hauge, R.H., Tour, J.M., and Smalley, R.E.: Science **301**, 1519,(2003)
58. Itkis, M.E., Niyogi, S., Meng, M.E., Hamon, M.A., Hu, H., and Haddon, R.C.: Nano Letters **2**, 155,(2002)
59. Kataura, H., Kumazawa, Y., Maniwa, Y., Umezu, I., Suzuki, S., Ohtsuka, Y., and Achiba, Y.: Synthetic Metals **103**, 2555,(1999)
60. Mintmire, J.W., and White, C.T.: Physical Review Letters **81**, 2506,(1998)
61. Ando, T.: Journal of the Physical Society of Japan **66**, 1066,(1997)
62. Kane, C.L., and Mele, E.J.: Physical Review Letters **90**,(2003)
63. Strano, M.S.: Journal of the American Chemical Society **125**, 16148,(2003)

64. Krupke, R., Hennrich, F., von Lohneysen, H., and Kappes, M.M. : Science **301**, 344,(2003)
65. Zheng, M., Jagota, A., Strano, M.S., Santos, A.P., Barone, P., Chou, S.G., Diner, B.A., Dresselhaus, M.S., McLean, R.S., Onoa, G.B., Samsonidze, G.G., Semke, E.D., Usrey, M., and Walls, D.J.: Science **302**, 1545,(2003)
66. Chattopadhyay, D., Galeska, L., and Papadimitrakopoulos, F. : Journal of the American Chemical Society **125**, 3370,(2003)
67. Zhou, W., Ooi, Y.H., Russo, R., Papanek, P., Luzzi, D.E., Fischer, J.E., Bronikowski, M.J., Willis, P.A., and Smalley, R.E. : Chemical Physics Letters **350**, 6,(2001)
68. Kavan, L., Rapta, P., and Dunsch, L.: Chemical Physics Letters **328**, 363,(2000)
69. Kavan, L., and Dunsch, L.: Chemphyschem **4**, 944,(2003)
70. Kavan, L., Rapta, P., Dunsch, L., Bronikowski, M.J., Willis, P., and Smalley, R.E.: Journal of Physical Chemistry B **105**, 10764,(2001)
71. Collins, P.G., Bradley, K., Ishigami, M., and Zettl, A. : Science **287**, 1801,(2000)
72. Heinze, S., Tersoff, J., Martel, R., Derycke, V., Appenzeller, J., and Avouris, P.: Physical Review Letters **89**,(2002)
73. Shim, M., and Siddons, G.P.: Applied Physics Letters **83**, 3564,(2003)
74. Zheng, M., Jagota, A., Semke, E.D., Diner, B.A., McLean, R.S., Lustig, S.R., Richardson, R.E., and Tassi, N.G.: Nature Materials **2**, 338,(2003)
75. Nakashima, N., Tomonari, Y., and Murakami, H.: Chemistry Letters, 638,(2002)
76. Djuve, J., Grant, L.M., Sjoblom, J., Goloub, T.P., and Pugh, R.J.: Langmuir **18**, 2673,(2002)
77. Brown, S.D.M., Jorio, A., Corio, P., Dresselhaus, M.S., Dresselhaus, G., Saito, R., and Kneipp, K.: Physical Review B **6315**, art. no.155414,(2001)
78. Reich, S., Thomsen, C., and Ordejon, P.: Physical Review B **65**, art. no.155411,(2002)

Part III

Optical Spectroscopy

7 Fluorescence Spectroscopy of Single-Walled Carbon Nanotubes

R.B. Weisman

An overview is presented of basic and applied aspects of the fluorescent photoluminescence from single-walled carbon nanotubes (SWNT). This fluorescence was first discovered in aqueous surfactant suspensions of SWNT that had been processed for enrichment in individual, unbundled nanotubes. Spectrofluorimetric measurements of emission intensity as a function of excitation and emission wavelengths revealed a rich pattern of peaks representing distinct (n, m) structural species. Careful analysis allowed each of these peaks to be assigned to a specific semiconducting (n, m) species. This spectral assignment provided a large body of precise optical transition energies for a significant range of tube diameters and chiralities. Important patterns of electronic structure emerged showing the related properties of nanotubes within "families" (sharing the same $n–m$ value) and "tribes" (sharing the same mod($n–m$,3) value). The results also allowed construction of an empirical "Kataura plot," useful for guiding experiments, that gives optical transition energies as a function of nanotube diameter for semiconducting species. In surfactant-suspended samples, optical transition energies are found to depend mildly on nanotube environment. Spectral line shapes reveal the predominant excitonic character of optical excitations in SWNT and provide information on environmental heterogeneity and on exciton dephasing rates. Nanotube fluorescence is quenched by aggregation, chemical derivatization, and by acidification in some aqueous suspensions. Fluorimetry offers a powerful method for determining the (n, m) composition of mixed nanotube samples. Instrumental methods for such fluorimetric analysis are discussed and compared. Finally, the unusual near-infrared emission from SWNT can be exploited to allow selective optical detection and imaging of nanotubes in complex environments. Early results are presented showing how this approach can be used to image the locations of nanotubes inside biological cells.

7.1 Introduction

When matter is electronically excited by thermal, optical, or electrical means, it commonly relaxes through the emission of light. This luminescence process has enormous scientific value because of the detailed information that spectral positions and intensities can reveal about a sample's electronic structure. In addition, luminescence has great practical value because it permits the sensitive qualitative and quantitative analysis of specific substances in complex mixtures. Single-walled carbon nanotubes (SWNT) represent an exciting new class of synthetic nanomaterials that hold great interest for their scientific novelty and for their potential utility in a wide range of applications [1,2]. The recent discovery of photoluminescence from SWNT offers a powerful tool for new basic and applied studies of nanotubes.

Absorption and emission of light at wavelengths from the ultraviolet to the near-infrared arise from excitations of the electrons in matter, and the wavelengths and strengths of these optical transitions reflect a sample's detailed electronic properties. SWNT are tubular nanostructures, typically about 1 or 2 nm in diameter and hundreds or thousands of nanometers long, consisting entirely of covalently bonded carbon atoms. Each of these carbons is bound to three neighboring atoms by sigma bonds, and its remaining carbon p-electron contributes to a delocalized pi-electron system. Electronic states of this system are delocalized along the translationally periodic, quasi-infinite axis of the nanotube, but are constrained in the transverse directions by an angular periodic boundary condition reflecting the tube's diameter and wrapping (chiral) angle.

A key feature of carbon nanotubes is the strong dependence of their electronic properties on physical tube structure [1,3]. Because SWNT are formed in a variety of discrete structures having distinct diameters and chiralities,

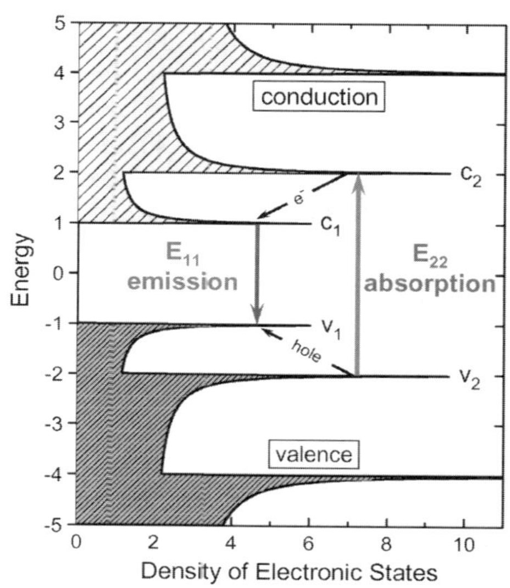

Fig. 7.1. Schematic density of electronic states for a semiconducting single-walled carbon nanotube. Van Hove singularities are labeled with "v" for valence band and "c" for conduction band, along with subscripts giving the sub-band index. Vertical arrows show intense optical transitions for light polarized along the tube axis

a corresponding variety of pi-electron band structures and excitations are found in mixtures of nanotubes. Each possible SWNT structure is uniquely described by a pair of integers, (n,m), that describe the length and orientation of that tubes's circumference vector when projected onto a graphene sheet. As a consequence of combining the roll-up boundary condition with graphene's unusual band structure, a SWNT is expected to be metallic if $n = m$, semi-metallic if the difference n–m is evenly divisible by 3, or semiconducting otherwise. Band gaps among the semiconducting nanotubes vary in approximate inverse proportion to the tube diameter. Another important feature of SWNT electronic structure is sets of sharp maxima, called van Hove singularities, in their electronic densities of states. These arise from the tubes' quasi-one-dimensional character. Nanotube optical spectra are dominated by strong dipole-allowed transitions, polarized along the tube axis, between van Hove singularities in matching valence and conduction sub-bands. These are referred to as E_{ii} transitions, with $i = 1, 2, 3$, etc. Figure 7.1 illustrates the idealized electronic state density and dominant low-energy optical transitions of a semiconducting SWNT [4].

7.2 Observation of Photoluminescence

All currently available SWNT production methods generate mixtures that contain many different (n, m) species. From the discussion above, one would expect the optical spectrum of such a mixture to show distinct, separated transitions for the various species at wavelengths reflecting their differing band structures. However, it was widely observed that the optical spectra of SWNT samples are relatively diffuse and uninformative. This situation changed with the discovery of sharpened spectral features in samples that had been processed to overcome the strong tendency of SWNT to aggregate into bundles bound by van der Waals attractions [5]. Raw, unpurified SWNT material from a HiPco reactor was mechanically dispersed in an aqueous solution of an ionic surfactant such as SDS (sodium dodecylsulfate). This was followed by intense ultrasonic agitation to de-bundle many of the nanotubes. Rapid coating by SDS of the individual nanotubes formed in this step prevented them from re-aggregating. Finally, the remaining slightly denser bundled nanotubes were separated from the individuals by extensive ultracentrifugation. The resulting decanted supernatant provided samples substantially enriched in individually suspended nanotubes. It was found that such samples show not only sharpened absorption spectra, but also photoluminescence in the near-infrared, with emission features evident at wavelengths from approximately 900 to 1600 nm in a sample of HiPco-grown nanotubes [5]. In fact, the observed photoluminescence spectrum was even more structured than the absorption spectrum. Each emission peak was located very near to an absorption peak in the set of lowest energy (E_{11}) transitions. The red-shift between corresponding absorption and emission peaks was only ca.

30 cm^{-1}. The emission was thereby identified as a set of band-gap photoluminescence transitions in a mixture of semiconducting SWNT having different E_{11} transition energies. In analogy to Kasha's Rule in molecular electronic spectroscopy [6], no emission was detectable in E_{22} or higher transitions.

Although precise measurements of luminescence quantum yields are hampered by a scarcity of reference standards in the near-infrared and by overlapping transitions of different SWNT species, it appears that only approximately 0.1% of optically excited semiconducting nanotubes in aqueous suspension relax by photon emission. The excited state lifetime is subnanosecond, as was deduced in the initial photoluminescence study [5] and later examined in detail by highly time-resolved spectroscopic investigations [7–9]. This combination of lifetime and emissive quantum yield indicates a radiative decay rate within the range of spin-allowed optical transitions. In molecular language, the emitting state is a spin singlet rather than a spin triplet. This identification allows the emission to be described as fluorescence, a term that we will favor in this chapter.

7.3 Deciphering the (n, m) Spectral Assignment

The discovery of nanotube band-gap fluorescence allowed a powerful tool to be applied to the centrally important spectral problem of mapping optical transitions to specific (n,m)-identified nanotube species. This additional experimental tool was two-dimensional spectrofluorimetry on samples of mixed SWNT [10]. Here the intensity of band-gap emission is measured as a function of two variables: the (E_{11}) emission wavelength and the (E_{22}) excitation wavelength. Figure 7.2 shows the dramatic result of such a measurement in the form of a surface plot. The x- and y-coordinates of each distinct peak in the main "mountain range" directly reveal the E_{11} and E_{22} transition energies of a single (n,m) nanotube species. Note that only semiconducting species were observed in this experiment, because the metallic and semi-metallic nanotubes are nonemissive. Assignment of observed transitions to specific (n,m) values was clearly a crucial goal. When the set of experimental E_{22}/E_{11} ratios was examined as a function of E_{22}, strong patterns became evident. By correlating these qualitative patterns with those found from model calculations, it was recognized that elements in the pattern represented "families" of nanotubes, defined as those species sharing the same value of n–m. It was also possible to deduce the changes in n and m values between adjacent spectral points within and between families. In this way, the entire network of relative (n,m) values was obtained for a large set of spectrofluorimetric data points [10].

The remaining step in spectral assignment was then "anchoring" the network by finding the absolute (n,m) identity of one peak. Taking into consideration a number of criteria, it was deduced that the anchoring choices were highly constrained, and that only a few possibilities were plausible. Each of

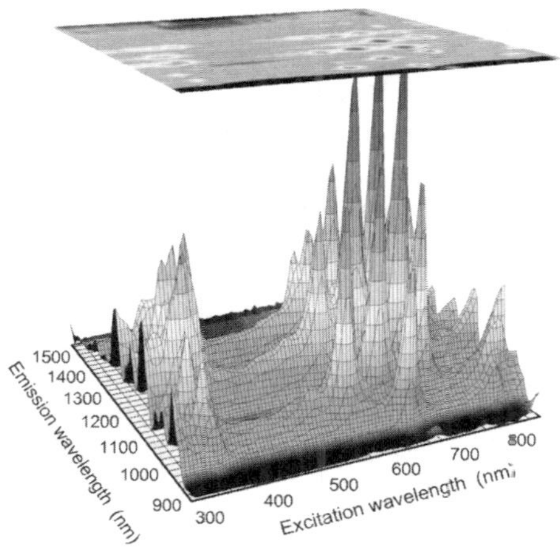

Fig. 7.2. Surface plot of photoluminescence (fluorescence) intensity measured as a function of emission and excitation wavelengths for a SWNT sample in aqueous SDS suspension. Each peak arises from a specific (n,m) semiconducting nanotube species (see Color Plates, p. 342)

these anchoring choices gave a candidate assignment. The correct assignment among these candidates was identified by measuring resonance Raman spectra using a variety of laser wavelengths in order to correlate radial breathing mode (RBM) frequencies with E_{22} wavelengths for a set of several different nanotube species. Each candidate assignment matched the set of observed RBM frequencies with a different set of (n, m) identities, and therefore, nanotube diameters. For each possible assignment, the plot of RBM frequencies versus inverse tube diameter was examined for a linear relationship, which was strongly expected on experimental and theoretical grounds. One assignment gave distinctly better linearity than the others. In accord with other criteria, it was identified as the correct assignment [10].

7.4 Implications of the Spectral Assignment

The completed assignment immediately provided a wealth of data on the precise optical excitation energies for a variety of identified semiconducting SWNT species spanning a wide range of diameters and chiralities. Figure 7.3 is a plot of measured E_{22}/E_{11} frequency ratios as a function of nanotube chiral angle. The data show strong family patterns, illustrated by the labeled lines drawn through points sharing the same value of n–m. The families also tend to divide themselves into two groups: those for which mod(n–m,3) equals 1 or 2. We will refer to these two subsets of semiconducting SWNT as "mod 1" and "mod 2" tribes. The measured E_{22}/E_{11} values extrapolate to a ratio near 1.8 in the limit of armchair (30°) chirality. The difference between

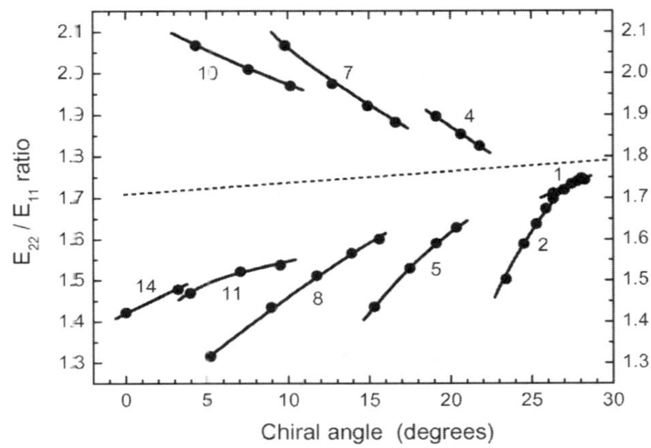

Fig. 7.3. Measured ratios of optical transition frequencies vs. chiral angle for semiconducting nanotubes. Points are experimental data and solid curves connect data within families of nanotubes sharing the same value of n–m. Curves are labeled with these values. The *dashed line* shows a plausible extrapolation of the average ratios to the "armchair" limit of 30° chiral angle

this value and the value of 2 predicted by many electronic structure models has been termed the "ratio problem" [11]. Recent theoretical research has attempted to explain this discrepancy through consideration of many-body interactions such as excitonic effects [11–13]. The spectroscopic data may also be used to display measured E_{11} and E_{22} transition wavelengths as a function of nanotube diameter. Figures 7.4 and 7.5 show such plots. These graphs reveal that a general linear increase of transition wavelength with diameter is combined with substantial systematic deviations that depend on chiral angle, diameter, and (mod 1 or mod 2) tribe. These deviations from a linear dependence are also found to be highly correlated between E_{11} and E_{22} transitions of a given (n, m) species, as shown in Fig. 7.6, which plots one of these deviations against the other for all observed species. If the deviations arose solely from trigonal warping, then the data in Fig. 7.6 would fall on a straight line with slope of −4 [14,15]. Instead, it appears that the influence of trigonal warping is superimposed with many-body effects to give curvature in the correlation graph. The set of (n, m) species with measured E_{11} and E_{22} spectral transitions has been extended through photoluminescence studies on samples grown by laser vaporization [16,17], a process that gives nanotubes of larger average diameter than the HiPco method.

Since their introduction by Kataura [18], plots of E_{ii} vs. nanotube diameter have provided valuable guidance for interpreting optical absorption and resonance Raman experiments on SWNT. Such plots typically display the results of simple tight binding calculations with a γ_0 scaling parameter of approximately 2.9 eV. The recent availability of (n, m)-resolved spectral

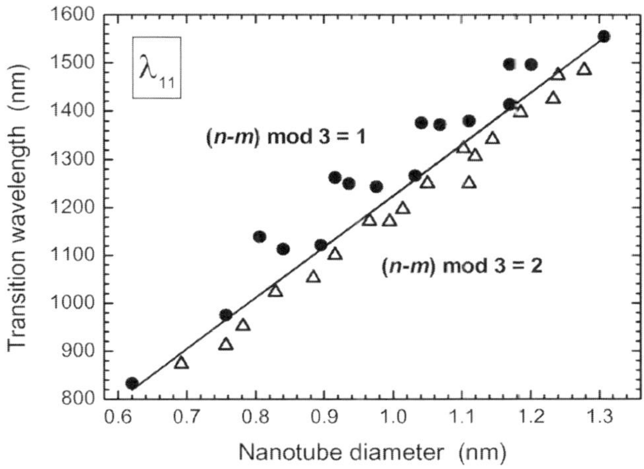

Fig. 7.4. Measured peak wavelengths of E_{11} transitions of SWNT in aqueous SDS suspension as a function of nanotube diameter. *Solid circles* show data for species with mod(n–m,3) = 1; *open triangles* show data for species with mod(n–m,3) = 2. The *solid line* indicates the boundary between the two sets of points. Deviations of points from the line reflect systematic features of SWNT electronic structure rather than experimental scatter

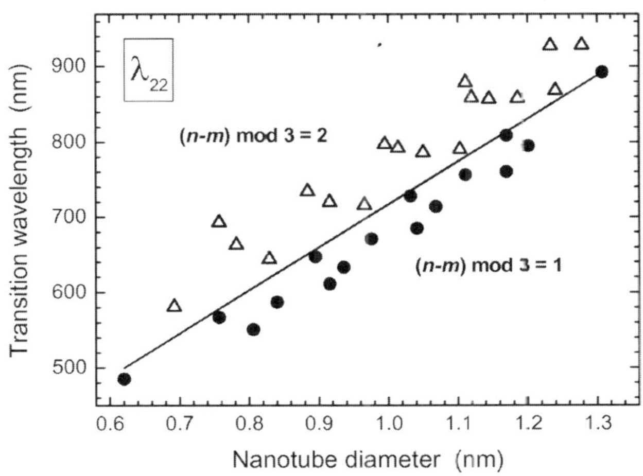

Fig. 7.5. Measured peak wavelengths of E_{22} transitions of SWNT in aqueous SDS suspension as a function of nanotube diameter. *Solid circles* show data for species with mod(n–m,3) = 1; *open triangles* show data for species with mod(n–m,3) = 2. The *solid line* indicates the boundary between the two sets of points. Deviations of points from the line reflect systematic features of SWNT electronic structure rather than experimental scatter

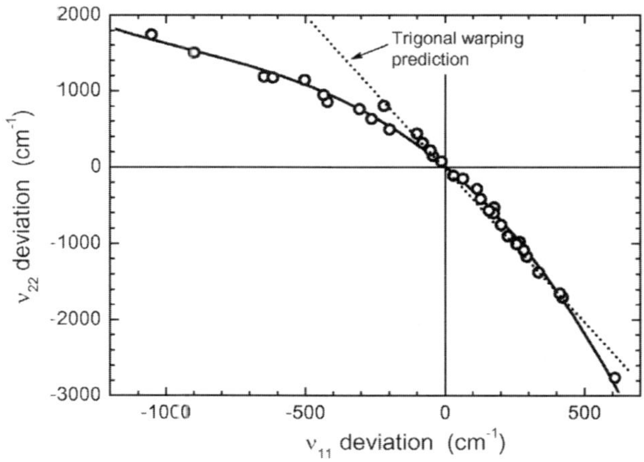

Fig. 7.6. Correlations between deviations of optical transition frequencies from the boundary lines shown in Fig. 7.4 and 7.5. The deviations in E_{222} transitions are plotted as a function of deviations in E_{11} transitions, with one experimental point plotted for each (n, m) species. The *solid curve* is a cubic fit through the data, and the *dotted line* shows the predicted effect from trigonal warping

data makes it possible to construct a model-free plot showing E_{11} and E_{22} vs. diameter as based solely on experimental findings [19]. For this purpose, the data for 33 (n, m)-species of SWNT in aqueous SDS suspension were fit to empirical functions of diameter and chiral angle. The resulting expressions accounted for the spectral peak positions with average errors of 1.3 meV (10 cm^{-1}) for E_{11} transitions and 5 meV (40 cm^{-1}) for E_{22} transitions. Note that these errors are less than 10 % of the transition line widths. The empirical fitting functions allowed extrapolation of the E_{11} and E_{22} transition energies from the measured set to all semiconducting nanotubes having diameters between 0.5 and 2.5 nm. Figure 7.7 shows the resulting empirical "Kataura plot," with the model-based values included for comparison. Significant differences are apparent. The observed E_{11} transition energies are as much as 25% higher than the model predictions, and the "scatter" arising from the dependence of transition energy on chiral angle is about twice as large as in the model plots. Note that the tight-binding model plot cannot be brought into agreement with experimental results by adjusting the γ_0 parameter.

The empirical plot shown in Fig. 7.7 is a valuable tool for guiding and interpreting absorption or emission spectroscopy experiments on SWNT in aqueous SDS suspension. It is also highly relevant to resonance Raman experiments, because it identifies the (n,m) species that will show resonantly enhanced Raman scattering with a given incident photon energy [20,21]. Nanotube researchers using Raman or electronic spectroscopies are therefore strongly advised to adopt the empirical plot in preference to the model-based version [18] because of its superior validity. However, a question of both

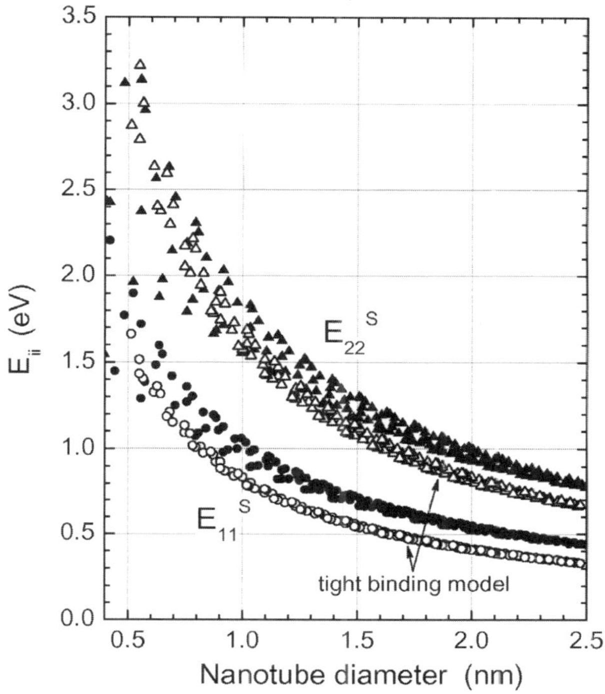

Fig. 7.7. Semiconducting SWNT optical transition energies for E_{11} and E_{22} as a function of nanotube diameter. The solid symbols show values found by extrapolation of empirical data. The open symbols show values computed using a simple tight binding model with $\gamma_0 = 2.90\,\text{eV}$. Adjustment of the γ_0 parameter cannot bring the model into agreement with the empirical values

fundamental and practical importance is whether the empirical findings are sensitive to nanotube environment: how do the transition energies differ for SWNT in other surfactants or in air? Recent work by Lefebvre et al. help to clarify this point [22–24]. These researchers measured photoluminescence emission and excitation spectra of SWNT suspended in air between silicon pillars above a silicon substrate. They report a pattern of transition energies very similar to that found in SDS-suspended samples, but with all transitions modestly shifted to higher energy. The tabulated transition energies show relative shifts between air and aqueous SDS environments of approximately 2 to 4% for E_{11} and 0.5 to 2% for E_{22} [23]. In addition, spectral studies of SWNT suspended in water by a variety of ionic, nonionic, and biopolymeric surfactants show that transition energies do depend somewhat on the surfactant identity. E_{11} transition energies in aqueous suspensions can vary by ca. 3%, with SDS giving the highest energies and protein coatings giving the lowest [25,26]. These findings provide confidence that the qualitative pattern of SWNT spectral transitions and the deduced assignment of transitions to (n, m) species reflect intrinsic properties of virtually isolated SWNT. Optical transition frequencies in the suspended samples are only mildly perturbed by the surroundings. However, for applications requiring precise values of E_{11} and E_{22} frequencies, it is necessary to adjust the plot shown in Fig. 7.7 by recalibrating the empirical parameters to reflect small spectral shifts induced by the particular environment of interest.

We note that preliminary investigations indicate that SWNT optical transition frequencies depend in a relatively complex way on sample temperature. It appears that the magnitudes and signs of the temperature-induced spectral shifts depend on a nanotube's diameter, chiral angle, and mod(n–m,3) tribe. These effects may dictate the control of sample temperatures in the most precise SWNT spectroscopic experiments. Because transitions of different (n, m) species shift differently as temperature changes, deliberate adjustment of sample temperature may also provide a useful research and analysis tool for disentangling overlapped spectral features in mixed samples, or for tuning optical transitions to desired frequencies.

7.5 Transition Line Shapes and Single-Nanotube Optical Spectroscopy

The discussion above is based only on spectral peak positions. Significant information is also revealed by the shapes and widths of optical transitions. For SWNT in aqueous surfactant suspension, analyses of emission spectra show that individual transitions are best represented by Voigt profiles, which are convolutions of Gaussian and Lorentzian shapes. The Lorentzian component reflects a finite dephasing lifetime in the transition's excited state, while the Gaussian component is associated with inhomogeneous broadening from variations in the central transition frequency among tubes having the same (n, m) species. Such inhomogeneous broadening may arise from a distribution of local environments. For SWNT samples in surfactants such as SDS or SDBS, the deduced emission full-widths at half-maximum are approximately 180 to 200 cm^{-1}, with comparable Lorentzian and Gaussian components. By contrast, E_{22} excitation spectral features show significantly broader profiles that appear almost entirely Lorentzian. Here, the more tightly bound v_2 states are probably less sensitive to environmental inhomogeneities, giving a smaller Gaussian contribution to the line shape that is masked by the much broader Lorentzian component arising from dephasing on the 10^{-14} s time scale.

The discovery of SWNT fluorescence opened the possibility of exploiting the high sensitivity of emission methods in order to study the electronic spectroscopy of individual nanotubes rather than bulk ensembles. The first such measurements were reported by Lefebvre et al. on unprocessed nanotubes suspended in air at room temperature between pillars over a silicon surface [22].

These workers reported emission linewidths of approximately $105\,\mathrm{cm}^{-1}$ from individual nanotubes. Hartschuh et al. spin-coated an aqueous suspension of SWNT onto a glass substrate to obtain a dilute sample of immobilized nanotubes and then recorded simultaneous Raman and fluorescence emission spectra of individual nanotubes of small diameter [27]. This experiment confirmed the matching between E_{11} transitions and Raman RBM frequencies that had previously been deduced in the (n,m) assignment of optical spectra [10]. In addition, the fluorescence emission spectra of individual nanotubes at room temperature were found to have Lorentzian shapes with widths of approximately $180\,\mathrm{cm}^{-1}$. However, there was significant variation (up to ca. $160\,\mathrm{cm}^{-1}$) in emission peak positions among different tubes with the same (n, m) identity. This variation is the source of the inhomogeneous line broadening that is evident in ensemble E_{11} spectra of bulk samples. By contrast, the air-suspended samples of Lefebvre et al. showed little evidence of spectral inhomogeneity, with individual nanotubes and ensembles of nanotubes displaying very similar peak widths and positions [24]. Typical observed widths were $120\,\mathrm{cm}^{-1}$ for E_{11} and $280\,\mathrm{cm}^{-1}$ for E_{22} transitions in that study. Another important finding from single-nanotube spectroscopy is the absence of blinking or variations in emission intensity on the time scale of 10^{-2} to at least 10^{2} seconds [27]. Finally, we note that special care may be needed to avoid unintended sample heating effects under the intense excitation conditions required in some single-nanotube spectroscopy experiments.

Several experimental results strongly suggest that the observed optical transitions have substantial excitonic character, reflecting strong spatial correlation between the conduction band electron and valence band hole created by light absorption. One clue is the "ratio problem" described above and the need to consider electron-hole interactions in theoretical attempts to explain it [12,13,28,29]. Another strong indication comes from the spectral line shapes observed in absorption, fluorescence, and fluorescence excitation spectra. In contrast to the sharply asymmetric profiles expected from the joint density of states [30], the observed spectra show Lorentzian line shapes for E_{22} transitions of bulk samples. E_{11} emission from individual nanotubes is also found to be Lorentzian in one report [27], although slightly asymmetric in another [24]. The widths of symmetric Lorentzians may be interpreted as varying inversely with the dephasing lifetime of the excitons involved in the transitions. Interestingly, dephasing times deduced from E_{22} line widths in bulk samples show a strong inverse dependence on E_{22} transition energy, ranging from ca. 13 to 31 fs for HiPco nanotubes. Optical excitations of semiconducting SWNT should certainly be viewed as more complex in character than is suggested by the simplified band structure scheme of Fig. 7.1.

7.6 Influence of Sample Preparation on Optical Spectra

Even after SDS-suspended SWNT samples have been ultrasonicated and centrifuged for enrichment in individual nanotubes, they still contain significant concentrations of nanotube bundles. SWNT in bundles may undergo at least two types of electronic perturbation. First, they experience an environment having a polarizability that is increased and that varies with position within the bundle. This effect will tend to red-shift and broaden observed electronic transitions. Second, electronic couplings within the bundle can enable rapid energy transfer from an optically excited nanotube to a neighbor having a smaller band gap. If the accepting species in this energy transfer process is a metallic nanotube, then fluorescence emission will be quenched. If the acceptor is instead a semiconducting nanotube with a smaller band gap, the fluorescence emission will be shifted to an unexpectedly long wavelength that may lie beyond the limit of detector response or solvent transparency and remain undetected. In samples containing the statistically expected 33% metallic (or semi-metallic) species, bundles of three or more nanotubes are likely to contain at least one metallic tube. This implies that samples of bundled suspended SWNT will be essentially non-emissive, in agreement with experimental observation. However, a sample containing a mixture of individual and bundled nanotubes will show emission characteristic of the individual species while its absorption spectrum shows a superposition of components from bundled and individual nanotubes.

Another factor that can apparently influence SWNT fluorescence efficiency is disruption of the pi-electron system through chemical derivatization or sidewall damage. As a general rule, the most strongly fluorescent samples consist of pristine nanotubes that have not undergone chemical processing [5,31]. Oxidative treatment with HNO_3 to remove residual catalyst impurities from raw nanotube material seems to cause enough conversion of sp^2 to sp^3 carbon sites to greatly suppress the fluorescence quantum yield. In fact, loss of fluorescence appears to be a more sensitive measure of SWNT chemical modification than the appearance of a D-band in the Raman spectrum. The chemical history of a sample must therefore be controlled when fluorimetric analysis is planned.

For aqueous suspensions of SWNT in ionic surfactants such as SDS or SDBS (sodium dodecylbenzenesulfonate), the solution acidity can also strongly affect nanotube fluorescence efficiency. It is found that nanotube fluorescence is strongly quenched at lower pH values [5]. However, unlike quenching by chemical derivatization, this effect is apparently fully reversible if the sample pH is subsequently increased through addition of a simple base such as NaOH [5,32]. Interestingly, different semiconducting SWNT species vary in their susceptibility to this acid fluorescence quenching, as shown in Fig. 7.8. The general pattern is that as a mixed sample's pH becomes lower, the long wavelength fluorescence features vanish before the short wavelength peaks. It has been suggested that this behavior is associated with differences

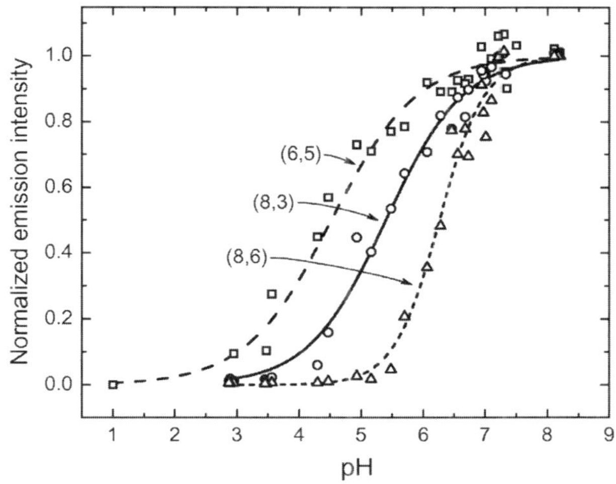

Fig. 7.8. Representative measurements of normalized fluorescence intensity as a function of sample pH for SWNT species in aqueous SDS suspension. Points show experimental data for the (6,5), (8,3), and (8,6) species, and the dashed, solid, and dotted curves are sigmoidal fits to those three data sets

in optical band gaps, which may predict the susceptibility of semiconducting nanotube species to protonation [32]. However, a subsequent study found evidence that the pH values at which different (n, m) species show fluorescence quenching correlates more closely with the nanotube diameter than with optical band gap [33]. Although the mechanism causing acidic quenching of SNWT fluorescence in aqueous ionic surfactants is not yet certain, the phenomenon is of great practical importance. In any experimental project that relies on SWNT fluorescence in such media to detect nanotubes or to analyze for their concentrations or (n, m) distributions, one must ensure that the pH is high enough to avoid quenching effects. It is possible that the (n, m)-dependent quenching effect may also be valuable as a method for monitoring local pH levels through the fluorescence spectra of nanotube mixtures.

7.7 Spectrofluorimetric Sample Analysis

Spectrofluorimetry has great potential as a tool for analyzing bulk samples of SWNT. Because the E_{22} absorption and E_{11} emission wavelengths are known as a function of (n, m) structure, one can readily use spectrofluorimetric data to obtain a qualitative assay identifying the specific semiconducting SWNT species present in a sample. A specimen to be analyzed must first be processed using a standardized procedure to obtain a reasonably well-dispersed, homogeneous bulk sample. Although ultrasonic dispersion followed by ultracentrifugation is needed to provide a sample highly enriched in individually suspended nanotubes [5], ultrasonic dispersion alone can give a substantial fraction of individual tubes [34]. Such a mixture of individual and bundled tubes will still show fluorescence emission from the individual tubes that may be adequate for analytical purposes. A full excitation-emission scan can be performed using a commercial spectrofluorometer to provide data such as shown in Fig. 7.2. However, because nearly all of the E_{11} emission falls beyond the spectral range of photomultiplier detectors, the instrument will need a near-infrared light detector such as an InGaAs or Ge photodiode. Such detectors normally need thermoelectric or cryogenic cooling to suppress dark current to usable levels. It is desirable to have an emission monochromator optimized for near-infrared wavelengths and an instrumental calibration function that reflects the overall wavelength-dependent sensitivity of the detection system. A Fourier transform infrared spectrometer can also be adapted to measure nanotube fluorescence spectra with high efficiency [17].

Although full two-dimensional scanned spectrofluorimetry provides very detailed information, it carries the disadvantages of relatively high instrumentation cost and slow data collection (hours per sample). Data acquisition can be expedited by measuring fluorescence intensity only at specific, predetermined combinations of excitation and emission wavelengths that correspond to known (n, m) species, rather than scanning a full two-dimensional matrix of wavelengths. There is also an alternative, simpler approach that reduces both instrumentation expense and data acquisition time. This exploits the relatively long Lorentzian tails in E_{22} absorptions (which have full widths at half-maximum of several hundred cm^{-1}) to achieve significant excitation of many different (n, m) species using a single excitation wavelength. An inexpensive red diode laser of suitable wavelength can be used to excite a sample, and the fluorescent emission can be rapidly analyzed by a multichannel array of InGaAs photodiodes mounted at the exit plane of a small spectrograph. High quality emission spectra can often be collected in seconds with such an instrument, and the resulting data can be analyzed as a superposition of fluorescence bands from specific (n, m) species, each having a previously determined center wavelength and shape. More than 20 (n, m) amplitudes can be deduced from one spectrum in this way. To enhance this one-dimensional approach, one can sequentially acquire several emission spectra excited by diode lasers or other monochromatic sources of different wavelengths that are selected to reveal separate subpopulations within the nanotube sample. Broadband light from a tungsten or xenon lamp can also be used to excite the full range of nanotubes. It should be noted that the number of SWNT structures having diameters within a narrow diameter range increases linearly with the center diameter of that range. This implies that the difficulty of resolving and identifying (n, m) species in a mixed sample grows greater as the average nanotube diameter in the sample increases. The limited near-infrared transparency of water poses another obstacle to fluorimetric analysis of larger diameter nanotubes, because most light beyond 1350 nm is blocked by a path length of several millimeters of H_2O. The substitution of D_2O for H_2O extends the useful range to approximately 1850 nm.

To date, quantitative determinations of (n, m) species concentrations through fluorimetry has not been possible, because the fluorescence intensity observed from a species depends not only on its mass concentration, but also on its absorptivity in the E_{22} transition and on its fluorescence quantum yield. It is not currently known how these quantities vary with diameter, chiral angle, and mod(n–m,3) tribe. However, it seems quite plausible that the diameter-dependence of optical properties is relatively mild over the rather limited range of tube structures contained in most SWNT specimens. If one makes this assumption, then spectrofluorimetry can quickly provide diameter distributions of bulk samples. Spectrofluorimetric analysis has already been used in this way to estimate the detailed composition of samples obtained through different growth processes [10,35–37]. One valuable and unexpected finding from these analyses is a strong preference for near-armchair chiral angles in SWNT grown via CVD with specific supported catalysts [35–37]. In Fig. 7.9, the semiconducting SWNT structures are marked by their (n, m) indices on a graphene sheet map. Here each species's diameter is proportional to the length of the line connecting its label to the vertex in the upper left corner, while its chiral angle is measured between that line and the axis labeled "zigzag". The fluorimetric intensities measured in a sample grown using a supported Co-Mo catalyst are indicated by the thickness of the hexagons drawn around corresponding (n, m) labels [35]. Clearly, the high spectral intensities are concentrated in species near (6,5) and (7,5) within a narrow range of diameters and at chiralities near 30 degrees. Performing such spectrofluorimetric assays is much less tedious than compiling histograms from TEM images of many individual nanotubes in a sample. Once SWNT absorptivities and fluorescence quantum yields have been determined as a function

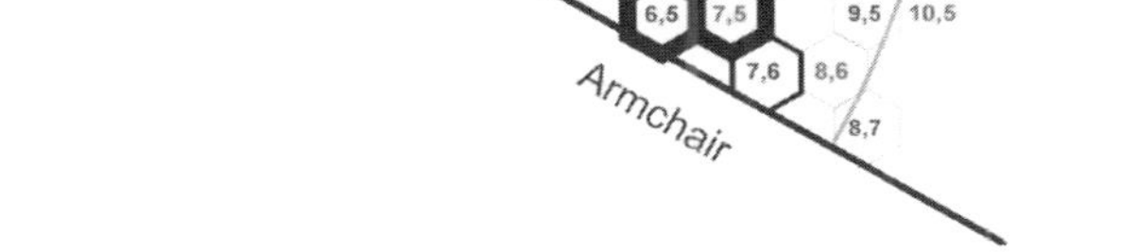

Fig. 7.9. Graphene sheet map showing the species-resolved fluorescence intensities measured in a SWNT sample grown from CO on a Co-Mo supported catalyst. The thickness of the hexagon surrounding each (n, m) label is proportional to the experimental emission intensity for that semiconducting nanotube species

of (n, m) structure, spectrofluorimetry will provide a calibrated method for quantitatively analyzing the composition of bulk nanotube samples. Even though metallic SWNT cannot be detected through fluorescence, this approach should prove essential in guiding the important continuing efforts to develop selective synthesis processes and effective separation schemes [35,38].

7.8 Detection, Imaging, and Electroluminescence

Fluorescence provides the most sensitive and selective spectroscopic method for detecting semiconducting SWNT. As is true for nearly all analytes, nanotubes can be observed through fluorescence at far lower concentrations than required for direct optical absorption spectroscopy. For bulk aqueous samples, SWNT concentrations in the range of parts per billion by mass can be readily detected using standard instrumentation. In fact, it is possible to study individual nanotubes in such samples by coupling a spectrometer or an infrared-sensitive camera to an optical microscope focused on a small observation volume [22–24,27]. Well before the discovery of SWNT fluorescence, vibrational spectroscopy of single nanotubes was achieved by investigators using resonance Raman spectroscopy of SWNT on clean solid surfaces [39]. However, samples containing low concentrations of SWNT in complex surroundings such as environmental specimens or biological media present special challenges to Raman methods. First, the resonant Raman signals from nanotubes may be masked by nonresonant Raman scattering from other species that are present at far higher concentrations. Small quantities of fluorescent compounds in the samples can also give background emission strong enough to obscure the nanotube Raman signals. By contrast, nanotube fluorescence spectroscopy offers an unusual level of background immunity because virtually no organic compounds emit light in the near-infrared emission range of SWNT fluorescence. In addition, resonant and nonresonant Raman backgrounds do not interfere with SWNT fluorescence detection because the frequency difference between fluorescence excitation and emission transitions far exceeds the Raman shift of any sample component. For these reasons, nanotubes can be detected and imaged with high contrast even when present at very low concentrations in surroundings as complex as biological cells and tissues.

Figure 7.10 illustrates the ability of fluorescence methods to detect nanotubes in biological environments [40]. It shows a near-infrared fluorescence image of a mouse macrophage cell that had been incubated for 24 hours in a growth medium containing suspended HiPco SWNT. The image was recorded with an optical microscope that had been modified in the following three ways. A red diode laser was mounted to provide sample excitation of E_{22} transitions. Dichroic optics were installed to reflect the excitation beam through the objective lens onto the sample and to block all light except near-infrared nanotube fluorescence from reaching the detector. Finally, the

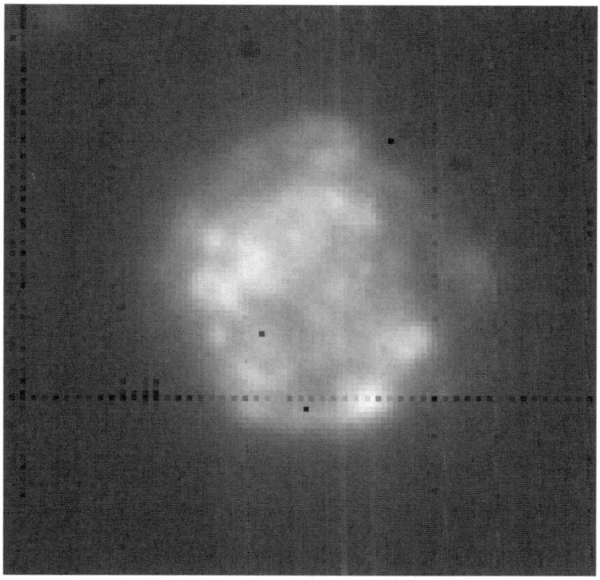

Fig. 7.10. Fluorescence micrograph of a single macrophage cell that had been incubated in a growth medium containing suspended SWNT. The cell was excited at 660 nm and nanotube emission was imaged only at wavelengths greater than 1125 nm. Bright regions (yellow in false color) show the highest nanotube concentrations; dark regions (blue in false color) show the lowest. The cell's diameter was approximately 20 μm. Black points and the horizontal dotted line arise from uncorrected defects in the InGaAs imaging camera (see Color Plates, p. 342)

normal camera was replaced with an InGaAs focal plane array sensitive in the range of SWNT E_{11} emission. The resulting image selectively depicts the spatial distribution of nanotubes inside the macrophage cell. It is clear that nanotubes are not spread uniformly through the cytoplasm, but instead have accumulated in localized structures. From this and other images and data, these structures were identified as phagosomes, indicating that the nanotubes had been actively ingested through the macrophages' normal biological function of phagocytosis. In addition, the nanotubes retain at least some fluorescent character despite the oxidative chemical environment inside the phagosomes. The promising results from this exploratory experiment suggest that near-infrared fluorescence methods will serve as powerful tools for detecting and imaging SWNT in biological cells, tissues, and organisms, and for enabling detailed studies of the interactions between SWNT and biological systems.

A final related topic of interest is electrically induced nanotube luminescence. It has been shown that the electron-hole pairs giving near-infrared nanotube emission in semiconducting nanotubes may be generated without optical excitation by electrically injecting charge carriers into the ends of an individual tube mounted as an FET device on a silicon substrate [41]. This effect may find application for (n, m)-identification of nanotubes in experimental electronic devices, and perhaps eventually as the basis for building nanoscale opto-electronic systems.

7.9 Conclusions

Band gap fluorescence of semiconducting single-walled carbon nanotubes is an important phenomenon that enables a wide variety of new experimental investigations. Fluorescence-based spectroscopic studies provide precise information about electronic excitation energies and their dependence on nanotube structure, while lineshape analyses and time-resolved emission experiments give insights into relaxation processes. The spectrofluorimetric ability to monitor in parallel a wide range of (n,m) species allows studies of processes in which nanotubes are chemically or physically manipulated in structure-dependent ways. Spectrofluorimetry will soon form the basis for rapid, non-destructive analytical methods that will reveal the qualitative and quantitative compositions of bulk nanotubes mixtures. Such methods will advance many areas of nanotube research by standardizing sample quality and by guiding efforts to selectively produce or sort nanotubes by structural type. Finally, the unusual near-infrared wavelengths of nanotube fluorescence provide a uniquely selective approach for detecting single-walled nanotubes in complex biological environments. Future applications of nanotube fluorescence in biomedicine may include studies of nanotube toxicity and biodistributions in organisms and the development of novel methods for noninvasive disease diagnosis.

Acknowledgements

The author gratefully acknowledges the vital contributions of his co-workers and collaborators, and the research support of the National Science Foundation (grants CHE-9900417, CHE-0314270, and CBEN center grant EEC-0118007) and the Welch Foundation (grant C-0807).

References

1. R. Saito, G. Dresselhaus, and M.S. Dresselhaus, Physical Properties of Carbon Nanotubes, Imperial College Press (London), 1998.
2. M.S. Dresselhaus, G. Dresselhaus, and Ph. Avouris, ed., Carbon Nanotubes: Synthesis, Structure, Properties, and Applications, Springer-Verlag (New York), 2001.

3. S. Reich, J. Janina, and C. Thomsen, Carbon Nanotubes: Basic Concepts and Physical Properties, Wiley (New York), 2004.
4. J.W. Mintmire and C.T. White: Phys. Rev. Lett. **81**, 2506 (1998)
5. M. O'Connell, S.M. Bachilo, C.B. Huffman, V. Moore, M.S. Strano, E. Haroz, K. Rialon, P.J. Boul, W.H. Noon, C. Kittrell, J. Ma, R.H. Hauge, R.B. Weisman, and R.E. Smalley: Science **297**, 593 (2002)
6. M. Kasha: Disc. Faraday Soc. **9**, 14 (1950)
7. Y.-Z. Ma, J. Stenger, J. Zimmerman, S.M. Bachilo, R.E. Smalley, R.B. Weisman, and G.R. Fleming: J. Chem. Phys. **120**, 3368 (2004)
8. J. Kono, G.N. Ostojic, S. Zaric, M.S. Strano, V.C. Moore, J. Shaver, and R.H. Hauge: Applied Physics A **78**, 1093 (2004)
9. A. Hagen, G. Moos, V. Talalaev, and T. Hertel: Applied Physics A **78**, 1137 (2004)
10. S.M. Bachilo, M.S. Strano, C. Kittrell, R.H. Hauge, R.E. Smalley, and R.B. Weisman: Science **298**, 2361 (2002)
11. C.L. Kane and E.J. Mele: Phys. Rev. Lett. **90**, 207401/1 (2003)
12. C.D. Spataru, S. Ismail-Beigi, L.X. Benedict, and S.G. Louie: Applied Physics A **78**, 1129 (2004)
13. C.D. Spataru, S. Ismail-Beigi, L.X. Benedict, and S.G. Louie: Phys. Rev. Lett. **92**, 077402/1 (2004)
14. S. Reich and C. Thomsen: Phys. Rev. B **62**, 4273 (2000)
15. R. Saito, G. Dresselhaus, and M.S. Dresselhaus: Phys. Rev. B **61**, 2981 (2000)
16. S. Lebedkin, F.H. Hennrich, T. Skipa, and M.M. Kappes: J. Phys. Chem. B **107**, 1949 (2003)
17. S. Lebedkin, K. Arnold, F.H. Hennrich, R. Krupke, B. Renker, and M.M. Kappes: New Journal of Physics **5**, 140.1 (2003)
18. H. Kataura, Y. Kumazawa, Y. Maniwa, I. Umezu, S. Suzuki, Y. Ohtsuka, and Y. Achiba: Synth. Met. **103**, 2555 (1999)
19. R.B. Weisman and S.M. Bachilo: Nano Lett. **3**, 1235 (2003)
20. R. Saito, A. Gruneis, G.G. Samsonidze, G. Dresselhaus, M.S. Dresselhaus, A. Jorio, L.G. Cancado, M.A. Pimenta, and A.G. Souza Filho: Applied Physics A, 1099 (2004)
21. S.K. Doorn, D.A. Heller, P.W. Barone, M.L. Usrey, and M.S. Strano: Applied Physics A **78**, 1155 (2004)
22. J. Lefebvre, Y. Homma, and P. Finnie: Phys. Rev. Lett. **90**, 217401/1 (2003)
23. J. Lefebvre, J.M. Fraser, Y. Homma, and P. Finnie: Applied Physics A **78**, 1107 (2004)
24. J. Lefebvre, J.M. Fraser, P. Finnie, and Y. Homma: Phys. Rev. B **69**, 075403-1 (2004)
25. V.C. Moore, M.S. Strano, E.H. Haroz, R.H. Hauge, and R.E. Smalley: Nano Lett. **3**, 1379 (2003)
26. P. Cherukuri, S.M. Bachilo, S.H. Litovsky, and R.B. Weisman: to be published (2004)
27. A. Hartschuh, H.N. Pedrosa, L. Novotny, and T.D. Krauss: Science **301**, 1354 (2003)
28. T.G. Pedersen: Phys. Rev. B **67**, 073401-1 (2003)
29. T. Ando: J. Phys. Soc. Jpn. **66**, 1066 (1997)
30. A.G. Souza Filho, A. Jorio, Hafner J.H., C.M. Lieber, R. Saito, M.A. Pimenta, G. Dresselhaus, and M.S. Dresselhaus: Phys. Rev. B **63**, 241404-1 (2001)

31. J.-S. Lauret, C. Voisin, G. Cassabois, P. Roussignol, C. Delalande, A. Filoramo, L. Capes, E. Valentin, and C. Jost: Physica E **21**, 1057 (2004)
32. M.S. Strano, C.B. Huffman, V.C. Moore, M.J. O'Connell, E.H. Haroz, J. Hubbard, M. Miller, K. Rialon, C. Kittrell, S. Ramesh, R.H. Hauge, and R.E. Smalley: J. Phys. Chem. B **107**, 6979 (2003)
33. R.B. Weisman, S.M. Bachilo, and D. Tsyboulski: Applied Physics A **78**, 1111 (2004)
34. M.F. Islam, E. Rojas, D.M. Bergey, A.T. Johnson, and A.G. Yodh: Nano Lett. **3**, 269 (2003)
35. S.M. Bachilo, L. Balzano, J.E. Herrera, F. Pompeo, D.E. Resasco, and R.B. Weisman: J. Am. Chem Soc. **125**, 11186 (2003)
36. S. Maruyama, Y. Miyauchi, Y. Murakami, and S. Chiashi: New Journal of Physics **5**, 149.1 (2003)
37. Y. Miyauchi, S. Chiashi, Y. Murakami, Y. Hayashida, and S. Maruyama: Chem. Phys. Lett. **387**, 198 (2004)
38. R.B. Weisman: Nature Mater. **2**, 569 (2003)
39. A. Jorio, R. Saito, J.H. Hafner, C.M. Lieber, M. Hunter, T. McClure, G. Dresselhaus, and M.S. Dresselhaus: Phys. Rev. Lett. **86**, 1118 (2001)
40. P. Cherukuri, S.M. Bachilo, S.H. Litovsky, and R.B. Weisman: J. Am. Chem. Soc. **126**, 15638 (2004)
41. J.A. Misewich, Ph. Avouris, R. Martel, J.C. Tsang, S. Heinze, and J. Tersoff: Science **300**, 783 (2003)

8 The Raman Response of Double Wall Carbon Nanotubes

F. Simon, R. Pfeiffer, C. Kramberger, M. Holzweber, and H. Kuzmany

Raman spectroscopy on carbon nanotubes (CNT) yields a rich variety of information owing to the close interplay between electronic and vibrational properties. In this paper, we review the properties of double wall carbon nanotubes (DWCNTs). In particular, it is shown that SWCNT encapsulating C_{60}, so-called peapods, are transformed into DWCNTs when subject to a high temperature treatment. The inner tubes are grown in a catalyst free environment and do not suffer from impurities or defects that are usually encountered for as-grown SWCNTs or DWCNTs. As a consequence, the inner tubes are grown with a high degree of perfection as deduced from the unusually narrow radial breathing mode (RBM) lines. This apostrophizes the interior of the SWCNTs as a nano-clean room. The mechanism of the inner nanotube production from C_{60} is discussed. We also report recent studies aimed at the simplification and industrial scaling up of the DWCNT production process utilizing a low temperature peapod synthesis method. A splitting of the RBMs of inner tubes is observed. This is related to the interaction between the two shells of the DWCNTs as the same inner tube type can be encapsulated in different outer ones. The sharp appearance of the inner tube RBMs allows an assignment of the tube modes to (n m) indexes and thus provides a precise determination of the relation between the tube diameter and the RBM frequencies.

8.1 Introduction

Carbon nanotubes have been in the forefront of the nanomaterial research since their discovery [1]. They are not only fundamentally interesting materials due to their appealing one-dimensional structure but several applications have been envisaged. Some of them have already been established such as scanning probe-heads [2] or field emission devices [3, 4]. There is an active ongoing work in these fields to exploit the properties of these materials better and to improve the device qualities. Furthermore, high expectations are related to their applications as building elements of electronics, composite reinforcing materials and many more.

Carbon nanotubes can be represented as rolled up graphene sheets, i.e. single layers of graphite. Depending on the number of coaxial carbon nanotubes, they are usually classified into multi-wall carbon nanotubes (MWCNTs) and single wall carbon nanotubes (SWCNTs). Some general considerations have been clarified in the last 13 years of nanomaterial research related to these structures. MWCNTs are more homogeneous in their physical properties as the large number of coaxial tubes smears out individual tube properties. This makes them suitable candidates for applications where their nanometer size and the conducting properties can be exploited. In contrast, SWCNT materials are grown as an ensemble of weakly interacting tubes with different diameters. The physical properties of similar diameter SWCNTs can change dramatically as the electronic structure is very sensitive on the rolling-up direction, the so-called chiral vector. Depending on the chiral vector, SWCNTs can be metallic or semiconducting [5]. This provides a richer range of physical phenomena as compared to the MWCNTs, however significantly limits the range of applications. To date, neither the directed growth nor the controlled selection of SWCNTs with a well defined chiral vector has been performed successfully. Thus, their broad applicability is still awaiting. Correspondingly, current research is focused on the post-synthesis separation of SWCNTs with a narrow range of chiralities [6–9] or on methods which yield information that are specific to SWCNTs with different chiralities. An example for the latter is the observation of chirality selective band-gap fluorescence in semiconducting SWCNTs [10].

A more recently discovered third class of CNTs are double-wall carbon nanotubes (DWCNTs). DWCNTs were first observed to form under intensive electron radiation [11] in a high resolution transmission electron microscope from C_{60} encapsulated in SWCNTs, so-called peapods [12]. Following the synthesis of C_{60} peapods in macroscopic amounts [13], bulk quantities of the DWCNT material are available using a high temperature annealing method [14]. Alternatively, DWCNTs can be produced with usual synthesis methods such as arc-discharge [15] or CVD [16] under special conditions. According to the number of shells, DWCNTs are between SWCNTs and MWCNTs. Thus, one expects that DWCNTs may provide a material where improved mechanical stability as compared to SWCNTs coexists with the rich variety of electronic properties of SWCNTs. There are, of course, a number of yet unanswered questions e.g. whether the outer tube properties are unaffected by the presence of the inner tube or whether the commensurability of the tube structures plays a role. These questions should be answered before the successful application of these materials.

In this contribution, we review Raman studies of DWCNTs. We show that the study of inner tubes, in particular those from C_{60} peapod based DWCNTs provides some unique insight into the physics of SWCNTs. Such studies enabled the observation of unprecedently sharp Raman modes, which evidence that the inside grown SWCNTs are highly perfect mainly due to the catalyst free nano-clean room interior of outer SWCNT reactor tubes. The sharp Raman features of inner tube Raman radial breathing modes (RBMs) enable the indexing of chiral vectors thus providing an alternative to the band-gap fluorescence method.

This review is organized as follows: we describe the experimental methods and the sample preparations that are used for the current study. We compare the properties of DWCNTs grown with different methods. We show that the C_{60} peapod based DWCNTs have unique properties which underline the nano-clean room conditions encountered in the inside of SWCNTs. We describe a novel method for the preparation of the C_{60} peapod precursor material that enables the large scale productions of DWCNTs. We present a detailed investigation of the electronic structure of the small diameter inner tubes. We also present the chiral vector assignment to such tubes thus refining the empirical parameters of the relation between the RBM frequencies and the tube diameters.

8.2 Experimental

SWCNT starting materials for the production of DWCNTs described herein were prepared by the laser ablation method. Their diameters were controlled in order to obtain efficient C_{60} encapsulation that results in high yield of inner nanotubes. The values of $d_{\mathrm{L}} = 1.39\,\mathrm{nm}$, $\sigma_{\mathrm{L}} = 0.1\,\mathrm{nm}$ were obtained for the mean diameter and the variance of the distribution for the different samples using a large number of exciting laser energies following [17]. The SWCNT materials were purified following [13]. Peapod samples were prepared by annealing SWCNTs with C_{60} in a quartz ampoule following [13]. The peapod filling fraction was close to 100% as evidenced previously on similar samples using Electron Energy Loss Spectroscopy (EELS) [18]. The peapod materials were transformed to DWCNTs using the high temperature annealing method of [14]. The samples in the form of bucky-paper are kept in dynamic vacuum and on a copper tip attached to a cryostat, which allows temperature variation in the 20–600 K temperature range. The Raman experiments were performed in a 180 degree backscattering geometry. A He/Ne, an Ar/Kr mixed gas, and a tunable Ti:sapphire laser pumped by an Ar laser were used for the excitation at 30 different laser lines. Multi frequency Raman spectroscopy was performed on a Dilor xy triple axis spectrometer in the 1.64–2.54 eV (755–488 nm) energy range and in a Bruker FT-Raman spectrometer for the 1.16 eV (1064 nm) excitation at 90 K. We operated the Dilor spectrometer in two modes, high and normal resolution. The high resolution uses the additive mode of the spectrometer and the spectral resolution as determined from the FWHM of the elastically scattered light was 0.4–0.7 cm^{-1} going from red to blue excitation. Similarly, spectral resolution in the normal mode was 1–2 cm^{-1} depending on the laser line. Measurements with the FT-Raman spectrometer were recorded with 1 cm^{-1} resolution. Raman shifts were calibrated against a series of spectral calibration lamps. *Ab initio*

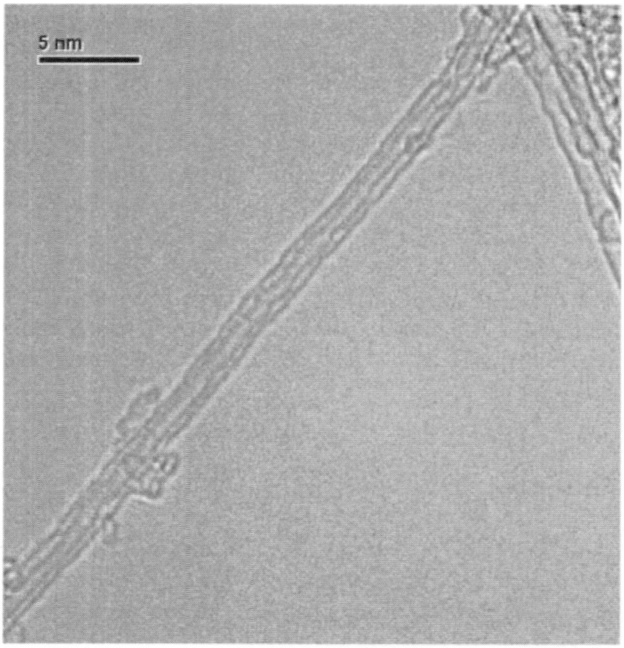

Fig. 8.1. High-resolution TEM micrograph of a C_{60} peapod based DWCNT sample

calculations were performed with the Vienna Ab Initio Simulation Package (VASP) [19].

8.3 Results and Discussion

8.3.1 Synthesis of Double-Wall Carbon Nanotubes

Double-wall carbon nanotubes can be classified into two groups depending on the method used for their production. DWCNT samples produced with usual preparation methods [15] have less controllable parameters such as their diameter distributions and, as shown below, their quality is inferior compared to the C_{60} peapod based DWCNTs. Following the discovery of DWCNTs from C_{60} peapods under intensive electron irradiation [11], it was demonstrated using HR-TEM that a 1200 °C heat treatment can also efficiently produce the inner nanotubes based on the C_{60} peapods [20]. The first characterization of bulk amounts of DWCNTs produced by this synthesis method was performed using Raman spectroscopy by Bandow et al. [14].

In Fig. 8.1 we show a typical HR-TEM micrograph of C_{60} peapod based DWCNT. The difference in the outer and inner tube diameter is thought to be close to twice the Van der Waals distance of graphite, 0.335 nm. Indeed, X-ray

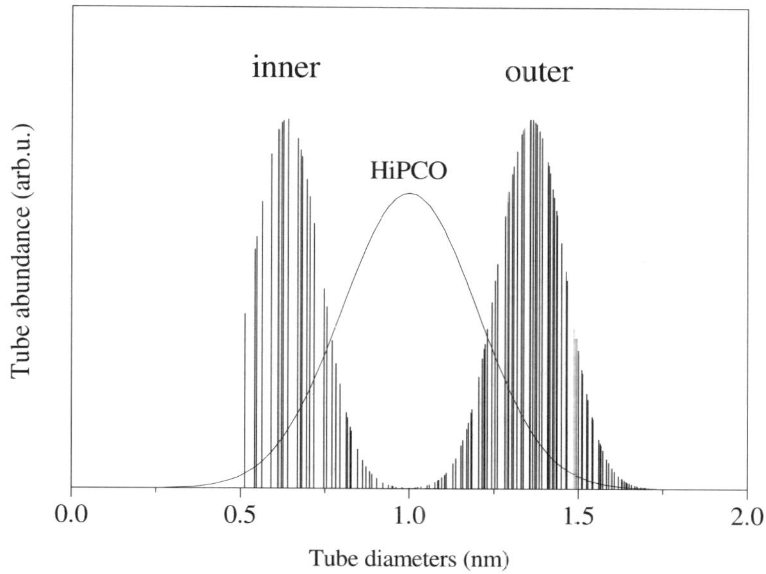

Fig. 8.2. Schematic diameter distribution of geometrically allowed tubes in the DWCNTs of the current study. The envelope of the approximate diameter distibution in a HiPco sample is shown for comparison

studies have indeed shown that the inner and outer tube diameter difference is only slightly larger, 0.72 ± 0.02 nm [21]. The corresponding situation for a DWCNT sample with outer tube distribution centered at the (10,10) tube is depicted in Fig. 8.2. The cut-off observed at small diameter inner tubes is given by the smallest outer tube diameter of 1.2 nm which allows the precursor C_{60} to enter [22–24].

The transformation from C_{60} peapods to DWCNTs can be conveniently followed by Raman spectroscopy. In Fig. 8.3 we compare the Raman spectra of pristine SWCNTs, C_{60}@SWCNT peapods, and DWCNTs based on the peapod material. The emergence of additional vibrational modes in the 200–450 cm^{-1} spectral range is evident in Fig. 8.3 for the DWCNT material as compared to the pristine and peapod materials. This emergence is accompanied by the disappearance of the modes of the encapsulated fullerene as denoted by circles in Fig. 8.3. The extra lines in the 200–450 cm^{-1} spectral range were identified as the radial breathing modes of the smaller diameter inner shell tubes. The disappearance of the fullerene peaks is evidence that the encapsulated C_{60} serves as carbon source for the internal tube formation [14, 25].

The inverse relationship between the RBM frequencies and tube diameters allows a tube diameter selective study of inner tube growth. In Fig. 8.4

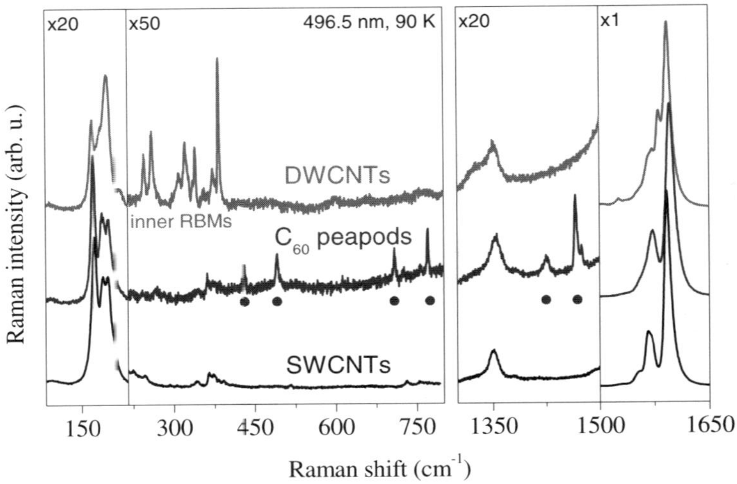

Fig. 8.3. Raman spectra of pristine SWCNT, C_{60} peapod and C_{60} peapod based DWCNT samples for $\lambda = 497$ nm laser excitation and 90 K. Bullets mark the positions of the modes of encapsulated C_{60}

we show the diameter selective growth of inner tubes. The topmost curve, A, shows the fully developed inner tube RBM spectrum after a 12 h annealing at 1280 °C annealing, whereas B and C correspond to 0.5 and 1 hour annealing. The comparison reveals the more rapid development of small diameter inner tubes, followed by the slower development of larger diameter tubes [26]. Similar diameter selective growth of inner tubes was found by Bandow et al. [27]. In the latter work, it was concluded from measurements at two laser excitations that small diameter inner tubes form first and are subsequently transformed to larger diameter inner tubes. This observation is somewhat different from the one reported here. The difference might be related to the photoselective property of the Raman experiments, as the data reported here are based on measurements on a larger number of laser lines. The formation of smaller tubes in the beginning, followed by the development of larger diameter tubes later on provides useful input for the theories aimed at explaining the inner tube formation. From computer simulation it was demonstrated that C_{60} peapod based DWCNTs are formed by Stone-Wales transformations from C_{60} dimer precursors formed at high temperature by cyclo-addition [28,29]. The free rotation of C_{60} molecules is a prerequisite for the dimer formation as it enables the molecules to have facing double bonds. It has been found experimentally that the ellipsoidal shaped C_{70} are present as both "standing" or "lying" peapod configurations i.e. with the longer C_{70} axis perpendicular or parallel to the tube axis [30]. In small diameter tubes the lying C_{70} configuration is preferred and the molecules have facing pentagons

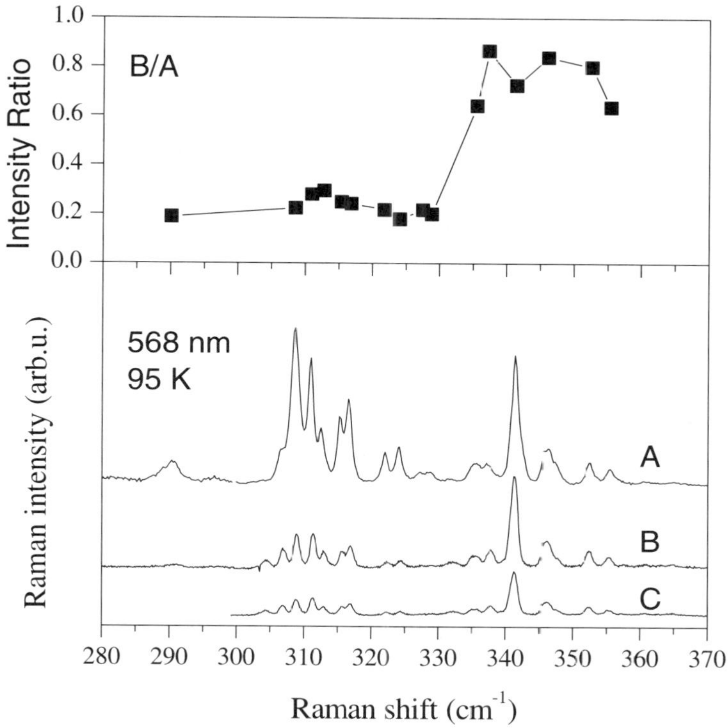

Fig. 8.4. Diameter selective growth of inner tubes in C_{60} peapod based DWCNT measured at $\lambda = 568\,\text{nm}$ laser excitation

and consequently no cyclo-additional double bond formation is possible in a linear chain. For such small diameter tubes, dimers may be formed in a canted C_{70} configuration. However, this structure has not been observed experimentally. Experiments with C_{70} peapod based DWCNTs have shown that inner tubes can indeed be formed from lying C_{70} peapods, which presents a challenge to the current theories [31]. Clearly, more theoretical work is required on the formation process of inner tubes from fullerene peapod samples.

Apart from the lack of full understanding of the inner tube growth, the production process of the precursor fullerene peapods is relatively complicated and its scaling up to larger amounts is difficult. The usual, vapor filling, process involves the annealing of the fullerene together with the opened SWCNT material sealed together in a quartz ampoule. Recently, a new fullerene encapsulation technique involving the refluxing of the SWCNTs and the fullerenes in solvents was presented [32]. In Fig. 8.5 we show the comparison of DWCNT samples based on vapor and solvent filled C_{60} peapods. Clearly, the yield and diameter distribution of the inner tubes are identical

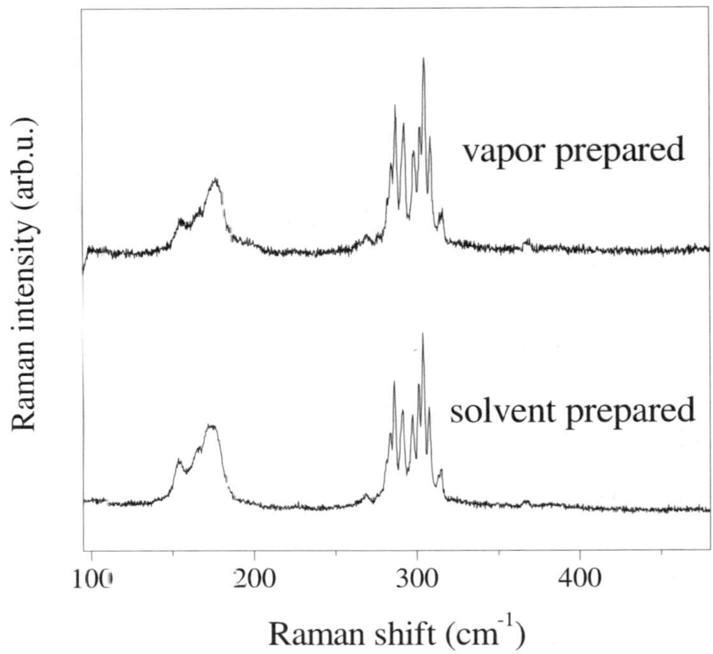

Fig. 8.5. Comparison of DWCNTs formed from vapor prepared (*upper curve*) and solvent prepared (*lower curve*) peapod samples at $\lambda = 676\,\mathrm{nm}$ laser excitation and room temperature

in the two kinds of samples which proves that the solvent filling method is indeed a simple alternative to the vapor filling method. In contrast to the latter, it can be easily scaled up to commercial quantities.

Production of DWCNTs has been reported by direct methods [15, 16]. Such DWCNTs are characteristically different from the tubes grown from the peapod precursors. In Fig. 8.6 we show the comparison of directly produced (C-DWCNT) and C_{60} peapod based DWCNT (60-DWCNT) materials. The C-DWCNTs were made from the catalytic decomposition of methane and was studied in detail previously [33]. There are clear similarities and differences in the comparison of the two kinds of DWCNT materials. Among the similarities, the Raman spectra of both compounds show two groups of bands. The lower frequency extending from 110–160 cm^{-1} for the C-DWCNT and from 165–200 cm^{-1} for the 60-DWCNT corresponds to the RBMs of the outer tubes. The modes seen at 190–250 for the C-DWCNT and at 250–370 cm^{-1} for the 60-DWCNT correspond to the RBMs of the inner tubes. The difference in the observed RBM Raman shifts is related to the different diameters of the inner and outer tubes in the C-DWCNT and 60-DWCNT compounds.

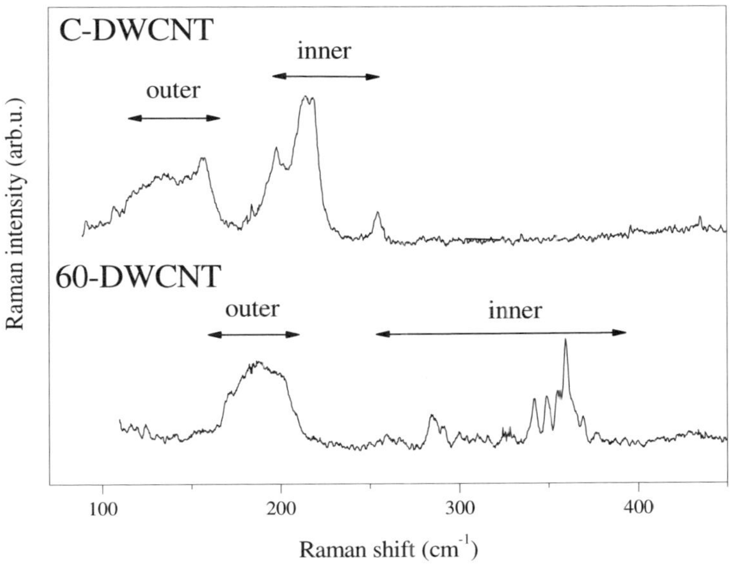

Fig. 8.6. Comparison of CVD grown (C-DWCNT) and C_{60} peapod based DWCNT (60-DWCNT) materials at $\lambda = 633$ nm laser excitation. Arrows indicate the Raman shift regions where RBMs of the outer and inner tubes are observed. The difference in inner and outer tube diameters in the two samples is apparent

It was found that C-DWCNT material shown in Fig. 8.6 has $d_{inner} = 1.52$ nm, $\sigma_{inner} = 0.6$ nm, and $d_{outer} = 2.26$ nm, $\sigma_{outer} = 0.4$ nm for the mean and the variance of the diameter distributions of the inner and outer tubes, respectively [16]. For the 60-DWCNT sample $d_{inner} = 0.67$ nm, $\sigma_{inner} = 0.1$ nm, and $d_{outer} = 1.39$ nm, $\sigma_{inner} = 0.1$ nm was found [25]. The most important difference between the two materials is the smaller inner tube RBM linewidths for the 60-DWCNT sample. In what follows, we discuss the properties of only the peapod based 60-DWCNT samples and will refer to these briefly as DWCNTs.

8.3.2 Energy Dispersive Raman Studies of DWCNTs

Electronic Structure of DWCNTs

The diverging behavior of the electronic density of states of SWCNTs gives rise to a significant resonant Raman enhancement and a photoselective Raman scattering. The photoselectivity, combined with the multi-frequency Raman method enables a detailed study of the electronic structure of the SWCNTs. The photoselective scattering also holds for the inner tubes of the DWCNTs. The laser energy dependent response of the inner tube RBMs is shown

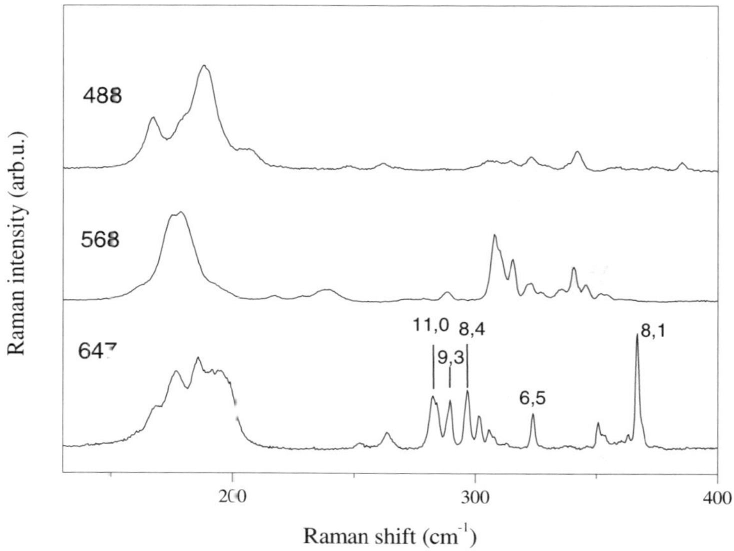

Fig. 8.7. Raman response of the RBM of a DWCNT sample when excited with different lasers. The spectra were recorded at 90 K. The indexing for some inner tube RBMs is shown for the $\lambda = 647$ nm laser energy

in Fig. 8.7. For the red excitation RBM linewidths as small as $0.35\,\text{cm}^{-1}$ were observed [25]. This is almost an order of magnitude smaller than reported previously on individual SWCNTs [34]. Related to the narrow linewidth of the vibrational modes, the resonant excitation for several lines is significant, resulting in Raman intensities almost 10 times larger than those from the outer tubes. This implies very sharp resonances between the Van Hove singularities. This issue is discussed in detail below. In addition to the sharp electronic resonances, the electron-phonon coupling is enhanced for small diameter tubes [35], which may also contribute to the observed signal enhancement of the inner tube vibrational modes [25].

The Kataura plot that is relevant for the current study is shown in Fig. 8.8. It describes the relation between the optical transitions and the tube diameters. It shows that when using red laser excitations (< 2 eV), nominally semiconducting inner tubes are expected to be observed. The metallic inner tubes are expected to appear only for higher energy excitations, above 2.5 eV.

The presence of metallic tubes, as e.g. the (9,3) inner tube at the 647 nm laser excitation in Fig. 8.7 is unexpected as they should only be observable at significantly larger excitation energy when resonance with the E_{11}^{m} transition occurs. This proves that smaller energy optical transitions are present for small diameter nanotubes that are absent in the simplest tight-binding calculations. Figure 8.9 compares the zone folded tight-binding and *ab ini-*

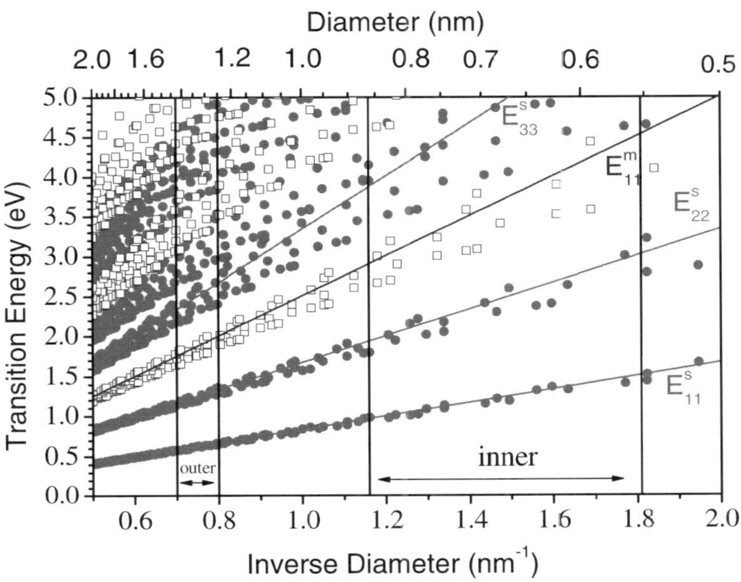

Fig. 8.8. Kataura plot with vertical lines indicating the diameter regions of inner and outer nanotubes that are relevant for the current study

tio derived density of states (DOS) for a (6,0) metallic tube. The *ab initio* calculations show the presence of some structures between the lowest energy optical transitions of the tight-binding model, giving rise to the experimentally observed resonance. The observed difference between the *ab initio* and the tight-binding theory results from the finite curvature of the small diameter nanotubes. This curvature induces the mixing of the σ and the inward pointing π orbitals.

The width of the Van Hove singularities of the DOS of SWCNTs is a measure of their one-dimensional character. It can be measured from high resolution energy dispersive Raman studies. The results in the 700–750 nm excitation energy range are shown in Fig. 8.10. Dashed lines mark the Raman shift where the RBMs of two selected inner SWCNTs were followed for several laser excitations. The FWHM of the energy dependent inner tube intensities and thus the FWHM of the Van Hove singularities was found to be 60 meV [36]. This value is very small and reflects the one dimensional character of the SWCNTs. However, this does not significantly differ from values obtained from CVD grown, individual SWCNTs in a previous study [37]. This shows that the high perfectness of the inner tubes as deduced from the phonon lifetimes has no influence on the width of the singularities in their DOS as compared to the non defect-free outer SWCNTs. It rather supports

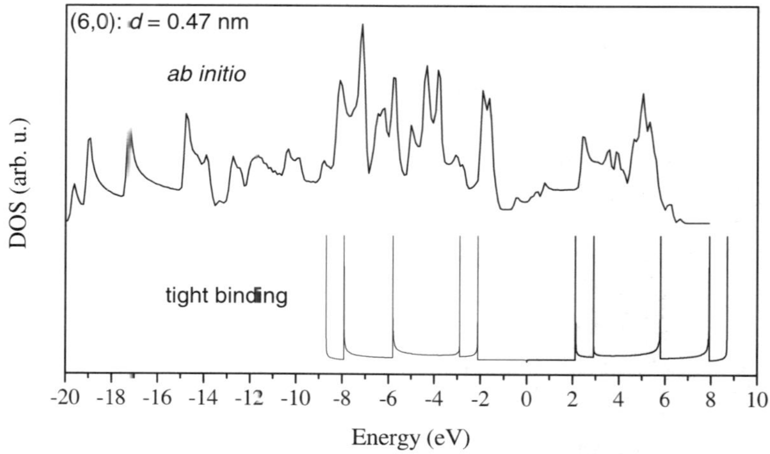

Fig. 8.9. Electronic density of states for a small diameter, metallic nanotube using *ab initio* and tight-binding techniques. The additional structure seen in the *ab initio* calculations between the two first Van Hove singularities of the tight-binding method gives rise to resonances at excitation energies lower than E_{11}^{m}

the enhancement of electron-phonon coupling for small diameter nanotubes as the origin for the signal enhancement. If, however, an enhanced electron phonon coupling is present for the small diameter tubes, one expects a range of interesting physical phenomena to arise for the small tube such as Peierls transition [38] or superconductivity [39]. Indeed, superconductivity has been observed with a critical transition temperature of $T_c = 15\,K$ in 0.4 nm diameter SWCNTs [40]. However, the relevance of such phenomena for the inner tubes embedded in DWCNTs is not yet settled and is currently being investigated.

Splitting of the Inner Tube RBM Response of DWCNTs

In addition to the well defined number of geometrically allowed inner tubes, a larger number of RBMs are observed. In Fig. 8.11 we show spectra measured in the high resolution mode at the 647 nm laser excitation at 90 K. The spectra after deconvolution with the resolution of the spectrometer are also shown. The resolution of the spectrometer contributes to an additional Gaussian broadening of the intrinsically Lorenztian RBM lineshapes. The width of the Gaussian was measured from the response of our apparatus to the exciting laser and was found to be 0.4–0.7 cm^{-1} depending on the laser energy. The presence of additional, split components is apparent in Fig. 8.11 for some tubes. Some RBMs split into even 3 and more components. This splitting is a natural consequence of the different number of geometrically allowed inner

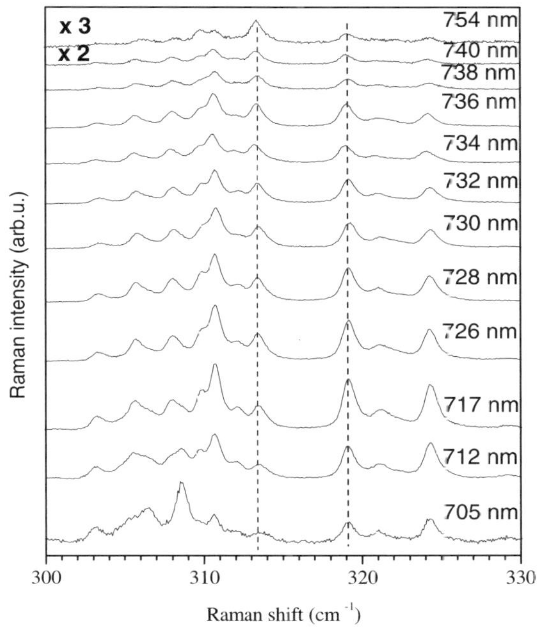

Fig. 8.10. Energy dispersive Raman spectra of the DWCNT sample in the 700–750 nm excitation energy range. Dashed lines indicate the same inner tube RBM followed for several laser excitations. The spectra are normalized to the incident power

and outer tubes and is related to the interaction between the two shells of the DWCNTs. As the diameters of both the inner and outer tubes are discrete sequences, some inner tubes can be grown in outer tubes with different diameters. Then the difference in inner-outer tube wall distance gives rise to a different interaction that causes the observed splitting of the lines. A rough estimate yields that for the DWCNTs studied here, 40 geometrically allowed outer tubes accommodate 20 geometrically allowed inner tubes. As a consequence, on average splitting into two components is expected. This estimate, however, does not account for the magnitude of the splitting that is currently being further investigated both experimentally and theoretically [41].

In addition to the two-wall related splitting of the RBMs of inner tubes, further peculiarities can be observed. In Fig. 8.12 we compare the inner tube RBMs of a DWCNT sample with the RBMs of a small diameter SWCNT sample. The latter was a HiPco sample with a mean diameter and a variance of $d = 1.05$ nm and $\sigma = 0.15$ nm, respectively [42]. The figure shows this comparison for 568 nm and 647 nm laser excitation. The larger number of RBMs in the DWCNT sample as compared to the HiPco sample and the

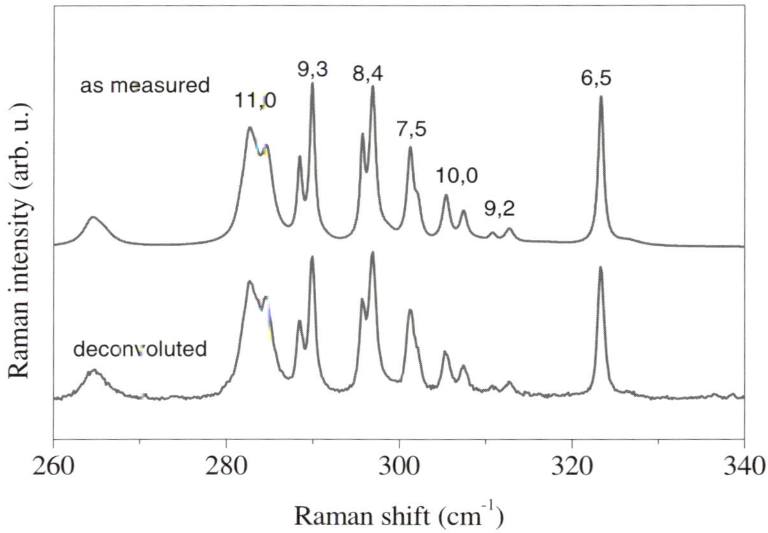

Fig. 8.11. High resolution Raman spectra of inner tube RBMs of DWCNT at $\lambda = 647\,\mathrm{nm}$ laser excitation and 90 K. We also show the corresponding spectrum after deconvolution with the resolution of the spectrometer

absence of splitting for the RBMs in the HiPco sample is observed. Broader RBM lines are observed in the HiPco sample, however this would not limit the observation of the splitting. As discussed above, the splitting is related to the two-shell nature of the DWCNT samples and thus its absence is natural in the HiPco sample. However, the absence of some RBMs corresponding to geometrically allowed tubes in the HiPco sample that are observed in the inner tube DWCNT spectrum is intriguing. The absence of geometrically allowed SWCNTs or the smaller number of optical transitions in the HiPco samples may explain for our observation. The dashed line in Fig. 8.12 shows an example for a tube RBM that is absent at 647 nm excitation from the HiPco spectrum, however appears at 568 nm excitation. Similar behavior was observed for other, missing HiPco RBM modes, i.e. the RBM modes of the SWCNTs of the HiPco samples are also present, although at much less number of laser lines. This clearly shows, that all the geometrically allowed SWCNTs are present both among the inner tubes and also in the HiPco sample, however, there is a significantly larger number of optical transitions for the inner tubes of the DWCNT sample. This is most probably related to a yet unexplained intricate interplay between the two shells of the DWCNT samples and calls for theoretical work on this issue. In addition to the larger number of lines observed in the inner tube RBM spectrum of DWCNTs, a slight downshift ranging from 2–3 cm^{-1} is also observed. This downshift is

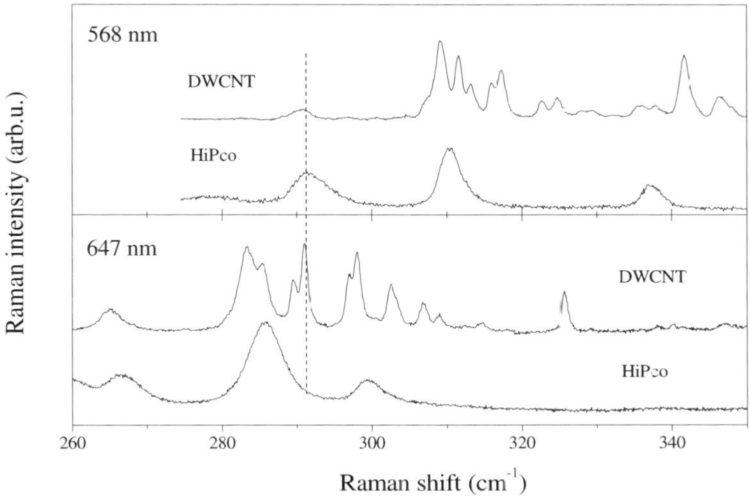

Fig. 8.12. Comparison of the inner tube RBMs of a DWCNT sample and the RBMs of a sample prepared by the HiPco process at $\lambda = 568\,\text{nm}$ and $647\,\text{nm}$ laser energy. Both samples were measured at 90 K in the high resolution mode

related to the different environment for an inner tube of the DWCNT and for an SWCNT in a HiPco sample. The earlier is surrounded by an outer tube, whereas the latter is embedded in bundles. These interactions give rise to a different value for the C_2 constant of the RBM mode frequencies. The broader linewidths observed in the HiPco sample underlines the highly perfect nature of the inner nanotubes.

Chiral Index Assignment for Inner Tubes

The reciprocal relation between the RBM frequencies and the tube diameters contributes to a significant spectral spread for the observed inner tube RBMs. This, together with the narrow linewidths observed for the inner tubes allows the (n, m) indexing of the tubes. In Fig. 8.13, the inner tube RBMs measured at several laser lines are shown. The observation of a large number of well defined RBMs allows the determination of RBM shifts corresponding to distinct inner tubes. The Raman shifts of observable inner tubes are summarized in Table. 8.1, together with the assigned chiral vectors, and tube diameters calculated with DFT methods. It was found that the diameters of small tubes, based on the lattice constant of graphene gives significant deviations as compared to the DFT calculated tube diameters [43]. More recent, energy dispersive Raman measurements have indicated an alternative indexing of the inner tubes [44].

Table 8.1. Inner tube RBM frequencies and tube diameters: (1) center of gravity line position of the RBM averaged from different laser excitations, (2) CNT chiral indices, (3) interpolated DFT determined tube diameters, (4) first and second tight-binding optical transition energies with $\gamma_0 = 2.9\,\text{eV}$, (5) theoretical RBM frequencies at the best fit with $C_1 = 233\,\text{cm}^{-1}\,\text{nm}$ and $C_2 = 14\,\text{cm}^{-1}$. ("n.i.": not identified), from [43].

ν_{RBM} expt. (cm^{-1})	Chirality (n,m)	d_{RBM} (nm)	E_{11}/E_{22} (eV)	ν_{RBM}theor. (cm^{-1})
246.1	(11,3)	0.997	0.84/1.59	247.8
252.4	(12,1)	0.978	0.86/1.60	252.1
n.i.	(10,4)	0.975	2.36/4.21	252.9
257.6	(9,5)	0.960	0.84/1.73	256.8
260.5	(8,6)	0.950	0.87/1.69	259.2
n.i.	(11,2)	0.947	2.40/4.24	260
n.i.	(7,7)	0.947	2.52/4.54	260
n.i.	(12,0)	0.937	2.40/4.25	262.6
265.3	(10,3)	0.921	0.87/1.84	267
270.5	(11,1)	0.901	0.85/1.91	272.5
274.7	(9,4)	0.901	0.82/1.75	272.5
n.i.	(8,5)	0.888	2.61/4.60	276.5
280.2	(7,6)	0.881	0.93/1.85	278.5
282.3	(10,2)	0.871	0.97/1.78	281.6
284.0	(11,0)	0.860	0.98/1.80	284.9
289.1	(9,3)	0.846	2.67/4.63	289.4
296.1	(8,4)	0.828	0.98/2.00	295.4
298.9	(10,1)	0.824	2.70/4.65	296.7
301.9	(7,5)	0.817	1.01/1.95	299.2
304.0	(6,6)	0.813	2.90/5.02	300.5
306.5	(9,2)	0.794	1.01/2.15	307.3
311.7	(10,0)	0.783	1.02/2.22	311.6
317.2	(8,3)	0.771	1.09/2.02	316.1
319.8	(7,4)	0.756	3.00/5.07	322.5
323.0	(9,1)	0.748	1.13/2.05	325.8
327.9	(6,5)	0.748	1.09/2.17	325.8
336.7	(8,2)	0.719	3.06/5.09	338.3
340.3	(9,0)	0.706	3.08/5.10	344.1
345.8	(7,3)	0.697	1.15/2.41	348.2
353.8	(6,4)	0.684	1.21/2.30	354.6
358.5	(5,5)	0.680	3.37/5.52	356.8
364.9	(8,1)	0.671	1.19/2.60	361.4
374.5	(7,2)	0.643	1.32/2.38	376.3
383.5	(8,0)	0.629	1.36/2.40	384.4
389.3	(6,3)	0.624	3.52/5.54	387.3
395.6	(5,4)	0.615	1.33/2.63	393.2
405.8	(7,1)	0.595	3.58/5.55	405.9
422.0	(6,2)	0.569	1.40/3.00	423.8
432.9	(7,0)	0.553	1.43/3.22	435.7
438.1	(5,3)	0.553	1.52/2.79	435.7
444.5	(4,4)	0.547	4.10/5.80	439.9
462.1	(6,1)	0.519	1.66/2.87	463.1

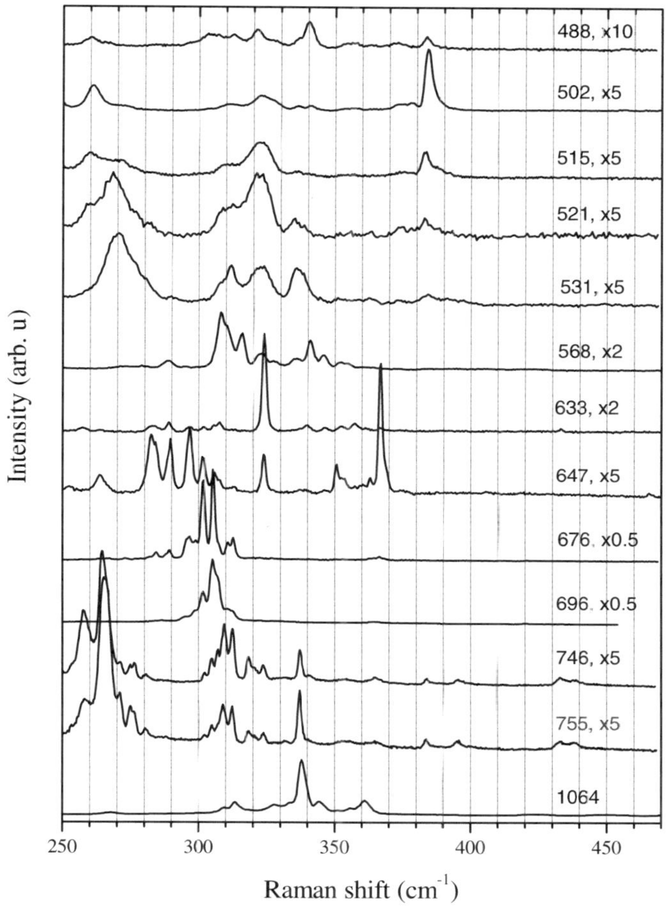

Fig. 8.13. DWCNT RBM spectra measured at 90 K for several laser excitations

In Fig. 8.14 we compare the inverse tube diameters calculated from the lattice constant of graphene and from a DFT calculation. A simple interpolation could be established: $1/d_{\mathrm{DFT}} = 1/d_{\mathrm{G}}$ - $\left(0.0050/d_{\mathrm{G}}^2 + 0.0013/d_{\mathrm{G}}^4\right)$, where d_{DFT} and d_{G} are the DFT and the graphene derived tube diameters, respectively, and d_{G} is expressed from the chiral indices as: $d_{\mathrm{G}} = 0.141\sqrt{3\,(m^2+n^2+mn)}/\pi$. The C_1 and C_2 constants of the RBM frequencies $\nu_{\mathrm{RBM}}{=}C_1/d_{\mathrm{DFT}} + C_2$ were determined from a linear regression and $C_1 = 233\,\mathrm{cm}^{-1}\,\mathrm{nm}$ and $C_2 = 14\,\mathrm{cm}^{-1}$ were found. Figure 8.15 shows that no linear relationship could be established between the RBM frequencies and the graphene derived inverse tube diameters. However, a very reliable fit is

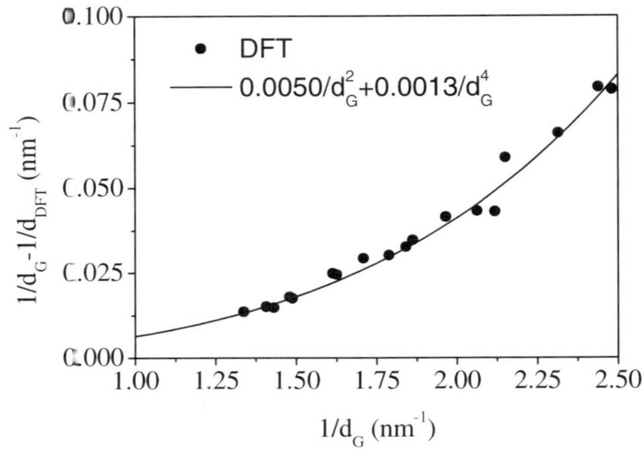

Fig. 8.14. Difference between graphene and DFT derived inverse tube diameters. The *solid line* is a polynomial interpolation

obtained with small discrepancies between calculated and measured RBM frequency values when the DFT optimized tube diameters were used. It establishes that the reciprocal relationship between the RBM frequencies and the tube diameters, when the latter is properly calculated, is valid down to the smallest observable inner tubes. The value determined for the C_1 constant is not restricted to small diameter tubes only, however studies on larger diameter tubes have been lacking the precision that could be obtained from the inner tube analysis of DWCNTs. Therefore a direct comparison with previously determined values can not be performed. It has been pointed out above that the different environment for an inner tube and for an SWCNT in a bundle prevents a comparison of the respective C_2 parameters.

In conclusion, the Raman studies of peapod based DWCNT materials have been reviewed. It was shown that this material has an unprecedently high perfectness related to the growth in the catalyst free environment. The growth mechanism is not fully understood and some alternative mechanisms have been discussed. A method is presented that enables the industrial scaling up of the DWCNT production. The properties of the DWCNT have been compared with a similar diameter SWCNT material, a HiPco sample. The geometrically allowed small diameter tubes are present in both materials. However, due to the smaller number of optical transitions in the HiPco sample, a smaller number of RBM lines are observed. A splitting of the geometrically allowed inner tube RBMs in the DWCNT sample was observed. It was explained by the different shell-shell distance of inner-outer tube pairs with varying diameters. The sharp appearance of the inner tube RBMs and their relatively larger spectral spread allowed a chiral index assignment for a broad spectral range and Raman shifts. It was found that a linear relation could

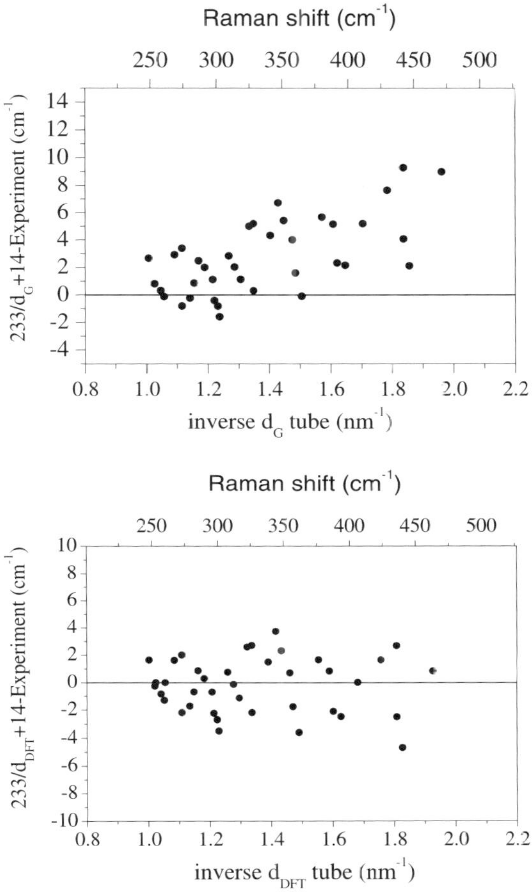

Fig. 8.15. Difference between the calculated and measured RBM Raman shifts when the graphene derived (upper panel) and DFT optimized (lower panel) tube diameters were used

be established between the RBM frequencies and DFT determined tube diameters. The empirical constants relating the RBM frequencies and the tube diameters have been refined. Nevertheless, a direct comparison with values determined on free-standing or bundled SWCNTs is not straightforward.

Acknowledgements

The authors gratefully acknowledge J. Kürti, and V. Zólyomi for many fruitful discussions and for the DFT calculations. H. Kataura and Y. Achiba are acknowledged for providing some of the C_{60} peapod materials. We thank

Hui-Ming Cheng for providing the CVD grown DWCNT sample. The authors gratefully acknowledge J. Bernardi for the TEM micrographs, A. Sen, Ch. Schaman, Th. Pichler, and Á. Kukovecz for their contributions to the Raman experiments. This work was supported by the Austrian Science Funds (FWF) project Nr. 14893 and by the EU projects NANOTEMP BIN2-2001-00580 and PATONN Marie-Curie MEIF-CT-2003-501099.

References

1. S. Iijima, Nature (London) **354**, 56 (1991)
2. J.H. Hafner, C.L. Cheung, C.M. Lieber, Nature **398**, 761 (1999)
3. G.Z. Yue, Q. Qiu, B. Gao, Y. Cheng, J. Zhang, H. Shimoda, S. Chang, J.P. Lu, and O. Zhou, Appl. Phys. Lett. **81**, 355 (2002)
4. A.N. Obraztsov, I. Pavlovsky, A.P. Volkov, E.D. Obraztsova, A.L. Chuvilin, V.L. Kuznetsov, J. Vac. Sci. and Techn. B **18**, 1059 (2000)
5. M.S. Dresselhaus, G. Dresselhaus, P.C. Ecklund: *Science of Fullerenes and Carbon Nanotubes*, Academic Press, San Diego 1996.
6. D. Chattopadhyay, L. Galeska, and F. Papadimitrakopoulos F, J. Am. Chem. Soc. **125**, 3370 (2003)
7. R. Krupke, F. Hennrich, H. von Lohneysen, and M.M. Kappes, Science **301**, 344 (2003)
8. Z.H. Chen, X. Du, M.H. Du, C.D. Rancken, H.P. Cheng, and A.G. Rinzler, Nano Lett. **3**, 1245 (2003)
9. M. Zheng, A. Jagota, M.S. Strano, A.P. Santos, P. Barone, S.G. Chou, B.A. Diner, M.S. Dresselhaus, R.S. McLean, G.B. Onoa, G.G. Samsonidze, E.D. Semke, M. Usrey, and D.J. Walls, Science **302**, 1545 (2003)
10. M.J. O'Connell, S.M. Bachilo, C.B. Huffman, V.C. Moore, M.S. Strano, E.H. Haroz, K.L. Rialon, P.J. Boul, W.H. Noon, C. Kittrell, J.P. Ma, R.H. Hauge, R.B. Weisman, and R.E. Smalley, Science **297**, 593 (2002)
11. B.W. Smith, M. Monthioux, and D.E. Luzzi, Chem. Phys. Lett. **315** 31 (1999)
12. B.W. Smith, M. Monthioux, and D.E. Luzzi, Nature **396**, 323 (1998)
13. H. Kataura, Y. Maniwa, T. Kodama, K. Kikuchi, K. Hirahara, K. Suenaga, S. Iijima, S. Suzuki, Y. Achiba, W. Krätschmer, Synth. Met. **121**, 1195 (2001)
14. S. Bandow, M. Takizaw, K. Hirahara, M. Yudasaka, and S. Iijima, Chem. Phys. Lett. **337** 48 (2001)
15. J.L. Hutchison, N.A. Kiselev, E.P. Krinichnaya, A.V. Krestinin, R.O. Loutfy, A.P. Morawsky, V.E. Muradyan, E.D. Obraztsova, J. Sloan, S.V. Terekhov, and D.N. Zakharov, Carbon **39**, 761 (2001)
16. Wencai Ren, Feng Li, Jian Chen, Shuo Bai, and Hui-Ming Cheng, Chem. Phys. Lett. **359**, 196 (2002)
17. H. Kuzmany, W. Plank, M. Hulman, C. Kramberger, A. Gruneis, T. Pichler, H. Peterlik, H. Kataura, and Y. Achiba, Eur. Phys. J. B **22**, (2001) 307.
18. X. Liu, T. Pichler, M. Knupfer, M.S. Golden, J. Fink, H. Kataura, Y. Achiba, K. Hirahara, and S. Iijima, Phys. Rev. B 65 (2002) 045419.
19. G. Kresse and D. Joubert, Phys. Rev. B **59** 1758 (1999)
20. B.W. Smith, and D.E. Luzzi, Chem. Phys. Lett. **321** 169 (1999)

21. M. Abe, H. Kataura, H. Kira, T. Kodama, S. Suzuki, Y. Achiba, K. Kato, M. Takata, A. Fujiwara, K. Matsuda, and Y. Maniwa, Phys. Rev. B **68**, 041405 (2003)
22. M. Melle-Franco, H. Kuzmany, and F. Zerbetto, J. Phys. Chem. B **109**, 6986 (2003)
23. S. Berber, Y-K. Kwon, D. Tománek, Phys. Rev. Lett. **88**, 185502 (2002)
24. A. Rochefort, Phys. Rev. B **67**, 115401 (2003)
25. R. Pfeiffer, H. Kuzmany, C. Kramberger, C. Schaman, T. Pichler, H. Kataura, Y. Achiba, J. Kürti, and V. Zólyomi, Phys. Rev. Lett. **90**, 225501 (2003)
26. M. Holzwebver, Ch. Kramberger, F. Simon, Á. Kukovecz, H. Kuzmany, and H. Kataura in the Proceedings of the XVII International Winterschool on Electronic Properties of Novel Materials, edited by H. Kuzmany, J. Fink, M. Mehring, and S. Roth, AIP Publishing, New York, 2003, p. 306.
27. S. Bandow, T. Hiraoka, T. Yumura, K. Hirahara, H. Shinohara, and S. Iijima, Chem. Phys. Lett. **384**, 320 (2004)
28. D. Tománek, private communication.
29. Y. Zhao, B.I. Yakobson, R.E. Smalley, Phys. Rev. Lett. **88**, 185501 (2002)
30. K. Hirahara, S. Bandow, K. Suenaga, H. Kato, T. Okazaki, H. Shinohara, and S. Iijima, Phys. Rev. B. **64**, 115420 (2001)
31. F. Simon, Á. Kukovecz, Ch. Kramberger, R. Pfeiffer, F. Hasi, H. Kuzmany, and H. Kataura, Phys. Rev. B in press, cond-mat/0403179.
32. F. Simon, H. Kuzmany, H. Rauf, T. Pichler, J. Bernardi, H. Peterlik, L. Korecz, F. Fülöp, and A. Jánossy, Chem. Phys. Lett. **383**, 362 (2004)
33. Feng Li, S.G. Chou, Wencai Ren, J.A. Gardecki, A.K. Swan, M.S. Ünlü, B.B. Goldberg, Hui-Ming Cheng, and M.S. Dresselhaus J. Mater. Res. **18**, 1251 (2003)
34. A. Jorio, R. Saito, J.H. Hafner, C.M. Lieber, M. Hunter, T. McClure, G. Dresselhaus, and M.S. Dresselhaus, Phys. Rev. Lett. **86**, 1118 (2001)
35. M. Machón, S. Reich, J.M. Pruneda, C. Thomsen, and P. Ordejón in the Proceedings of the XVII International Winterschool on Electronic Properties of Novel Materials, edited by H. Kuzmany, J. Fink, M. Mehring, and S. Roth, AIP Publishing, New York, 2003, p. 427.
36. A. Sen, Ch. Kramberger, Ch. Schaman, R. Pfeiffer, H. Kuzmany, and H. Kataura in the Proceedings of the XVII International Winterschool on Electronic Properties of Novel Materials, edited by H. Kuzmany, J. Fink, M. Mehring, and S. Roth, AIP Publishing, New York, 2003, p. 314.
37. A. Jorio, A.G. Souza, G. Dresselhaus, M.S. Dresselhaus, R. Saito, J.H. Hafner, C.M. Lieber, F.M. Matinaga, M.S.S. Dantas, and M.A. Pimenta, Phys. Rev. B **63**, 245416 (2001)
38. George Grüner: Density waves in solids, Addison-Wesley Publishing Company, 1994.
39. Michael Tinkham: Introduction to superconductivity, Krieger Publishing Company, 1975.
40. Z.K. Tang, L.Y. Zhang, N. Wang, X.X. Zhang, G.H. Wen, G.D. Li, J.N. Wang, C.T. Chan, and P. Sheng, Science **292** 2462 (2001)
41. R. Pfeiffer, Ch. Kramberger, F. Simon, H. Kuzmany, V.N. Popov, and H. Kataura, Eur. J. Phys. B **42**, 345 (2004)
42. Á. Kukovecz, Ch. Kramberger, V. Georgakilas, M. Prato, and H. Kuzmany, Eur. Phys. J. B **28**, 223 (2002)

43. Ch. Kramberger, R. Pfeiffer, H. Kuzmany, V. Zólyomi, and J. Kürti, Phys. Rev. B **68**, 235404 (2003)
44. R. Pfeiffer et al. unpublished

Part IV

Transport and Electromechanical Applications

9 Carbon Nanotube Electronics and Optoelectronics

Ph. Avouris, M. Radosavljević, and S.J. Wind

In this chapter, we review progress to date in carbon nanotube electronics and optoelectronics. We discuss the underlying physics of CNT-FETs, highlighting the similarities and differences relative to conventional silicon metal-oxide-semiconductor field-effect transistors (MOSFETs), and we examine how these affect CNT-FET electrical characteristics. As device scaling is the key technology driver in today's semiconductor technology, we explore how CNT-FETs behave when scaled to smaller dimensions and the impact this scaling behavior may have on their suitability for technological insertion. We look at results achieved to date on simple CNT-based circuits, and we consider the requirements of more complex architectures. Finally, we discuss the optoelectronic properties of CNTs and show that CNT-FETs can also be used as light emitting and light decoding devices.

9.1 Introduction

The discovery of the integrated circuit by Jack Kilby at Texas Instruments has lead to four decades of dizzying advances in silicon-based electronic technology. This progress has been achieved primarily through the sustained scaling of physical dimensions in the metal-oxide-semiconductor field-effect transistor (MOSFET), which has lead to many successive generations of devices with increased transistor performance and density. Most experts agree that by continuing on this scaling path, the community will reach technological, economic and, most importantly, fundamental physical limits as soon as the end of this decade. Therefore it becomes crucial to start preparing for technologies that will enable continued implementation of increasingly higher performance devices.

In general, two distinct approaches are taken to address these issues. In one case, revolutionary technologies are proposed based on totally new concepts, for example two terminal molecular devices, spintronics, or quantum computing may be pursued. While some of these technologies may one day be implemented into products, they are obviously targeted for the far future. We focus on another, more evolutionary approach that is based on the well established three terminal transistor concept, but utilizes different materials, specifically carbon nanotubes (CNTs) that may address many of the problems present in aggressively scaled silicon devices. Indeed, the silicon electronics industry has long experimented with progressively more radical alternative materials (such as SOI, SiGe and high-k dielectrics to name a few) in order to improve device performance. In this vein, single-walled CNTs (SWCNTs) have some unique advantages including their nanoscale dimensions as well as the electronic and optical properties which we will describe in this chapter. Specifically, we will describe the progress on carbon nanotube field-effect transistors (CNT-FETs) used in electronics as switches and opto-electronics as light-emitters/detectors. While current state-of-the-art devices contain individual CNTs, the results are applicable to CNT arrays for larger current or light signals, provided that synthesis of homogenous CNT samples is developed.

9.2 Electronic Structure and Electrical Properties of Carbon Nanotubes

A single-walled carbon nanotube can be thought of as a graphene sheet (i.e., a single layer of graphite) rolled up into a cylinder, as in Fig. 9.1a. Consequently, the electronic properties of single-walled carbon nanotubes can be derived directly from those of graphene [1–3], with unique characteristics resulting from the precise arrangement of the carbon atoms of the rolled up graphene sheet. This "rolling up" can be described in terms of the "chiral vector," **C** (see Fig. 9.1a), which connects two sites of the two-dimensional graphene sheet which are crystallographically equivalent: $\mathbf{C} = n\hat{\mathbf{a}}_1 + m\hat{\mathbf{a}}_2$, where $\hat{\mathbf{a}}_1$ and $\hat{\mathbf{a}}_2$ are the unit vectors of the hexagonal graphene lattice. The unit cell of the nanotube is defined as the rectangle formed by **C** and the one-dimensional translational vector, **T**, as shown in Fig. 9.1a. Thus, the geometry of any nanotube can be described by the integer pair (n, m) which determine the chiral vector.

The band structure of graphene is quite unusual. It has states crossing the Fermi level at six points in k-space, as shown in Fig. 9.1b. The folding of the graphene layer to form a CNT introduces an additional level of quantization due to the confinement of electrons around the circumference of the nanotube. The component of the wave-vector $\mathbf{k}_c$ is constrained by the condition $\mathbf{k}_c \cdot \mathbf{C} = 2\pi j$, where j is an integer. Thus, in going from a 2D graphene sheet to a 1D CNT, each graphene band is split into a number of 1D subbands indexed by j, with allowed energy states corresponding to slices through the graphene band structure. Slices which pass through a K-point of the graphene Brillouin zone result in a metallic nanotube; all other slices yield semiconducting CNTs. Figs. 9.1c and 9.1d illustrate the contrasting cases of a metallic $(5, 5)$ tube and a semiconducting $(5, 0)$ tube, with their respective quantization in k-space and density of states in a π tight-binding description.

The general rules for determining whether a CNT is metallic or semiconducting based on the chiral indices (n, m) are: when $n = m$, the tube is

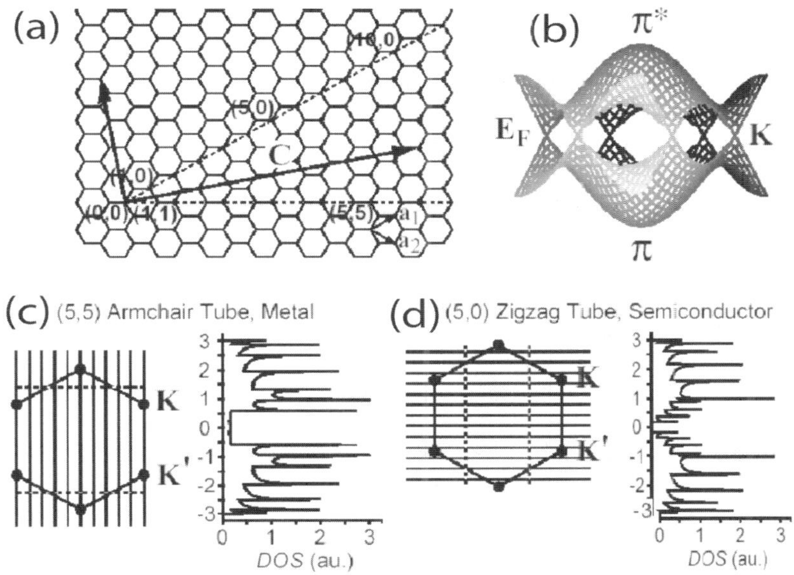

Fig. 9.1. (**a**) Definition of the chiral and translation vectors and the (n,m) notation of carbon nanotubes. (**b**) Band structure of the graphene layer. (**c**) and (**d**) Allowed 1D states as cuts of the graphene Brillouin zone for a (5,5) metallic (c) and a (5,0) semiconducting (d) nanotube, along with their associated density of states

metallic; when $n - m = 3i$, where i is a nonzero integer, the tube is a small-gap semiconductor; and all other combinations of (n, m) are true semiconductors. The band gap of semiconducting CNTs is approximately inversely proportional to the tube diameter. According to the tight-binding description of the electronic structure, the band gap, E_{gap} is given by $\gamma(2a/\sqrt{3}d_{CNT})$, where γ is the hopping matrix element, $a = \sqrt{3}d_{C-C}$, with d_{C-C} being the carbon-carbon bond distance, and d_{CNT} is the nanotube diameter [1–3].

The two-terminal conductance of a metallic single-walled CNT is given by the Landauer-Buttiker formula for one-dimensional conductors [4]: $G = (2e^2/h) \cdot \sum_i^N T_i$, where $2e^2/h$ is the quantum unit of conductance, and T_i is the transmission of the ith conducting channel. When $T_i = 1$, corresponding to the case of no scattering inside the nanotube or at the contacts, a metallic nanotube has a resistance, $R = 1/G = h/(4e^2)$, because there are 2 contributing conductance channels near the Fermi energy [5].

In the case of scattering within the CNT, an effective scattering length, λ_{eff}, can be used to describe the scattering. Multiple processes may contribute to scattering, and thus, $\lambda_{eff}^{-1} = \lambda_{el}^{-1} + \lambda_{ac}^{-1} + \lambda_{op}^{-1}$, where λ_{el} is the mean free path for elastic scattering, and λ_{ac} and λ_{op} are the mean free paths for scattering by acoustic and optical phonons, respectively. Due to the one-dimensional confinement of electrons on the surface of the nanotube, which allows only forward and backward motion, and the requirements for energy and momentum conservation, the available phase space for scattering is drastically constrained. The result is weak elastic scattering, with $\lambda_{el} \geq 1\,\mu\text{m}$, in the bulk of pure metallic CNTs. Acoustic phonons contribute weakly to inelastic scattering as well, with $\lambda_{ac} \approx 1\,\mu\text{m}$ [6]. Thus, transport in metallic CNTs at low energies is ballistic over distances of a micron or so. Optical phonons can scatter efficiently once carriers exceed optical phonon energies ($\sim 180\,\text{meV}$), with $\lambda_{op} \approx 20$–$30\,\text{nm}$ [7–9]. This can result in current saturation at elevated bias [7] and can lead to nanotube breakdown [10]. Thus far, there is somewhat less information on scattering in semiconducting CNTs, however, there are indications that at low energies, λ_{eff} is of order a few hundred nanometers [11–13].

For the case of a CNT device in which a semiconducting nanotube is connected to metallic leads, there are other sources of contact resistance in addition to the Landauer resistance mentioned above: Schottky barriers form at the metal/nanotube junctions, through which carriers must tunnel [14]. These Schottky barriers have a profound effect on the function and performance of CNT-based transistors, as will be discussed below. Additional sources of series resistance, which are not intrinsic to the device itself, can in some instances make an overwhelming contribution to the total measured CNT resistance, as well.

9.3 Potential and Realized Advantages of Carbon Nanotubes in Electronics Applications

Having discussed the structure and properties of CNTs in Sect. 9.2 it is useful to understand the general material-based benefits that may translate in additional performance enhancements for CNT devices. Some of these attributes have already been demonstrated in experiments, while others have yet to be realized [15].

In terms of transport, the 1D nature of CNTs severely reduces the phase space for scattering, allowing CNTs to realize maximum possible bulk mobility of this material. The low scattering probability and high mobility are responsible for high "on" current (in excess of 1mA/μm) in semiconductor CNT transistors. Furthermore, the chemical stability and perfection of the CNT structure suggests that the carrier mobility at high gate fields may not be affected by processing and roughness scattering as in the conventional semiconductor channel. Similarly in metallic CNTs, low scattering together with the strong chemical bonding and extraordinary thermal conductivity allows them to withstand extremely high current densities (up to $\sim 10^9\ \text{A/cm}^2$).

Electrostatics is improved in these devices as well. The fact that there are no dangling bond states at the surface of CNTs allows for a much wider choice of gate insulators beyond the conventional SiO_2. This improved gate control without any additional gate leakage becomes very important in the scaled devices with effective SiO_2 thickness below 1 nm. Also, the strong 1D electron confinement and full depletion in the nanoscale diameter of the SWCNTs (typically 1–2 nm) should lead to a suppression of short-channel effects in transistor devices.

The combined impact of transport and electrostatic benefits together with the fact that semiconducting CNTs are, unlike silicon, direct-gap materials, suggests applications in opto-electronics as well.

As far as integration is concerned, semiconducting CNTs benefit from their band structure which gives essentially the same effective mass to electrons and holes. This should enable similar mobilities and performance in n-type and p-type transistors which is necessary for a complementary metal-oxide semiconductor (CMOS)-like technology. Finally, since CNTs can be both metallic and semiconducting, an all-nanotube electronics can be envisioned. In this case, metallic CNTs could act as high current carrying local interconnects while semiconductors would form the active devices. The most important appeal of this approach is an ability to fabricate one of the critical device dimensions (the CNT diameter) reproducibly using synthetic chemistry.

9.4 Fabrication and Performance of Carbon Nanotube Field-Effect Transistors

The first carbon nanotube field-effect transistors (CNT-FETs) were reported only a few years after the initial discovery of the CNT [16, 17]. These early devices, shown schematically in Fig. 9.2a, were relatively simple in structure: Noble metal (gold or platinum) electrodes were lithographically patterned atop an oxide-coated, heavily doped silicon wafer, and a single-walled CNT was deposited atop the electrodes from solution. The metal electrodes served as the source and drain, and the CNT was the active channel. The doped substrate served as the gate electrode, separated from the nanotube channel by the thick ($\sim$ 100–200 nm) oxide layer. These devices displayed clear p-type transistor action, with gate voltage modulation of the drain current over several orders of magnitude. The devices displayed high parasitic resistance ($\gtrsim$ 1 MΩ), low drive current, low transconductance, $g_m \sim 1nS$, high subthreshold slope, $S = [d(log_{10}I_d)/dV_g]^{-1} \sim$ 1 V/decade, and no current saturation. Due to the thick gate dielectric, these devices required large values of gate voltage (several volts) to turn on, making them unattractive for practical applications.

Following these initial CNT-FET results, advances in CNT-FET device structures and processing yielded improvements in their electrical characteristics. Rather than laying the nanotube down upon the source and drain electrodes, relying on weak Van der Waals forces for contact, the CNTs were first deposited on the substrate, and the electrodes were patterned on top of the CNTs, as in Fig. 9.2b. In addition to Au, Ti and Co were used [18–20],

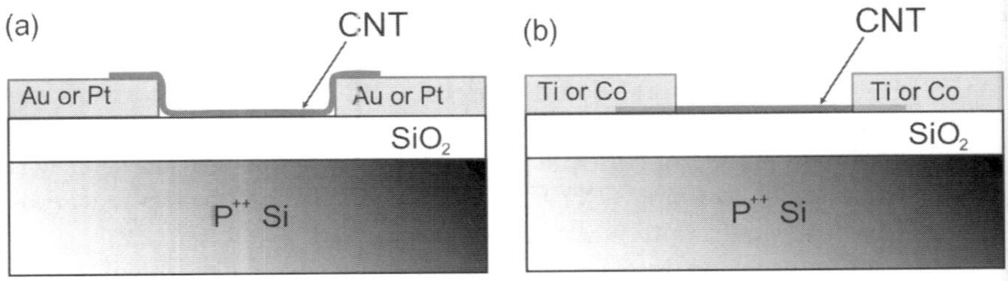

Fig. 9.2. (**a**) Schematic structure of first CNT-FETs, with CNT draped over metal electrodes. (**b**) "Improved" CNT-FET structure, with metal electrodes deposited upon the CNT, followed by thermal processing to improve contact. The substrate serves as the gate for both device structures, and it is separated from the CNT by a thick oxide layer

with a thermal annealing step to improve the metal/nanotube contact. In the case of Ti, the thermal processing leads to the formation of TiC at the metal/nanotube interface [19], resulting in a significant reduction in the contact resistance – from several MΩ to $\sim 30\,\mathrm{k\Omega}$. On-state currents $\sim 1\,\mu\mathrm{A}$ were measured, with a transconductance of $\sim 0.3\,\mu\mathrm{S}$, an improvement of more than 2 orders of magnitude relative to the Van der Waals contacted devices. This CNT-FET device configuration can be found extensively in the literature (see, e.g., McEuen et al. [21]). More recently, it has been found that Pd forms low resistance contact to CNTs for p-type devices [12]. It is speculated [12] that Pd offers improved wettability of the CNT surface relative to other metals, as well as good Fermi level alignment relative to the nanotube band edge. This point will be explored further in the following section.

As mentioned above, early CNT-FETs were p-type in air, i.e., hole conductors. The role of the ambient on CNT-FET conduction will be discussed in detail below, however, it was found that n-type (i.e., electron) conduction could be achieved by doping using an alkali (electron donor) source [22–24] or by thermal annealing in vacuum [19, 25]. In addition, it is possible to achieve an intermediate state, in which both electron and hole injection are allowed, resulting in ambipolar conduction [19, 25]. The ability to controllably fabricate both p-type and n-type CNT-FETs is critical to the formation of complementary metal-oxide-semiconductor (CMOS) logic circuits. Such CNT-FET circuits will be discussed later in this chapter.

Early experiments on CNT-FETs were built upon oxidized silicon wafers, with the substrate itself serving as the gate and a thermally grown oxide film, typically $\sim 100\,\mathrm{nm}$ or more thick, serving as the gate dielectric. The thick gate oxide required relatively high gate voltages ($\sim 10\,\mathrm{V}$) to turn on the devices, and the use of the substrate as the gate implied that all CNT-FETs must be turned on and off together, precluding the implementation of complex circuits. A more advanced CNT-FET structure [26] is shown in Fig. 9.3. The device comprises a top-gate separated from the nanotube channel by a

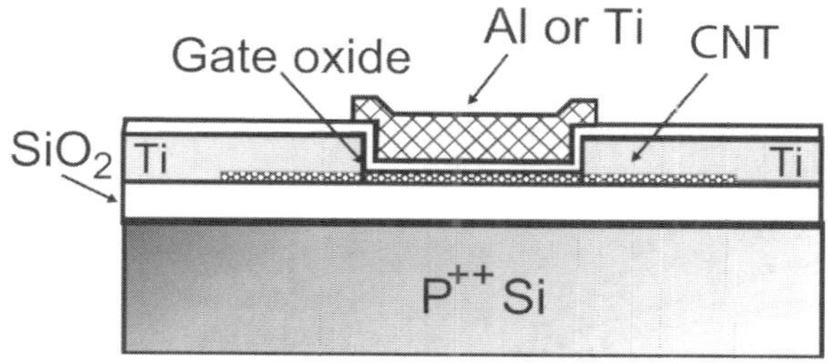

Fig. 9.3. Schematic cross-section of top-gate CNT-FET

thin gate dielectric. The top-gate allows independent addressing of individual devices, making it more amenable to integration in complex circuits, while the thin gate dielectric improves the gate-channel (CNT) coupling, enabling low voltage operation. In addition, the reduction of the capacitance due to gate-to-source and gate-to-drain overlap suggests that such a device structure would be appropriate for high frequency operation. Such a CNT-FET can also be switched using the conductive substrate as a bottom-gate, allowing for direct comparison between top and bottom-gate operation. Comparison of the output characteristics for top and bottom-gate operation of the device in Fig. 9.3 are shown in Fig. 9.4a and 9.4b, respectively. Operating the device with the top-gate yields distinctly superior performance relative to bottom-gate operation, with a lower threshold voltage ($-0.5\,\mathrm{V}$ vs. $-12\,\mathrm{V}$) and higher transconductance ($g_m = 3.25\,\mu\mathrm{S}$ vs. $0.1\,\mu\mathrm{S}$). Figure 9.4c shows superior sub-threshold behavior for top-gate operation with an improvement in subthreshold slope of more than a factor of 10 (130 mV/decade vs. 2 V/decade).

In order to gauge whether or not CNT-FETs have potential for future nanoelectronic applications, it is important to compare their electrical performance to those of advanced silicon devices. Wind et al. [26] demonstrated that although the device structure is far from optimized, the electrical characteristics, such as the "on current" and the transconductance, g_m, of the device shown in Fig. 9.3 exceeds those of state-of-the-art silicon MOSFETs. Further enhancements to CNT-FET structures, such as the use of high dielectric constant gate insulators [27, 28], and additional improvements in the metal/nanotube contact resistance at the source and drain [12] have lead directly to improved CNT-FET performance. Such improvements are not restricted to p-type CNT-FETs. Figure 9.5 shows the transfer characteristics of an n-type CNT-FET with a thin (5 nm) SiO_2 gate dielectric fabricated by potassium doping of a p-type device [29].

It is not sufficient to fabricate individual CNT-FETs. Future applications will require vast numbers of devices on a single substrate. In an effort to test the parallel fabrication of many CNT-FETs, Collins et al. [30] deposited mixtures of individual and ropes of CNTs upon an oxidized silicon wafer and formed a regular array of source, drain and gate electrodes. Although

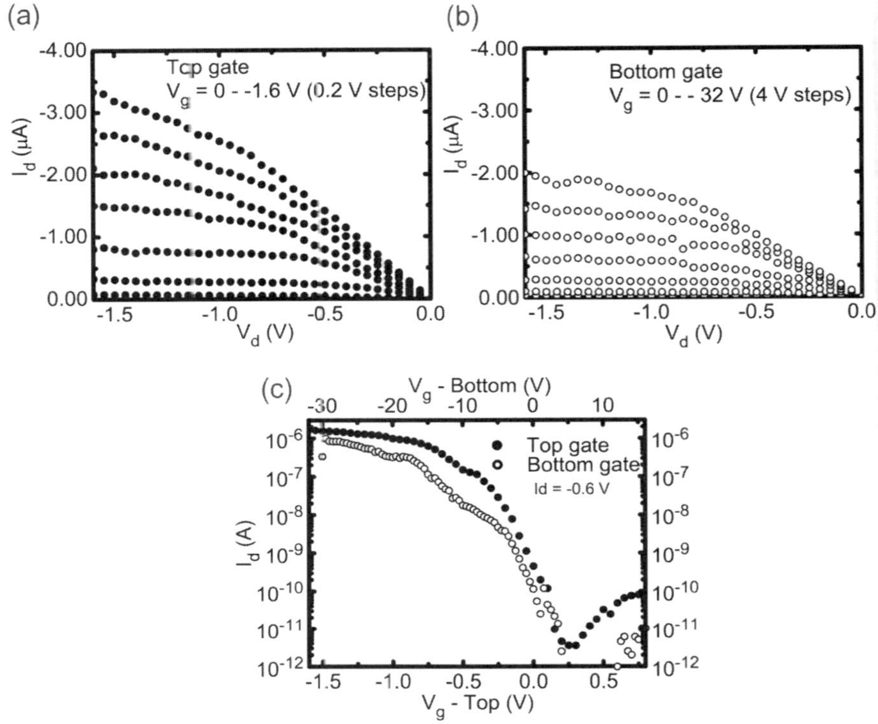

Fig. 9.4. Electrical characteristics of the CNT-FET shown in Fig. 9.3 for both top-gate and bottom-gate operation. The oxide thicknesses for the top-gate and the bottom-gate are 15 nm and 120 nm, respectively. (**a**) and (**b**) Output characteristics. (**c**) Transfer charactersitics for top- and bottom-gate operation

the deposited mixture of nanotubes contained random mixtures of semiconducting and metallic CNTs, functional FETs could be formed by selectively destroying the metallic nanotubes in a controlled manner via application of a sufficiently high current. The semiconducting tubes were preserved by using the gate electrode to deplete them of carriers during the destruction of the metallic tubes, leaving behind a large array of functional CNT-FETs [30]. In another approach, CNT-FETs were fabricated upon an existing silicon NMOS decoder circuit using CVD growth from metal catalysts located on the silicon devices [31]. The decoder was used for rapid electrical characterization of the CNT-FETs. CNTs are also being explored for low cost, low performance applications, such as thin film transistors (TFTs). Snow et al. [32] formed large area transistors by depositing random arrays of nanotubes between metal electrodes. They found that conduction between the electrodes could proceed via percolation through the network of CNTs. Their devices had field-effect mobilities $\sim 10\,\text{cm}^2/\text{V-s}$, which is significantly higher

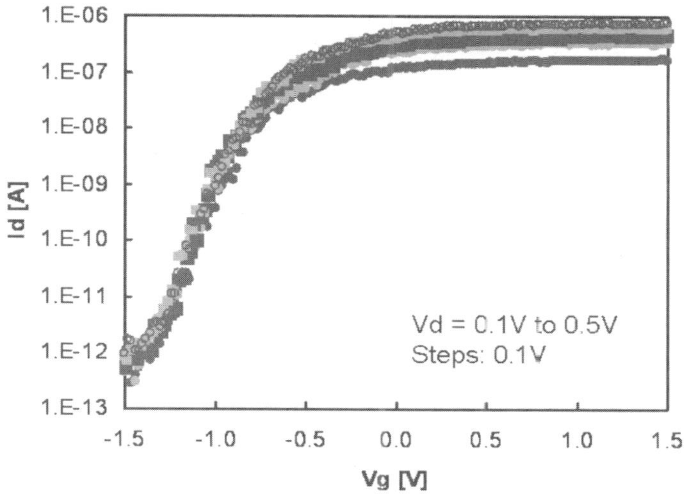

Fig. 9.5. Transfer characteristics of an n-type CNT-FET on 5 nm thick gate oxide (from [29])

than TFTs made from amorphous silicon [32]. The off-state current in these devices is limited by the presence of metallic nanotubes in the percolation network. In order for CNT-based TFTs to be practical, transport through these metallic tubes must be reduced.

9.5 Carbon Nanotube Transistor Operation in Terms of a Schottky Barrier Model

While the performance of CNT-FETs shows a great deal of promise, further improvements and optimization, especially in scaled devices requires a detailed understanding of the switching mechanism. So far we have made an implicit assumption that the conventional or bulk switching of silicon devices is applicable to CNT-FETs [33]. However, this notion may not be entirely correct in all situations, since in general the charge transfer must take place at the metal-nanotube interface leading to band-bending in CNTs and the creation of a Schottky barrier (SB) [34]. While this barrier is present at the metal-semiconductor contact, its height is dependent on a number of material parameters such as workfunction difference, as well as the interface quality. For this reason, significant amount of effort has been and continues to be made in order to search for the right material and process combination that yields a barrier free junction in CNT devices [12,35]. Such CNT-FETs would show minimal effect of the SB on transport and consequently approach a more traditional bulk switching, CMOS-like behavior. However, evidence for

Schottky barriers which affects the performance of CNT-FETs in most device geometries is still clearly present. For example, a CNT-FET with titanium-carbide contacts show equal hole and electron currents depending on the sign of the applied gate bias, V_g, so called ambipolar conduction [19]. This suggests the existence of two barriers, one for electrons and one for holes, of approximately equal height implying that each must be about half the bandgap ($E_{gap}/2 \sim 300\,\text{meV}$). Applying conventional semiconductor analysis indeed yields similar thermal activation barriers but on the order of 10 meV [19]. This finding suggested that thermionic contribution alone can not account for the observed current levels, which is supported by modeling results showing that Schottky barriers in 1D are much thinner than their planar analogues [14]. Consequently, carrier tunneling through these thin barriers becomes dominant conduction mechanism and cannot be neglected when quantifying the barrier height [36].

Similar conclusions can be drawn from attempting to understand the subthreshold behavior of CNT-FETs, in particular as a function of gate oxide thickness. The switching of a MOSFET is described by the inverse subthreshold slope, S, defined in the previous section, is approximated by $S \simeq k_BT/q \cdot \ln(10) \cdot (1 + C_D/C_g)$ where C_D and C_g are the depletion and gate capacitance, respectively. In the case of a fully depleted device, C_D is zero and, therefore, S depends only on the temperature, having a value of 65 mV/decade at room temperature. The original CNT-FETs with thick gate oxides in back-gated geometry had unexpectedly high S values of approximately 1,000 mV/decade. Moreover, similar high values of S are observed for both charge carriers types in unipolar and ambipolar devices, eliminating the influence of oxide traps as a possible explanation. On the other hand, when devices are fabricated using thinner oxides, such as the top-gated CNT-FET in Fig. 9.3, the value of S dropped significantly into the range of 100–150 mV/decade [26]. Such a dependence of S is not consistent with bulk switching mechanism which should give 65 mV/decade in the long channel limit. Instead, this unorthodox scaling of the subthreshold slope with oxide thickness is compatible with existence of sizeable Schottky barriers at the metal/CNT junctions, and theoretical modeling showed that the gate field impact on at this junction is responsible for the improvement in S observed [37, 38].

Further evidence of the presence of Schottky barriers in the CNT-FET devices is found in local gating experiments. In one study, the on-current in CNT-FETs is shown to increase significantly by an application of a local potential from a metal coated scanning probe tip only at the positions above the metal/CNT junction [39]. Similarly, the impact of Schottky barriers in the subthreshold characteristics of the CNT-FET is clearly observed in transistors with multiple, lithographically patterned top-gates [11]. In this case, local gates over the metal/CNT junction are used to electrostatically thin the

Schottky barriers and reduce the S value closer to that of the bulk switching device [11] (see Fig. 9.10).

More recently, different groups of researchers have identified material combinations that decrease the barrier height and improve the on-state performance of CNT-FETs. For example, by using larger diameter (smaller bandgap) CNTs, in combination with the metal of the right workfunction one can hope to reduce the barrier for one charge carrier type. This lowering and effective removal of the barrier is seen for CNTs with diameter, $d_{CNT} > 2\,\text{nm}$ and Pd electrodes. In that case, nearly ohmic p-type CNT-FETs are produced, while the barrier still remains for the same devices when operated as n-type transistors [12]. However, even the p-type CNT-FETs with Pd electrodes show significant barriers if d_{CNT} is well below 2 nm. Other groups have also reported nearly ohmic contacts for similar or larger d_{CNT} nanotubes and Au/Cr or pure Au electrode combinations [13,40]. In addition, the effect of the Schottky barriers can be reduced by studying very long (in excess of 100 μm) channel CNT-FETs, where the bulk nanotube resistance is mainly responsible for the device performance [40]. This and other dependences of the device performance on the material parameters of the CNT will be discussed in the next section.

9.6 The Role of Nanotube Diameter and Gate Oxide Thickness

The electronic properties of CNTs are strongly dependent upon the tube diameter, d_{CNT}, and chirality (given by the chiral indices m and n), as described earlier in this chapter. The CNT bandgap, E_{gap}, is inversely proportional to d_{CNT}, as is the effective mass, m^*, for electrons and holes. Thus, at a given temperature, large diameter CNTs will have a larger free carrier concentration than smaller diameter nanotubes, and they will have a lower effective mass. In addition, because of the smaller E_{gap} of large diameter CNTs, the band lineup at the metal/nanotube interface will likely result in lower Schottky barriers at the source and drain. Tunneling through these barriers will be easier as well because of the smaller m^*. As a result, CNT-FETs comprising large diameter tubes should be capable of higher I_{ON} than those with smaller d_{CNT}. This is borne out in the literature, as CNT-FETs carrying large on currents are generally fabricated with tubes grown by CVD [12, 41], which typically have larger diameters ($\sim$ 2–4 nm) than those grown by laser ablation [42] or HiPCO [43] techniques ($\sim$ 0.8–1.5 nm). Certainly, a high on-state current is a most desirable property and is a requirement for many device applications, however, a low off-state current is just as important, particularly for low power applications. I_{ON}/I_{OFF} ratios of at least 10^4–10^5 are the norm for advance silicon technology [44]. In the case of CNT-FETs, those factors that tend to lead to higher on-state currents (smaller E_{gap}, lower Schottky barriers, small m^*) also make it difficult to achieve acceptably low I_{OFF}.

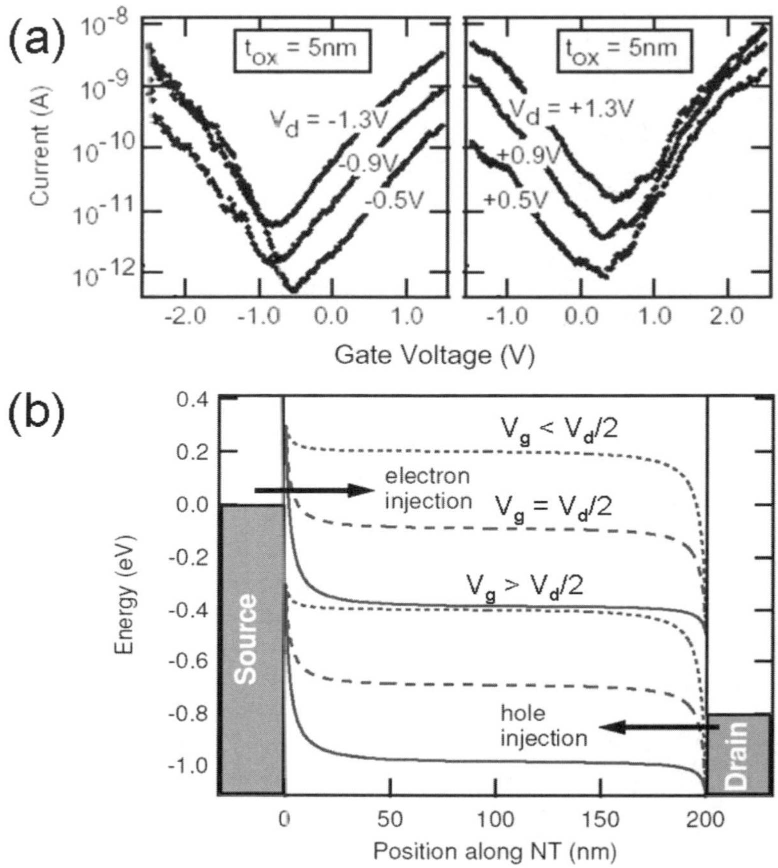

Fig. 9.6. (**a**) Transfer characteristics of ambipolar CNT-FET with ultrathin gate oxide. (**b**) Band bending situation for (a) (from [73])

This problem is exacerbated for the case of extremely scaled gate insulators. Reducing the gate dielectric thickness while maintaining a constant V_{gs} results in thinning of the Schottky barriers at both the source and the drain electrodes, increasing the probability of drain current leakage. Figure 9.6 illustrates this effect for an ambipolar CNT-FET with ultrathin gate oxide (the transfer characteristics are shown in Fig. 9.6a, while the theoretical band bending situation [45] is shown in Fig. 9.6b.) This drain leakage current increases exponentially with increasing drain voltage, resulting in a reduced I_{ON}/I_{OFF} ratio. It may be possible to alleviate this problem by designing an asymmetric CNT-FET in which the gate field at the source and drain are different, resulting in different injections rates for electrons and holes. Such an asymmetric device, with a narrow trench designed below the CNT in the

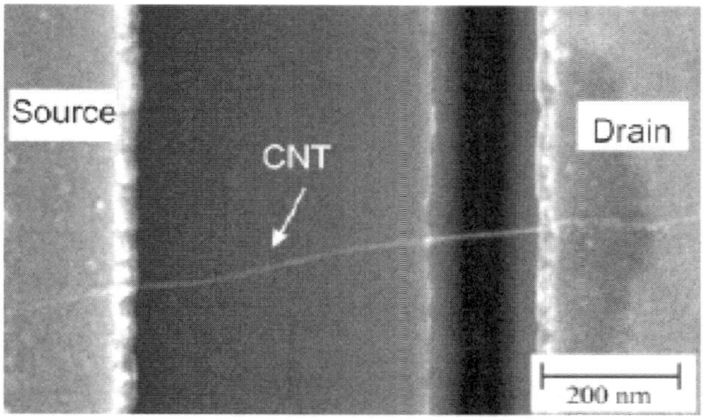

Fig. 9.7. CNT-FET with asymmetric trenches near the drain designed to decouple Schottky barriers effects from the CNT bulk (from [46])

vicinity of the drain electrode, is shown in Fig. 9.7 [46]. Incorporating such a feature in a CNT-FET may be an undesirable complication, and it may be unfeasible for very thin gate dielectrics. Thus, practical considerations dictate that the appropriate choice of CNT diameter and gate dielectric thickness be a trade-off between achieving a high I_{ON} at the highest tolerable I_{OFF}.

9.7 Environmental Influences on the Performance of CNT-FETs

The effect of ambient air on the performance and functionality of CNT-FETs can be also understood within the framework of the Schottky barrier model of conduction. In particular, this model helps clarify and separate the effects due to the bulk of the nanotube channel from those arising from the effects at the contact between the metal electrode and the nanotube.

As has already been discussed in the introduction, the original back-gated CNT-FETs operating in air function as p-type transistors without any explicit doping of the CNT [16, 17]. Similar CNT devices operated in two-terminal configuration (without the backgate) showed an increase in the electrical resistance and sign change of the Seebeck coefficient upon removal of air [30]. This is not surprising since in most cases nanotubes in these devices are open to air and no effort is made to protect them from the environmental influences. Further study of specific gases identified that most of this effect was associated with oxygen, and thus a proposal was made that the effect is due to electron transfer between CNTs and atmospheric O_2; in particular that CNTs are hole doped in the ambient and n-doped or intrinsic in vacuum.

These results were extended by investigation of the impact of gases on three-terminal CNT-FETs. As shown in Fig. 9.8a, a CNT-FET which has

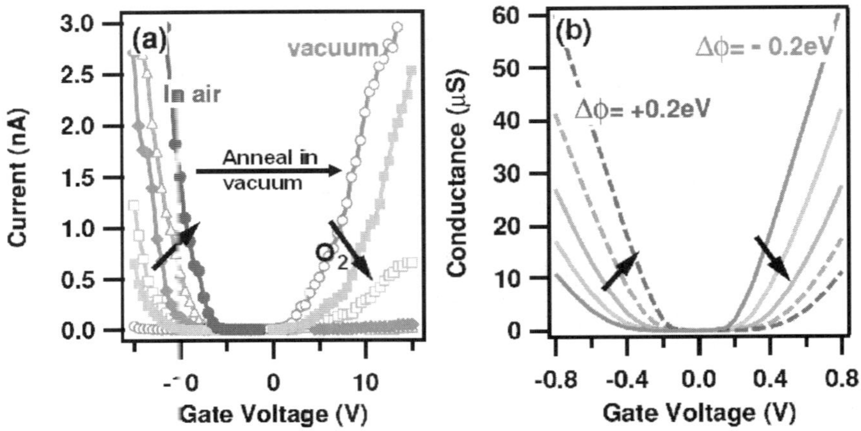

Fig. 9.8. Effect of environment on carbon nanotube FETs. (**a**) A Au-contacted p-FET is converted to n-FET by annealing in vacuum. Slow reintroduction of oxygen, as denoted by arrows, simultaneously suppresses electron conduction at positive V_g and increases the hole conduction at negative V_g without changing the threshold voltage of the transistor. (**b**) Simulations assuming only the change in metal workfunction with change in oxygen pressure show qualitative agreement with data (from [25, 38])

been annealed in vacuum behaves as an n-type transistor [25]. The slow re-introduction of dry O_2 shows that the original, p-type character can be recovered at atmospheric pressure with intermediate stages of ambipolar conduction also readily observable. This behavior is consistent with the previous reports that p-type character of CNT-FETs is a result of the interaction with O_2, but it also provides additional information which together with other experiments [47], and theoretical modeling [38] helps clarify the nature of this interaction. In particular, it is apparent that this treatment has no effect on the threshold voltage, which shifts quite significantly when nanotubes are doped by potassium, which is a well-established dopant.

Thus, the joint conclusion of these studies is that the effect of O_2, rather than doping is that it adsorbs at the metal/CNT junction and modifies the Schottky barriers that control the switching of the CNT-FET. As we have already discussed earlier in this chapter, the bonding of a CNT to a metal electrode results in charge transfer and creation of the Schottky barrier at the CNT/metal interface. The actual band line-up and charge-transfer may be strongly affected by the co-adsorption of other species such as gaseous oxygen at the junction. The role of the adsorbates is to locally change the surface potential (local work function), or directly interact with the metal/CNT interface [38, 47]. Modeling results in Fig. 9.8b of the surface potential change with oxygen presence result in curves without any threshold shift as seen in the experiment [38]. In fact, direct observation of this modification has been

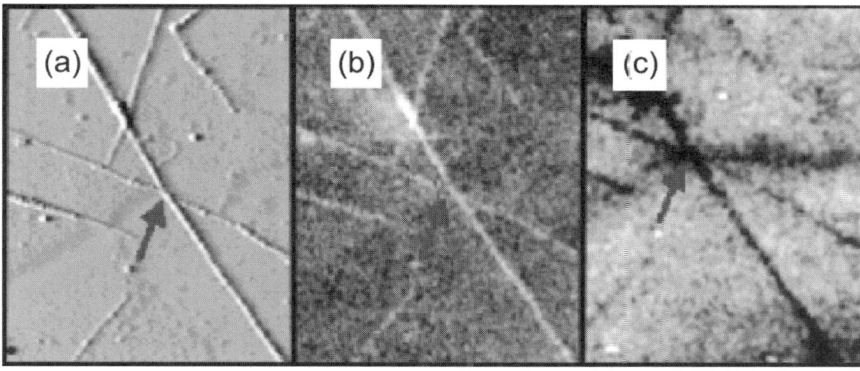

Fig. 9.9. In situ Scanning Kelvin probe microscopy (SKPM) of CNTs on gold in controlled environment. (**a**) Atomic force microscope image of CNTs on flame annealed (111) Au. (**b**)–(**c**) SKPM of the same area in air and ultra-high vacuum (UHV), respectively. Nanotubes appear brighter which implies a lower vacuum level than gold in air and vice versa in UHV. Arrows indicate same position in all three images (from [47])

achieved by scanning probe techniques [47]. Figure 9.9 shows scanning Kelvin probe images of a few CNTs dispersed on a gold surface in the ambient (b) and following a mild anneal in vacuum (c). (The contrast between CNTs and the Au substrate changes as the sample is cycled between air and vacuum indicating a reversal in the charge transfer direction between CNT and Au, which in turn facilitates injection of holes and electrons, respectively into a CNT device.) These findings may also have technological implications, as the sensitivity of the Schottky barriers to other co-adsorbed species may be applicable for engineering of the CNT-FET characteristics [47].

9.8 Scaling of CNT-FETs

The incredible growth of the silicon microelectronics industry has been based upon the ability to scale transistor features to smaller and smaller dimensions. Transistor scaling rules are well understood and are based upon straighforward electrostatic considerations [34]. Because the CNT-FET geometry differs from that of a planar silicon FET, it would be expected to follow different scaling rules. For example, in the case of a planar Si MOSFET, the current is inversely proportional to the gate dielectric thickness. However, the geometry of a CNT-FET is quite different; it corresponds more closely to that of a small conducting cylinder and a large planar electrode (the gate). In this case, for bulk switching, simple electrostatics would dictate a weak inverse logarithmic dependence of the current on insulator thickness [20]. Furthermore, as mentioned above, the basic switching in most CNT-FETs measured

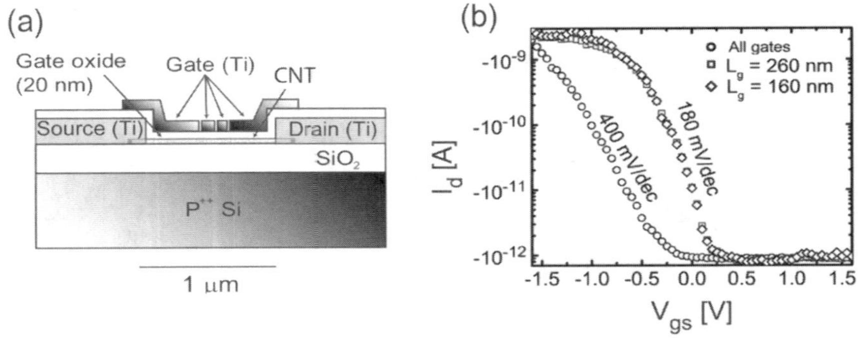

Fig. 9.10. Schematic cross-section of segmented top-gate CNT-FET. (**b**) Transfer characteristics showing the difference between Schottky barrier modulation achieved by operating all gate segments together (*open circles*) and bulk switching achieved by independent operation of different combinations of interior gate segments (*open squares* and *diamonds*)

to date involves the gate field-induced modulation of Schottky barriers at the source and drain electrodes. This mechanism is sensitive to the precise device geometry, i.e., CNT-FET structures which produce a high electric field at the metal/nanotube junction produce the most effective switching [48]. Heinze et al. [48] found unexpected geometry-dependent vertical scaling laws for the subthreshold slope and transconductance for back-gated Schottky barrier CNT-FETs with variable gate oxide thickness. The lateral scaling of CNT-FETs is affected by the fact that at low excitations, the elastic mean free path, λ_{el}, is relatively long, $\gtrsim 1\,\mu\text{m}$. Thus, for relevant device channel lengths, transport is essentially ballistic. Using a segmented gate device shown in Fig. 9.10a to decouple switching in the CNT bulk from Schottky barrier switching, Wind et al. [11] observed length-independent transport in the bulk regime (Fig. 9.10b), providing evidence for ballistic transport in CNT-FETs. Javey et al. [12] also reported evidence for ballistic transport based on measured on-state conductance approaching G_0. Thus, for channel lengths of a few hundred nanometers or less, length scaling does not apply to CNT-FETs, at least for *dc* operation (*ac* transport will be affected by time-of-flight effects). For much longer nanotubes, scattering is indeed evident [40], although these devices still exhibit unusually high mobilities ($> 100,000\,\text{cm}^2/\text{V-sec}$).

9.9 Prototype Carbon Nanotube Circuits

The promising characteristics of individual CNT-FETs have lead to initial attempts at integration of these devices into useful structures of several CNT-

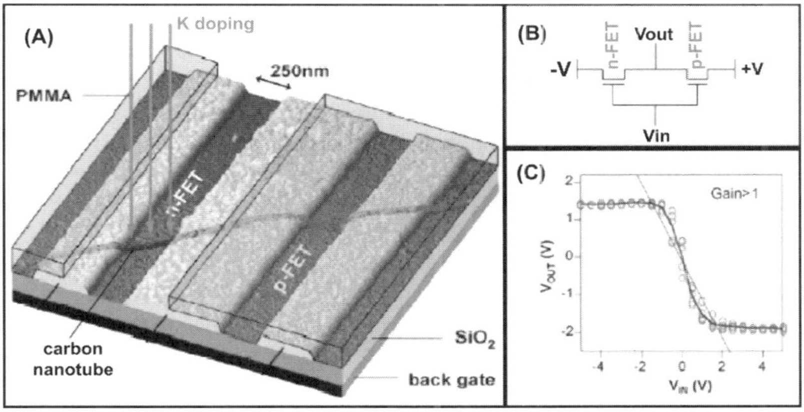

Fig. 9.11. Logical NOT gate fabricated on a individual CNT-FET. (**a**) Sample geometry showing an individual semiconducting nanotube, laying across three electrodes. One of the two FETs is protected by PMMA polymer and remains p-type. The other is exposed to potassium doping and is converted to n-type. (**b**) Schematic of the circuit for the voltage inverter. Common backgate is used as V_{in}, common contact is used as V_{out}, while other two contacts have $\pm V$ applied. (**c**) Inverter characteristics showing gain larger than 1 (*solid black line*) (from [23])

FETs that can perform a logic operation, function as memories [49] or sensors [50]. In the following, we limit our discussion to advances in logic circuitry.

The nanotube logic gates have been, in most cases, based on a complementary technology analogous to silicon CMOS, which is important as it may ease integration of CNTs onto this well established technology. The first complementary (CMOS-like) logic gates were reported by Derycke et al. [23] In this work, two different techniques were used to produce n-type devices for complementary logic – resulting in multiple transistor logic gates fabricated from adjoining nanotubes, or on the same tube. Inter-nanotube voltage inverter ("NOT" gate) is created by combining two CNT-FETs: a p-type device in the ambient and a vacuum annealed n-type transistor. A more compact and integrated approach uses potassium doping to convert one of two CNT-FETs on the same nanotube to n-type. The masking of the other transistor which remains p-type is accomplished by photoresist. The completed device, an intra-nanotube CMOS inverter is shown in Fig. 9.11. The circuit shown has a voltage gain of about 2, suggesting that integration, without signal degradation, of many devices along a single nanotube can be accomplished. It is important to note that these initial and indeed all other logic circuits to date have been built using unoptimized CNT-FETs, which suggests that their performance will only improve with further advances in design and fabrication of individual devices as well as development of novel integration schemes.

Shortly thereafter, Bachtold et al. [51] used p-type CNT-FETs along with resistors to build prototype logic gates based on an older transistor-resistor scheme. They went a step further in complexity and wired three such inverter gates to form a ring oscillator. Unfortunately, the large parasitics severely degraded the performance of the circuit which oscillated at only about 5 Hz. Later Javey and co-workers [52] used another scheme for converting into n-type CNT-FETs to wire up CMOS inverters with gains in excess of 10 and CMOS ring oscillators with frequencies in the 100 Hz range.

However, these reported frequencies are well below the expected *ac* response of CNT-FETs which is difficult to assess directly because of the relatively small current signals in these devices. Most recently, the non-linear current-voltage characteristics of CNT-FETs were used to demonstrate that CNT-FETs *dc* characteristics are not affected by *ac* fields at least up to 0.5 GHz [53].

9.10 Optoelectronic Properties of Carbon Nanotubes

Semiconducting nanotubes, unlike silicon, have a direct band-gap in momentum space [1]. As a result, no momentum transfer is needed to induce interband transitions and direct light absorption and emission is possible. At the same time, the 1D confinement of the electron (e) and hole (h) is expected to lead to a strong e-h interaction leading to the formation of 1D excitons [5, 54–57]. Early optical absorption studies of CNTs involved bulk mixtures of metallic and semiconducting CNTs and showed broad spectral features [58, 59]. It was later found that CNTs can be dispersed effectively in a liquid medium using surfactants. Semiconducting CNTs encapsulated inside micelles showed relatively narrow absorption and photoluminescence excitation and emission spectra [60–62]. In addition to photoluminescence from CNTs in micelles in the liquid state, laser-excited photoluminescence from CNTs in micelles deposited on solid substrates [63], as well as spectra from suspended CNTs have been reported [64]. As an analytical tool, the photoluminescence studies of dispersed CNTs have proven to be extremely valuable because, in conjunction with Raman experiments, they permit the identification of the (n, m) indices of the CNTs in a given sample.

Most experimental papers have discussed the excitations of CNTs in terms of band-to-band transitions, i.e. transitions between Van Hove singularities, and have used tight binding theory (TBT) to interpret the spectra. However, a single-particle description of the excited CNT states neglects electron–electron and electron–hole interactions, which should be particularly strong in a 1D system [5,54]. The latter interaction is expected to bind the photoexcited electron and hole to form Wannier-like excitons [5,54–57]. Indeed, first-principles electronic structure calculations that are based on a many-body Green's function approach in which both the quasiparticle (single-particle) excitation spectrum and the optical (electron-hole excitation) spectrum were

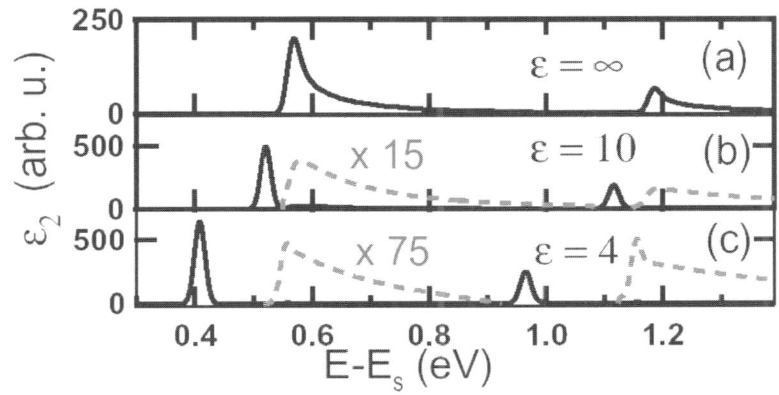

Fig. 9.12. Computed absorption spectra ε_2 of a $(19, 0)$ carbon nanotube in different dielectrics. (**a**) $\varepsilon = \infty$, equivalent to no electron-hole interaction. (**b**) $\varepsilon = 10$, and $\varepsilon = 4$. E_s is the unknown self-energy shift. (from [57])

computed, predict the formation of strongly bound CNT excitons. In fact, exciton formation with a binding energy approaching 100 meV was predicted even in the case of the small metallic $(3, 3)$ CNT [56]. For the semiconducting $(8, 0)$ CNT a binding energy of ~ 1 eV was obtained. The LDA value for the band-gap of this CNT was 0.6 eV at the Γ-point, but after quasiparticle corrections opened up to 1.75 eV [56]. While expected to be quite accurate such calculations are very laborious and thus are limited to a few small systems. For this reason Perebeinos et al. [57] recently performed approximate calculations based on TB wavefunctions, but included both the direct and exchange e–h interactions using the Bethe-Salpeter equation [65]. These calculations allowed them to derive general scaling rules for the properties of CNT excitons as a function of their diameter/chirality and the dielectric constant of the environment. It was found that the exciton binding energy, E_b, scales with the CNT radius R, effective mass m, and dielectric constant of the environment, ε, as [57]:

$$E_b = A_b (R/a_B)^{\alpha-2} \cdot (m/m_e)^{\alpha-1} \varepsilon^{-\alpha}. \tag{9.1}$$

In (9.1), a_B is the Bohr radius, m_e is the free electron mass, while A_b and α are fitting parameters, with $A_b = 24.1$ eV and $\alpha = 1.4$. Most importantly, they found a strong transfer of intensity from the interband transition to the exciton. When the binding is large, all absorption intensity is concentrated in the exciton (see Fig. 9.12) [57]. Recent time-resolved studies of the excited states of CNTs support the exciton picture on the basis of the observation of strong transitions where forbidden interband transitions are expected, and on the measurement of polarization memory in photo-induced absorption and photo-bleaching bands that lasts on a relatively long (ps) time scale [66].

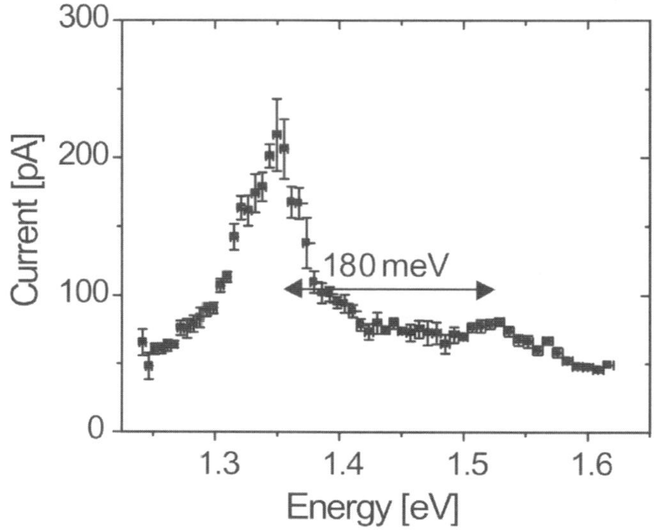

Fig. 9.13. Photoconductivity spectrum of a single semiconducting nanotube. The main peak at $\sim 1.35\,\mathrm{eV}$ corresponds to the zero-phonon line of the second allowed electronic transition of the nanotube, while the side peak at $\sim 180\,\mathrm{meV}$ higher energy involves the simultaneous excitation of one quantum of a C-C stretch optical phonon. (From [67])

Other evidence comes from the observation of vibronic side bands in the photoconductivity spectra of single CNTs (see discussion below) [67]. The lineshapes and the relative intensities of the zero-phonon line and vibronic side-band can be accounted for only within the excitonic picture [57].

For optoelectronic applications, solid-state, electrically driven CNT light-emitting devices and light detectors are desirable. Photogenerated currents from CNTs have been reported by several groups [68–70]. Early studies involved bundles of semiconducting and metallic nanotubes. Recently, however, photocurrents and photovoltage from individual CNTs in ambipolar and unipolar CNT-FETs were also reported [71]. As shown in Fig. 9.13, scanning the laser in the range of the second exciton state of a laser-ablation CNT produces a well-defined peak in the photocurrent spectra at $1.35\,\mathrm{eV}$ and a side band $\sim 180\,\mathrm{meV}$ higher energy. The peak corresponds to the energy of the zero phonon line of the second exciton, while the side band is due to a C–C stretching optical mode of the CNT [1]. The estimated quantum yield of photon to e–h pair conversion in the ambipolar CNT-FETs is $\sim 10\%$ [71]. Thus, CNT-FETs can act as efficient photodetectors with the added capability of polarization detection.

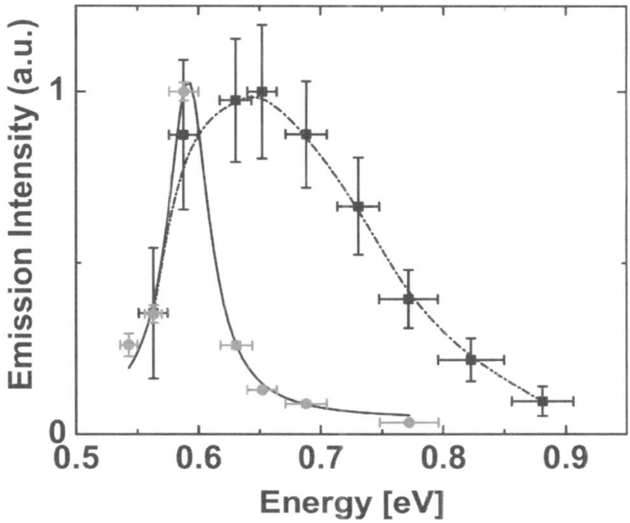

Fig. 9.14. Electroluminescence (i.e. emission due to radiative e–h recombination) spectra of a short ($\sim$ 300 nm) and a long ($\sim$ 50 μm) nanotube incorporated as channels of field-effect transistors. (From [71])

Ambipolar CNT-FETs can be used to produce a single gate-controlled light source [72]. In the case of Schottky barrier ambipolar devices with a symmetric source and drain structure, it was shown that about equal currents of electrons and holes are injected from opposite ends of the CNT when $V_g = V_d/2$ [73]. The electrons and holes are confined by the 1D CNT structure; where they meet, a fraction of them, determined by momentum and spin selection rules and other factors, may radiatively recombine. Indeed, IR emission was observed from ambipolar CNT-FETs [72]. This emission was polarized along the CNT axis and, most importantly, the emission intensity was found, as predicted, to be the highest when $V_g = V_d/2$. The peaks of the electroluminescence emission spectra [74], were in accord with the expected band-gaps of the CNTs used in the experiment. However, the electroluminescence peak shape was found to depend on the length of the CNT channel. Long CNTs ($\gtrsim$ 5 μm) gave narrow, nearly symmetric peaks; an example is shown in Fig. 9.14. Short CNTs ($\sim$ 200–300 nm), on the other hand, produced broad asymmetric spectra [74]. These observations were interpreted in terms of a complete or a partial relaxation of the hot carriers during their residence time in the long, or short CNTs channels, respectively. From the broad lineshapes the energy distributions of the carriers were deduced and interpreted in terms of fast relaxation through coupling with optical phonons followed by a slower relaxation via the acoustic phonons. Very recent spatially-resolved studies of the light emission from CNT-FETs involving very long (e.g., $\sim$ 50 μm) CNTs, as a function of the applied gate and drain

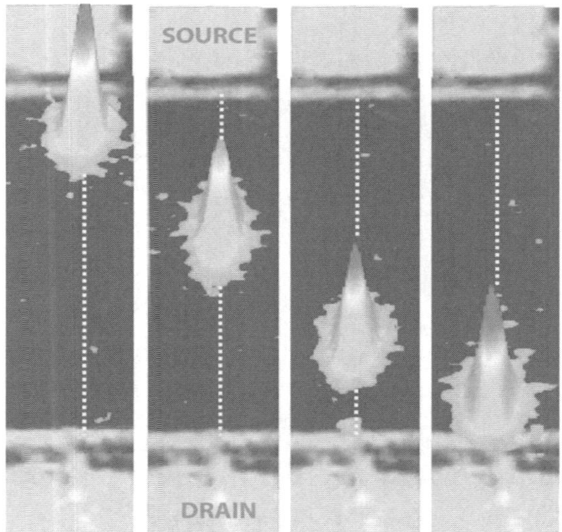

Fig. 9.15. Electroluminescence from a long ($\sim 50\,\mu m$) carbon nanotube field effect transistor. The emission spot can be translated along the axis of the nanotube by varying the gate bias under constant current ($18\,\mu A$) conditions. (From [75]) (see Color Plates, p. 343)

biases, have provided unique insights on electrical transport processes in a CNT [75]. Unlike the case of LEDs where chemical doping defines the location of the light emission, the ambipolar CNTs are undoped and the position of the light emission is controlled by the gate as shown in Fig. 9.15. The emitting spot gives the location where the electron- and hole-current meet inside the CNT, while its shape allows the determination of the e–h recombination lengths. Local sources of e–h pair creation at high fields such as Zener tunneling and hysteresis effects can also be visualized [75].

9.11 Summary

In this chapter, we have discussed the properties of CNTs that make them ideally suited for electronic and optoelectronic applications. Already CNT-FET transistors that outperform their silicon counterparts have been demonstrated. Moreover, the CNT-FET, depending on the biasing conditions, can not only be as an electrical switch, but also as a novel light source and a light detector. Finally, CNTs provide an ideal model system to study the electrical transport properties of 1D nanostructures and molecules.

Acknowledgement

SJW acknowledges the support of the Nanoscale Science and Engineering Initiative of the National Science Foundation under NSF Award Number CHE-0117752 and by the New York State Office of Science, Technology, and Academic Research (NYSTAR).

References

1. *Carbon Nanotubes*, M. Dresselhaus, G. Dresselhaus, and Ph. Avouris, Eds., Springer-Verlag (Berlin, 2001)
2. M.S. Dresselhaus, G. Dresselhaus, and R. Saito, Phys. Rev. B **45,** 6234 (1992)
3. J.W. Mintmire, B.I. Dunlap, and C.T. White, Phys. Rev. Lett. **68**, 631 (1992)
4. R. Landauer, Philos. Mag. **21**, 863 (1970)
5. T. Ando, J. Phys. Soc. Japan **66**, 1066 (1996)
6. D. Mann, A. Javey, J. Kong, Q. Wang, and H. Dai, Nano Lett. **3**, 1541 (2003)
7. Z. Yao, C.L. Kane, and C. Dekker, Phys. Rev. Lett., **61**, 2941 (2000)
8. J.-Y. Park, S. Rosenblatt, Y. Yaish, V. Sazonova, H. Ustunel, S. Braig, T.A. Arias, D.W. Brouwer, and P.L. McEuen, Nano Lett. **4**, 517 (2004)
9. A. Javey, J. Guo, M. Paulson, Q. Wang, D. Mann, M. Lundstrom and H. Dai, Phys. Rev. Lett. **92**, 106804-1 (2004)
10. P.G. Collins, M. Hersam, M. Arnold, R. Martel, and Ph. Avouris, Phys. Rev. Lett. **86**, 3128 (2001)
11. S.J. Wind, J. Appenzeller, and Ph. Avouris, Phys. Rev. Lett. **91**, 058301 (2003)
12. A. Javey, J. Guo, Q. Wang, M. Lundstrom, and H.J. Dai, Nature **424**, 654 (2003)
13. Y. Yaish, J.Y. Park, S. Rosenblatt, V. Sazonova, M. Brink, and P.L. McEuen, Phys. Rev. Lett. **92**, 046401 (2004)
14. F. Leonard, and J. Tersoff, Phys. Rev. Lett. **83**, 5174 (1999)
15. Ph. Avouris, J. Appenzeller, R. Martel, and S.J. Wind, Proc. of the IEEE **91**, 1772 (2003)
16. S.J. Tans, A.R.M. Verschueren, and C. Dekker, Nature **393**, 49 (1998)
17. R. Martel, T. Schmidt, H.R. Shea, T. Hertel, and Ph. Avouris, Appl. Phys. Lett. **73**, 2447 (1998)
18. H.T. Soh, C.F. Quate, A.F. Morpurgo, C. marcus, J. Kong, and H. Dai, App. Phys. Lett. **75**, 627 (1999)
19. R. Martel, V. Derycke, C. Lavoie, J. Appenzeller, K.K. Chan, J. Tersoff, and Ph. Avouris, Phys. Rev. Lett. **87**, 256805 (2001)
20. R. Martel, H.-S.P. Wong, K. Chan, and Ph. Avouris, in *International Electron Devices Meeting 2001: IEDM Technical Digest* (IEEE, Piscataway, N.J., 2001), p. 159.
21. P.L. McEuen, M.S. Fuhrer and H. Park, IEEE Trans. Nanotechnol. **1**, 78 (2002), and references therein.
22. M. Bockrath, J. Hone, A. Zettl, P.L. McEuen, A.G. Rinzler, and R.E. Smalley, Phys. Rev. B **61**, R10606 (2000)
23. V. Derycke, R. Martel, J. Appenzeller, and Ph. Avouris, Nano Lett. **1**, 453 (2001)

24. X. Liu, C. Lee, C. Zhou, and J. Han, Appl. Phys. Lett. **79**, 3329 (2001)
25. V. Derycke, R. Martel, J. Appenzeller, and Ph. Avouris, Appl. Phys. Lett. **80**, 2773 (2002)
26. S.J. Wind, J. Appenzeller, R. Martel, V. Derycke, and Ph. Avouris, Appl. Phys. Lett. **80**, 3817 (2002)
27. J. Appenzeller, J. Knoch, R. Martel, V. Derycke, S.J. Wind, and Ph. Avouris, IEEE Trans. Nanotechnol. **1**, 184 (2002)
28. A. Javey, H. Kim, M. Brink, Q. Wang et al. Nature Materials **1**, 241 (2002)
29. M. Radosavljevic, J. Appenzeller, Ph. Avouris, and J. Knoch, Appl. Phys. Lett. **84**, 3693 (2004)
30. P.G. Collins, M.S. Arnold, and Ph. Avouris, Science **292**, 706 (2001)
31. Y. -C. Tseng, P. Xuan, A. Javey, R. Malloy, Q. Wang, J. Bokor, H. Dai, Nano Lett. **4**,123 (2004)
32. E.S. Snow, J.P. Novak, P.M. Campbell, and D. Park, Appl. Phys. Lett. **82**, 2145 (2003)
33. A. Rochefort, M. Di Ventra, and Ph. Avouris, Appl. Phys. Lett. **78**, 2521 (2001)
34. S.M. Sze, Physics of Semiconductor Devices, 2nd edition (John Wiley & Sons, 1981)
35. A. Javey, Q. Wang, W. Kim, and H.Dai, in *International Electron Devices Meeting 2003: IEDM Technical Digest* (IEEE, Piscataway, N.J., 2003), p. 741.
36. J. Appenzeller, M. Radosavljevic, J. Knoch, and Ph. Avouris, Phys. Rev. Lett. **92**, 048301 (2004)
37. J. Appenzeller, J. Knoch, V. Derycke, R. Martel, S.J. Wind, and Ph. Avouris, Phys. Rev. Lett. **89**, 126801 (2002)
38. S. Heinze, J. Tersoff, R. Martel, V. Derycke, J. Appenzeller, and Ph. Avouris, Phys. Rev. Lett. **89**, 106801 (2002)
39. M. Freitag, M. Radosavljevic, Y.X. Zhou, A.T. Johnson, and W.F. Smith, Appl. Phys. Lett. **79**, 3326 (2001)
40. T. Durkop, S.A. Getty, E. Cobas, and M.S. Fuhrer, Nano Lett. **4**, 35 (2004)
41. S. Rosenblatt, Y. Yaish, J. Park, J. Gore,V. Sazanova, and P.L. McEuen, Nano Lett. **2**, 869 (2002)
42. A. Thess, R. Lee, P. Nikolaev. H. Dai, P. Petit, J. Robert, X. Chunhui, L.Y. Hee, K. Seong Gon, A.G. Rinzler, and D.T. Colbert, Science **273**, 483 (1996)
43. P. Nikolaev, M.J. Bronikowski, R.K. Bradley, F. Rohmund, D.T. Colbert, K.A. Smith, and R.E. Smalley, Chem. Phys. Lett. **313, 91** (1999)
44. International Technology Roadmap for Semiconductors, 2003 edition. (See also http://public.itrs.net)
45. S. Heinze, J. Tersoff and Ph. Avouris, Appl. Phys. Lett. **83**, 5038 (2003)
46. Y.-M. Lin, J. Appenzeller and Ph. Avouris, Nano Lett. **4**, 947 (2004)
47. X.D. Cui, M. Freitag, R. Martel, L. Brus, and Ph. Avouris, Nano Lett. **3**, 783 (2003)
48. S. Heinze, M. Radosavljevic, J. Tersoff and Ph. Avouris, Phys. Rev. B **68**, 235418 (2003)
49. T. Rueckes, K. Kim, E. Joselevich, G.Y. Tseng, C.L. Cheung, and C.M. Lieber, Science **289**, 94 (2000)
50. J. Kong, N.R. Franklin, C.W. Zhou, M.G. Chapline, S. Peng, K.J. Cho, and H.J. Dai, Science, **287**, 622 (2000)
51. A. Bachtold, P. Hadley, T. Nakanishi, and C. Dekker, Science **294**, 1317 (2001)

52. A. Javey, Q. Wang, A. Ural, Y.M. Li, and H.J. Dai, Nano Lett. **2**, 929 (2002)
53. D.J. Frank, and J. Appenzeller, IEEE Electr. Device Lett. **25**, 34 (2004)
54. T.G. Pedersen, Phys. Rev. B **67**, 073401 (2003)
55. C.L. Kane and E.J. Mele, Phys. Rev. Lett. **90**, 207401 (2003)
56. C.D. Spataru, S. Ismail-Beigi, L.X. Benedict, and S.G. Louie, Phys. Rev. Lett. (2004)
57. V. Perebeinos, J. Tersoff, and Ph. Avouris, Phys. Rev. Lett. **92**, 257402 (2004)
58. M. Ishida, S. Mizuno, T. Yoshihino, Y. Saito, and A. Nakamura, J. Phys. Soc. Jpn. **68**, 3131 (1999)
59. R. Saito and H. Kataura in M. Dresslahus, G. Dresselhaus and Ph. Avouris, Eds., Carbon Nanotubes, Springer-Verlag (Berlin, 2001), pages 213–246.
60. M.J. O'Connell, S.M. Bachilo, C.B. Huffman, V.C. Moore, M.S. Strano, E.H. Haroz, K.L. Rialon, P.J. Boul, W.H. Noon, C. Kittrell, J. Ma, R.H. Hauge, R.E. Smalley, and R.B. Weisman, Science **297**, 2361 (2002)
61. S.M. Bachilo, M.S. Strano, C. Kittrell, R.H. Hauge, R.E. Smalley, and R.B. Weisman, Science **298**, 2361 (2002)
62. A. Hagen and T. Hertel, Nano Lett. **3**, 383 (2003)
63. A. Hartschuh, H.N. Pedrosa, L. Novotny and T.D. Krauss, Science **301**, 1354 (2003)
64. J. Lefebvre, Y. Homma, and P. Finnie, Phys. Rev. Lett. **90**, 217401 (2003)
65. G. Strinati, Phys. Rev. B **29**, 5718 (1984)
66. O.J. Korovyanlo, C.-X Sheng, Z.V. Vardeny, A.B. Dalton, and R.H. Baughman,. Phys. Rev. Lett. **92**, 017403 (2004)
67. X. Qiu, M. Freitag, V. Perebeinos, and Ph. Avouris, Nano Lett. **5**, 749 (2005)
68. A. Fujuwara, Y. Matsuoka, H. Suematsu, N. Ogata, et al. Jpn. J. Appl. Phys., Part 1 **40**, L1229 (2001)
69. Y. Yamada, N. Naka, N. Nagasawa, Z.M. Li, and Z.K. Tang, Physica B **323**, 239 (2002)
70. Y. Zhang and S. Iijima, Phys. Rev. Lett. **82**, 3472 (1999)
71. M. Freitag, Y. Martin, J.A. Misewich, R. Martel, and Ph. Avouris, Nano Lett. **3**, 1067 (2003)
72. J.A. Misewich, R. Martel, Ph. Avouris, J.C. Tsang, S. Heinze, and J. Tersoff, Science **300**, 783 (2003)
73. M. Radosavljevic, S. Heinze, J. Tersoff and Ph. Avouris, Appl. Phys. Lett. **83**, 2435 (2003)
74. M. Freitag, J. Chen, A. Stein, T. Tsang, J. Misewich, R. Martel, V. Perebeinos, and Ph. Avouris, Nano Lett. **4**, 1063 (2004)
75. M. Freitag, J. Chen, J. Tsang, Q. Fu, J. Liu and Ph. Avouris, Phys. Rev. Lett. **93**, 076803 (2004)

10 Carbon Nanotube–Biomolecule Interactions: Applications in Carbon Nanotube Separation and Biosensing

A. Jagota, B.A. Diner, S. Boussaad, and M. Zheng

We describe a DNA-carbon nanotube hybrid material that has proven useful for dispersion and structure-based separation. Single-stranded DNA binds strongly to carbon nanotubes, rendering them dispersable in water as charged colloidal particles. These can be subjected to separation techniques. Certain sequences form hybrids that allow separation on the basis of the properties of the core nanotube material. Carbon nanotubes dispersed by a non-ionic surfactant can also be separated by an alternative technique. Molecular models suggest that the hybrid structure consists of helical wrapping of the DNA around the nanotube, in good agreement with AFM data. We propose that the separation mechanism relies on modulation of the electric field of the DNA phosphate charge by interactions with the nanotube core. Finally, we show an example of the use of carbon nanotube-biomolecule interactions for detection using a CNT-based field effect transistor.

10.1 Introduction

The interaction of carbon nanotubes with biological molecules is of significant interest for at least two reasons. Firstly, as shown by examples in this chapter, biological molecules can and have been harnessed to solve difficult problems in the manipulation and sorting of carbon nanotubes. Secondly, understanding interaction between the two is critical for the development of carbon nanotube-based sensor devices for biological molecules. This chapter focuses on our studies of carbon nanotube interaction with biomolecules. In the first part, we present a summary of the development of a DNA-nanotube hybrid material, and the use of the hybrid in structure-based carbon nanotube separation. The second part describes an alternative separation method for carbon nanotubes dispersed in non-ionic surfactant, providing a contrast to the DNA-based separation method. In the third part, we present an analysis of DNA-nanotube hybrid structures and the mechanism of DNA-based nanotube separation. The last part explores the use of carbon nanotube-based field-effect transistors in sensing charged proteins adsorbed onto the side wall of nanotubes.

10.2 DNA-Assisted Dispersion and Separation of Carbon Nanotubes

DNA is a natural-occurring polymer that plays a central role in biology. Many unique properties of DNA have inspired a search for its non-biological applications. Molecular recognition between complementary strands of double-stranded DNA has been used to construct various geometric objects at the nanometer scale [1], and to organize the assembly of colloidal particles [2,3]. The π-stacking interaction between bases in DNA has prompted the exploration of its electronic properties for possible use in molecular electronics [4]. Less utilized are the potential inorganic substrate-binding properties of DNA, in contrast to recently demonstrated efficacy of oligopeptides for this purpose [5,7].

A large molecular library can be formed by single-stranded DNA (ssDNA), which offers an intriguing possibility for carbon nanotube binding: depending on its sequence and structure, aromatic nucleotide bases in ssDNA may be exposed to form π-stacking interactions with the side-wall of carbon nanotubes. To search for DNA sequences that bind carbon nanotubes, we followed a well-established in vitro evolution procedure [8]. We discovered that DNA binding to carbon nanotubes is extremely effective and facile. Specific conditions for dispersion depend on the source of carbon nanotubes. For purified HiPco nanotubes, we found that practically any ssDNA would work in the presence of a denaturant, with mild sonication. The primary role of denaturant appears to be to disrupt base-pairing, which can also be prevented by avoiding dG(guanine):dC(cytosine) and dA(adenine):dT(thymine) baseparing interactions in the sequence design. For as-produced HiPco nanotubes, in addition to ssDNA, short double-stranded DNA and total RNA extracted from *Saccharomyces cerevisiae* and *Escherichia coli* can also disperse carbon nanotubes. In this case, however, vigorous sonication is needed for effective dispersion. DNA-CNT solutions are stable for months at room temperature. Removal of free DNA by either anion exchange column chromatography or nuclease digestion does not cause nanotube flocculation, indicating that DNA binding to carbon nanotubes is very strong. Even though there is no reliable way to determine the percentage of individually dispersed nanotubes in solution, all of our evidence suggests that DNA converts bundled CNT into individually dispersed tubes. The electronic absorption spectra of DNA-CNT solutions show well-resolved structures and systematic dependence on pH. We have also observed strong near-IR fluorescence from DNA-CNT solutions. These features are characteristic of individually dispersed nanotube solutions, obtained with surfactants after intense sonication treatment [9]. Atomic force microscopy measurements show that DNA-CNT's have diameters ranging from 1 to 2 nm. The diameters are larger than the 0. 7 to 1.1 nm range expected for HiPco tubes, but are consistent with DNA coating of nanotubes (Fig. 10.1).

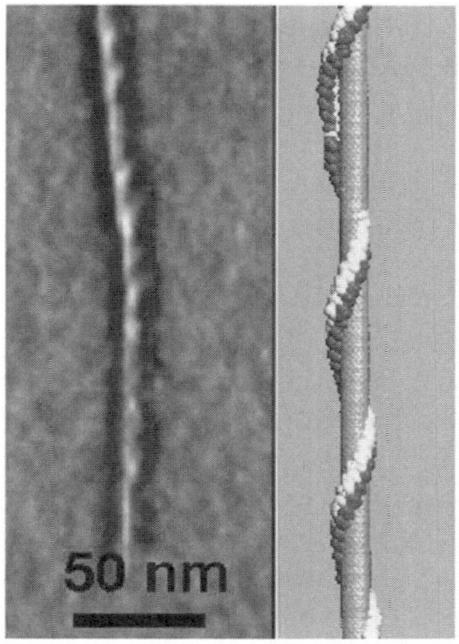

Fig. 10.1. AFM image of random polyd(CT) wrapped around a carbon nanotube, in good agreement with the proposed molecular wrapping model (see Sect. 3 of this chapter)

To explore the influence of the DNA sequence and length on dispersion efficiency, we tested some simple patterns of sequence with defined lengths. The dispersion efficiency was measured by the optical absorption intensity of dispersion solution at 730 nm, which is relatively invariant with respect to pH changes. Among fixed length (60–mer) homo-polymers that can be made by solid-phase synthesis, we found that poly d(T) has the highest dispersion efficiency. Among the four different lengths (60-, 30-, 21- and 15-mer) of poly d(T) examined, we found that d(T)30 gave the highest yield.

In comparison with other polymers that also disperse carbon nanotubes [10], DNA appears to be much more efficient. In a typical experiment, 1 mg of DNA can disperse equal amount of as-produced HiPCO CNT in 1 mL volume, yielding 0.2 to 0.4 mg/mL CNT solution after removal of non-soluble material by centrifugation. Such obtained CNT solution can be further concentrated by ten-fold to give a soluble CNT concentration as high as 4 mg/mL. DNA chain flexibility and backbone charge may all contribute to such high dispersion efficiency. In addition, DNA offers the advantage of defined length and sequence, and well-developed chemistries for functionalization. The latter provides a way to functionalize the carbon nanotube surface without modifying it covalently. As a demonstration, we made DNA-CNT using an oligonucleotide that was modified at one end with biotin. We observed biotin-dependent deposition of nanotubes onto streptavidin-coated agarose beads. This demonstration also illustrates a way to control the placement of carbon nanotubes on a solid substrate.

Carbon nanotube (CNT) separation is an enabling step for many potential applications and fundamental studies that require defined nanotube structures and properties. CNT's can be classified into two categories based on their electronic structures: metallic and semiconducting. The latter can be further classified by tube diameters, since the band gap of a semiconducting tube, a critical parameter that needs to be controlled for nanoelectronic applications, is inversely proportional to its diameter. Reports in the literature indicate that it is possible to separate metallic from semiconducting tubes by taking advantage of differences in their physical or chemical properties [11–14]. Diameter-based separation is more difficult, because differences in the physical/chemical properties caused by diameter changes are smaller, and variations in tube length could be a dominant factor in physical based separation methods. We have identified an oligonucleotide sequence that self-assembles into a highly ordered structure on CNT, allowing not only metal/semiconducting separation, but also diameter-dependent separation. We found that anion-exchange chromatography provides a macroscopic means to assay for electrostatic properties of nanoscale DNA-CNT hybrids. More specifically, we found that the outcome of anion-exchange based DNA-CNT separation, as measured by optical absorption spectral changes from fraction to fraction, is strongly dependent on the DNA sequence. To explore this dependence, we conducted a systematic but limited search of the huge ssDNA library under identical chromatographic conditions. We found that sonication effectively cuts CNT's in the presence of DNA, and that short CNT's increase sample recovery from an anion exchange column. In a typical dispersion experiment, a DNA/CNT mixture was kept in an ice-water bath and sonicated (Sonics, VC130 PB) for 120 min at a power level of 8 W. The average length of CNT's after this level of sonication is $\approx$ 140 nm as measured by AFM. We tested simple homo-polymers of dA, dC and dT, and sub-libraries composed of random combinations of two of the four nucleotides (dG, dA, dT, dC): poly d(A/C), poly d(A/G), poly d(A/T), poly d(C/T), poly d(C/G) and poly d(G/T). Among these, poly d(G/T), poly d(G/C) showed the largest variation in the optical absorption spectra from fraction to fraction. To narrow down the choice of sequence, we then tested representative sequences in the poly d(G/T) and poly d(G/C) sub-library. We found that the best separation was obtained with a sequence of repeats of alternating G and T, d(GT)n, with total length ranging from 20 to 90 bases (n = 10 to 45).

In a report published earlier [15], we provided evidence showing diameter as well as electronic property based separation of HiPco carbon nanotubes dispersed by d(GT)n sequences. Using carbon nanotubes synthesized by the "CoMoCat" process as the starting material [16], we found that individual (n,m) type enriched carbon nanotubes can be obtained by our separation process. In this experiment, unpurified single wall carbon nanotubes from Southwest Nanotechnologies (SWeNT, Norman, OK) and single-stranded DNA

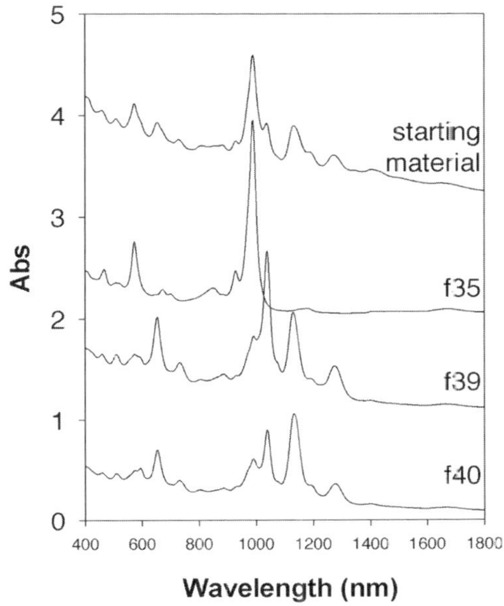

Fig. 10.2. Absorption spectra of fractions from anion exchange chromatography separation of d(GT)30-dispersed CoMoCat tubes

d(GT)30 were used. Anion exchange chromatography separation was done as described earlier [15]. The SWeNT tubes have two major types of semiconducting tubes (6,5) and (7, 5), as was characterized in the literature [16]. In Fig. 10.2, optical absorption profiles of eluted fractions 35, 39, 40 along with that of the starting material are shown. The absorption spectrum of f35 has two dominant peaks at 990 nm and 574 nm. These are close to the literature assignment of 975 nm (E11) and 567 nm (E22) for the (6, 5) tubes, but are red-shifted [16]. Similarly, peaks at 928 nm and 702 nm can be assigned to (9, 1) tubes (same diameter as (6, 5)); peaks at $\sim$ 970 nm (overlapping with 990 nm peak) and 674 nm can be assigned to (8, 3). By intensity comparison, it is clear that f35 is largely enriched with (6, 5) tubes.

To better understand the role of the d(GT)n sequence we selected for CNT separation, we used atomic force microscopy (AFM) to study the DNA assembly on CNT. Molecular modelling suggests that ssDNA can adopt many different modes of binding to CNT, with little difference in binding free energies. These modes include helical wrapping with different pitches, consistent with our observations. In contrast, d(GT)n exhibits a qualitatively different binding to CNT than most of the other ssDNA sequences. AFM measurements show that d(GT)n-CNT hybrids have a much more uniform periodic structure with a regular pitch of $\sim$ 18 nm. Whatever structure d(GT)n has on CNT, it appears to be very sensitive to minute changes in the structure

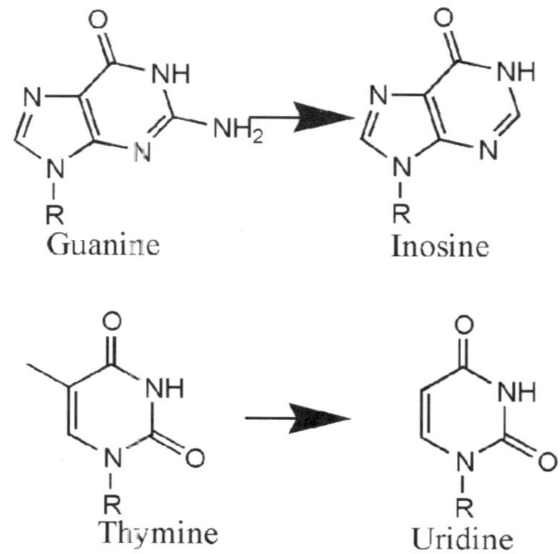

Fig. 10.3. Structures of G, T and their homologues

of the nucleotide bases (Fig. 10.3 mutation data). Replacement of G or T by their homologues, inosine or uridine, respectively, decreases the degree of separation on the anion exchange column. Replacing d(GT)n with d(GGTT)n/2 that has identical chemical composition also decreases the extent of the separation.

It is known that GT-rich sequences can self-assemble into supramolecular structures that play important roles in telomere replication [17]. More relevantly, GT-rich sequences have been observed to form nano-wires involving hydrogen-bonding interactions among different strands [18]. As a model for the d(GT)n-CNT structure, we propose that two anti-parallel d(GT)n strands interact with each other through hydrogen bonds to form a double-stranded strip, which then wraps around the CNT with close-packed bases resembling molecular tiles lying on the side-wall of the nanotube [15]. Such a double-helical structure is built on the unique hydrogen bonding network between two d(GT)n strands, and is expected to be more rigid and to have fewer allowed conformations than a single-helical structure. A rigid DNA structure that creates an identical charge distribution within a given type of nanotube is necessary for successful separation. This is probably why most sequences disperse CNT's but do not give good separation.

10.3 Separation of Carbon Nanotubes Dispersed by Non-ionic Surfactant

In addition to DNA-based separation, we have also explored alternative methods to separate carbon nanotubes dispersed in non-ionic surfactants. Strano and collaborators [19] have reported a marked sensitivity of the absorption spectra, the Resonance Raman spectra and the quantum yield of fluorescence emission to pH in HiPco carbon nanotubes dispersed in sodium dodecyl sulfate. The oscillator strength of the semiconductor E_{11} electronic transitions decreases with decreasing pH with the largest diameter tubes (small band gap) showing the highest pKas and the smallest diameter nanotubes (large band gap) the lowest. Similar behavior for the same E_{11} transitions was observed by Zheng et al. in DNA-wrapped carbon nanotubes [14]. Phonon coupling to the electronic continuum of the metallic nanotubes as reflected in the Fano line of the Raman spectra shows the highest pKa of all [19].

Strano and coworkers [19] showed that the loss of the oscillator strength of the electronic transitions was dependent on the presence of O_2. UV irradiation of carbon nanotubes under an atmosphere of N2 resulted in the restoration of the absorption bands at pH 3.5. While the role of O_2 is still unclear, it has been proposed that it is necessary to allow protonation to occur on the nanotube sidewall. Possible mechanisms include the formation of an oxygen adduct to which the proton binds and a lowering of an energetic barrier to direct nanotube protonation [19], or direct oxidation of the nanotube (Zheng and Diner, unpublished results).

There has been considerable discussion around how significant the presence of oxygen is in its modulation of the electronic properties of nanotubes with Collins et al. [20] reporting a decrease of the bandgap of semiconducting nanotubes upon exposure to O_2. The importance of bulk O_2 doping in nanotube electronic behavior remains controversial, however, with the claim by Derycke et al. [21] that the principal effect of O_2 on carbon nanotube field-effect transistors is at the metal-semiconductor contacts.

Several groups have used density functional theory combined with various approximations to calculate the energetics of the interaction of oxygen with carbon nanotubes by different mechanisms. Jhi et al. [22] have calculated modest binding energy for O_2 chemisorption (0.25 eV) with an electron transfer of 0.1 e^- per O_2 molecule. Sorescu et al. [23] have argued that the physisorption or chemisorption of O_2 to the nanotube can result in a partial electron transfer to the adsorbed diatomic molecule but that the O_2 binding is energetically less favorable than estimated by Jhi et al. In the case of physisorption, the extent of electron transfer was calculated to be extremely small (0.01 e^-) with weak binding (0.70 kcal/mole) and a high activation barrier. In the case of chemisorption, the electron transfer was calculated to be substantial (0.28 e^-) upon formation of a singlet state bridged structure on the outside of the nanotube but the free energy for the formation of this state was found to be endothermic (−10.2 kcal/mole). Formation of epoxides is considerably more favorable energetically, particularly in the smaller diameter carbon nanotubes. Epoxide formation suffers, however, from the high activation energy associated with the breakage of the O-O bond. It is possible, however, that the binding energies might be substantially enhanced by an inclusion of dispersion forces in the case of physisorption and of surface defects in the case of chemisorption. Despite the marginality of the energetics calculations, strong experimental evidence does exist for nanotube oxidation. NEXAFS and vibrational spectroscopy have provided strong evidence for both carboxyl and ether functionalities in nanotubes fabricated by both plasma laser vaporization and by the HiPco process. Of the functional groups mentioned above, those subject to protonation in the pH range described by Strano et al. (pKa ≈5.5 in SDS dispersed nanotubes) include carboxylates and superoxide anions.

If an anionic oxygen adduct were to be bound to the sidewalls of the carbon nanotubes, then the protonation of such a site would likely result in the loss of the anionic charge. If the presence of O_2 were to facilitate nanotube protonation, then such protonation would generate a cationic charge. Nanotube oxidation by oxygen (Zheng and Diner, unpublished results) may also give rise to cationic charge. The presence of these charges and their dependence on the electronic structure provide an opportunity for diameter (bandgap)- dependent fractionation of nanotubes. Electrophoresis, isoelectric focussing and ion exchange chromatography are all separatory techniques that allow the fractionation of charged materials based on net charge, pKa and localized charge density, respectively. In order to allow the intrinsic charge of the nanotubes to determine their behavior using these fractionation methods, the nanotubes need to be individually dispersed in non-ionic detergents.

HiPco carbon nanotubes were dispersed in Triton X-405 by extensive sonication, centrifuged at 140,000 g to pellet incompletely dispersed aggregates and concentrated by ultrafiltration. The absorption spectra were well resolved, consistent with individually dispersed nanotubes. Isoelectric focussing indicated that the Triton X-405 dispersed nanotubes were anionic in charge, migrating toward the anode to the most acidic fraction (pH < 3.7). It was also found that the dispersed carbon nanotubes bound to anion but not to cation exchange media. In the former case, the nanotubes were fully bound at pH 8 and could be progressively eluted by way of pH gradient elution down to pH 1.5. Alternatively the nanotubes could be incubated at a particular intermediary pH and selectively bound and subsequently eluted by way of a salt gradient. Lowering the pH preincubation from 8 to 2 resulted in an increasing fraction of the nanotubes that did not adhere to the anion exchange column. All of these observations are consistent with the nanotubes having an intrinsic negative charge, the density of which is diminished by lowering the pH. Fractions were eluted from the column by either the pH gradient method or the salt gradient method at fixed pH (3.1). Both of these gave rise to an early fraction pink in color and enriched in metallic nanotubes absorbing in the 450–600 nm range and a major fraction eluting later that was green in color and enriched in semiconducting nanotubes with E_{11} transitions in the 1100–1300 nm range (Fig. 10.4). Preliminary Resonance Raman spectra

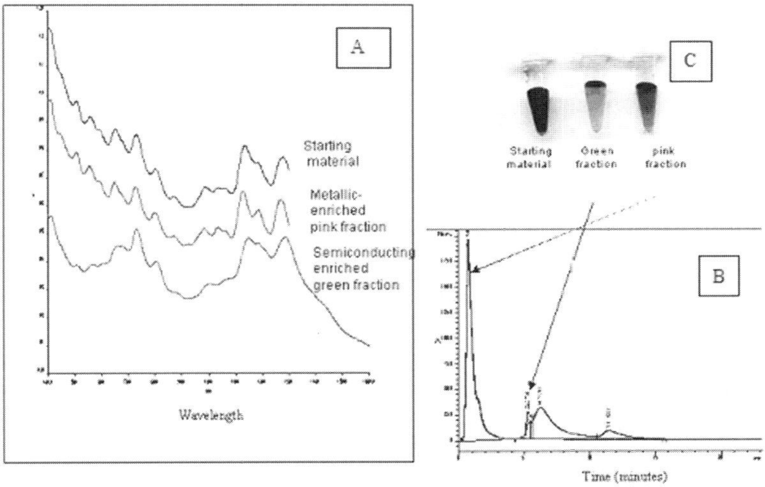

Fig. 10.4. Chromatographic separation of Triton X-405 dispersed carbon nanotubes using anion exchange chromatography. (**a**) absorption spectra of the fractions indicated in chromatogram (**b**); (**c**) vials containing fractions

of the latter fraction indicated an enrichment in nanotubes of intermediary diameter (radial breathing mode, RBM $= 230$–$250\,\mathrm{cm}^{-1}$, courtesy of Shin Grace Chou). This selectivity based on electronic structure and diameter appears to be consistent with the pH titration data of Strano [19]. The metallic tubes would be expected to elute early in a basic to acidic pH gradient and to be neutral or positively charged at pH 3.1. In fact, pH titration of several absorbance bands associated with semiconducting nanotubes show apparent pKa's of ≈ 3 in 1% Triton X-405. For the semiconducting tubes, the diameter of the eluted tubes would initially be expected to decrease with decreasing pH, or with higher salt concentration at a fixed pH, a reflection of the decrease in pKa with decreasing diameter. Consistent with this description, the dominant green nanotube fraction appears enriched in nanotubes of intermediate size. However, as shown for the DNA-wrapped nanotubes with periodic charge, the linear charge density increases with diameter, resulting in stronger binding to the column. Consequently, the two effects, pKa and linear charge density may work against each other later in the column run, reducing the column resolution for the smaller diameter nanotubes.

The selection based on selective protonation provides an alternative means of nanotube fractionation compared to that provided by DNA wrapped nanotubes where the selection is based on linear charge density and on electronic structure. These approaches are complementary and potentially provide two of a series of complementary methods that will ultimately contribute to the ability to fractionate nanotubes to individual chiral forms.

10.4 Structure and Electrostatics of the DNA/CNT Hybrid Material

In this section we describe our models for the structure and electrostatics of the DNA/CNT hybrid material, and relate them to observations of the efficacy of DNA as a dispersant for carbon nanotubes and for their separation. Considerations of the structure and the electrostatics are fairly general and can be applied to understanding how other molecules interact with carbon nanotubes.

Fundamentally, we view the driving force for the formation of the DNA/CNT hybrid to be favourable base-nanotube stacking interactions coupled with the solubilizing influence of the exposed, charged, phosphate groups in aqueous solution. A complete model of this structure would need to self-consistently include the influence of the aqueous medium, charges on the DNA backbone and its interaction with the nanotube, deformability of DNA and the nanotube, and the effect of dissolved counter-ions. We have tackled the problem here in parts. First we present results of molecular modeling (performed *in vacuo*) that yield a structure consisting of a carbon nanotube wrapped helically by strands of DNA, which is shown to be in good agreement with AFM observations. Separately, using electrostatics and treating the material and its surroundings as a continuum, we address questions about the behaviour of the hybrid material related to the charged phosphate groups and the surrounding counter-ions.

10.4.1 Structure of the DNA/CNT Hybrid

Because the backbone of single-stranded DNA has five torsionally labile bonds [24], it is quite flexible. Indeed, this flexibility allows it to find suitable low-energy conformations around the relatively rigid carbon nanotube. Molecular simulations to study these conformations, described next, were conducted using commercial molecular modeling software [25]. Candidate hybrid structures were constructed by building a strand of DNA in the vicinity of a carbon nanotube by sequential addition of single nucleotide residues to a previously constructed hybrid. Between each addition, the structure was equilibrated by a molecular dynamics run for 1ns at 300 K, followed by energy minimization.

Initial simulations were with ssDNA strands consisting only of pyrimidine bases on a (10,0) carbon nanotube. This process resulted in the formation of a variety of structures, of varying binding enthalpy and regularity. By systematically varying the five backbone torsions, we discovered a set of conformations that resulted in regular, low-energy hybrid structures with bases oriented approximately parallel to hexagons on the nanotube surface, and the sugar-phosphate backbone stretched to accommodate the twist of the bases.

Figure 10.5 shows the lowest energy structure we found by this procedure for polyd(CT) wrapped around a (10,0) carbon nanotube. Broadly, as

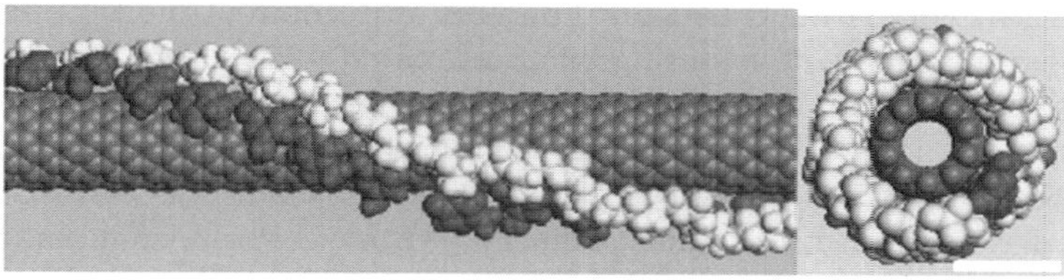

Fig. 10.5. Low-energy structure of Poly d(CT) on a (10,0) carbon nanotube. The sugar-phosphate backbone has the lightest shading, whereas the nucleotide bases have the darkest shading

anticipated, the structure consists of bases stacking parallel to the surface of the nanotube, and a backbone exposed to the surroundings, allowing its easy hydration. It is instructive to compare this structure with that of B-DNA. The most striking difference is that the normal to the base, which is approximately aligned with the helical axis in B-DNA, is orthogonal to the helical (and nanotube) axis in CNT-DNA. This is accomplished, not surprisingly, by adjustment of the torsional angles. Torsion angles in our model for CNT-DNA were found to belong to one of two sets. In the notation of Saenger [24] they were either

$$(\chi, \alpha, \beta, \gamma, \delta, \epsilon, \zeta) = (-138 \pm 7, 168 \pm 13, 152 \pm 11, 57 \pm 7, 99 \pm 9, 57 \pm 20, 128 \pm 18) \quad (10.1)$$

or

$$(\chi, \alpha, \beta, \gamma, \delta, \epsilon, \zeta) = (180 \pm 6, 168 \pm 13, 84 \pm 11, 162 \pm 7, 99 \pm 9, 113 \pm 4, 73 \pm 8) \quad (10.2)$$

By comparison, in B-DNA, typical values are [24]

$$(\chi, \alpha, \beta, \gamma, \delta, \epsilon, \zeta) = (-102, -41, 136, 38, 139, -133, -157) \quad (10.3)$$

The main difference is in torsions α, ϵ, and ζ, confirming that the base-phosphate unit itself is relatively rigid and that flexibility in backbone torsions allows the base to swing into a stacking conformation.

The enthalpy of formation of the hybrid (polyd(T)/(10,0)CNT) was computed to be 1.17 eV/nm of the nanotube. This value is for wrapping by a single strand. We observed in AFM some variability in helical pitch of the wrapping for the same helical angle, suggesting wrapping of multiple parallel strands. Clearly, in this case the enthalpy would increase correspondingly. In comparison, two (10,0) carbon nanotubes brought together resulted in enthalpy release of 1.2 eV/nm. Because poly(U) (and presumably poly(T)) does not stack separately in solution, we estimated the entropic penalty for wrapping by comparing constrained degrees of freedom of the wrapped chain to those in solution [10], resulting in an estimated 0.15 eV/nm at 300 K. This shows that the binding enthalpy dominates the entropic penalty. Poly(C), in comparison, has a known tendency for base stacking in solution, implying a reduced driving force for hybrid formation, which was consistent with observed reduction in dispersion efficacy. We have observed in simulations systematic differences due to DNA sequence and nanotube type, which will be described elsewhere in more detail.

10.4.2 Electrostatics of Elution of the DNA/CNT Hybrid

The investigation of the DNA-CNT structure presented so far has ignored the effect of the surrounding aqueous medium, charges on the phosphate groups, counter-ions, and differences in electronic structure of the nanotubes. The role of charges is particularly critical as it differentiates hybrids from each other based on the electronic properties of the nanotube core.

We assume that the principal charges are those on the phosphate groups. According to our structure, these are separated from each other by about 7 Angstroms, and lie about 4 Angstroms above the nanotube surface. From geometry, we note that linear charge density can be computed assuming helical wrapping by

$$Q_l = -n_{max} q_l \tan(\alpha)\sqrt{1+\frac{1}{\tan^2(\alpha)}} \tag{10.4}$$

where Q_l is the linear charge density along the nanotube axis, q_l is the linear charge density along the ssDNA chain ($\approx 1/7$ electrons per Angstrom), n_{max} is the number of strands ($\approx$5), and α is the helical angle with respect to the tube axis, $\approx \pi/4$. Therefore, we may expect as much as one electronic charge per Angstrom distance along the nanotube axis.

It is well-known [26,27] that if the linear charge density of a polyelectrolyte exceeds a critical value, then a fraction of the charge, θ, is neutralized by the condensed counter-ions, the neutralized fraction being independent of ionic strength. The critical charge density is defined in terms of the dimensionless Manning parameter, $\xi > 1$, which, based on equation (10.4), is

$$\xi = \frac{e^2}{4\pi\epsilon\epsilon_o k_b T b} \approx 6.2 \quad \text{for 1 nm tube saturated with ssDNA} \tag{10.5}$$

$$\theta = 1 - \frac{1}{\xi} \approx 0.83 \quad \text{for monovalent counterions} \tag{10.6}$$

where e is the electronic charge, ϵ and ϵ_o are the dielectric constant of water (≈ 80) and permittivity of space, respectively, k_b is Boltzmann's constant, T is absolute temperature, and b is the linear separation between charges along the nanotube axis. As a consequence, in suspension the DNA/CNT hybrid is expected to be surrounded by a Manning condensate of closely associated counter-ions that neutralize the surface charge significantly.

The effective linear charge density is affected by factors that depend on the type of embedded nanotube. Each phosphate group brings a negative charge in the vicinity of the carbon nanotube. Because the nanotube is polarizable, the negative charge will induce a charge on the nanotube surface. To indicate approximately the nature of this charge, we treat the nanotube surface as an infinite body with uniform and isotropic dielectric properties, and the exterior with properties representing water.

The dielectric properties of carbon nanotubes are anisotropic [28]. The dielectric constant for all types of nanotubes along the circumferential direction is about the same. In contrast, along the tube axis, the dielectric constant for metallic nanotubes is an extremely large number (formally infinite for metals), while for semiconductors it is on the order of 4.0. An effective isotropic dielectric constant can be obtained by equating any of the invariants of the 2nd rank dielectric tensor between the aniostropic and isotropic forms [29,30]. Because the difference in the dielectric properties along the nanotube axis is so large, it dominates the difference in effective dielectric properties. Physically, the inability of the metal to support a field in any single direction causes it to have image charges very different from a semiconductor that can support fields in any direction.

For the simplest case, consider the nanotube/water system as a bi-material interface with discontinuous dielectric properties. The combined field of the negative and induced charges in the space outside the nanotube can be represented by an image charge q' placed inside the nanotube. The value of the image charge [31] depends on the relative dielectric constants of the external media ϵ_{ext} (80 for water, 1 for vacuum) and the nanotube ϵ_{nt}. Specifically, for two half-spaces,

$$\frac{q'}{q} = -\left(\frac{\epsilon_{nt} - \epsilon_{ext}}{\epsilon_{nt} + \epsilon_{ext}}\right) \tag{10.7}$$

Because the effective dielectric constant depends on the type and diameter of the nanotube, the effective charge of the DNA-CNT hybrid is modulated significantly by the type of nanotube embedded in the DNA wrap. This effect is convoluted with changes in linear charge density due to the number of DNA strands wrapped, which increases with nanotube diameter. Together these two factors alter the effective charge density, which affects critical salt concentration for elution sufficiently to provide the basis for charge-based discrimination between different types of carbon nanotubes.

For metals, in using (10.7), it has been assumed that there will be net positive charging of the nanotube due to the presence of the negatively charged phosphate group. This implies that the nanotube, over the long time-scales of the experiment, acts as an electrically grounded material so that the compensating electrons hop off the nanotube. If, on the other hand, it is assumed that the nanotube is electrically isolated, there is no net change in linear charge density of the nanotube/DNA hybrid, only the local nature of electric

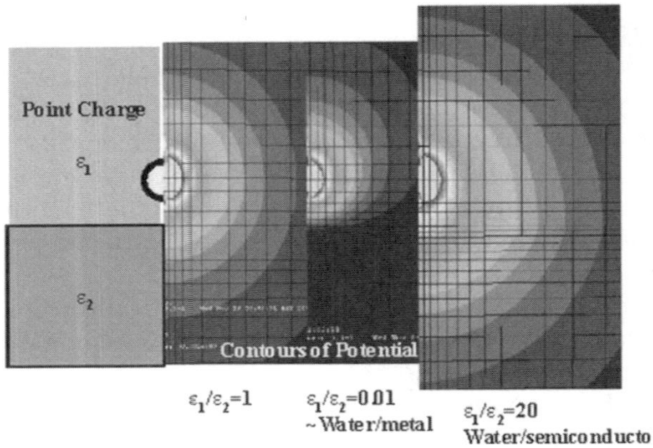

Fig. 10.6. Electrostatic potential (a.u.) surrounding a point charge near a bimaterial interface. A comparison of the homogenous case ($\epsilon_1 = \epsilon_2$), water/metal, and water/semiconductor interfaces shows that the presence of the metal strongly attenuates the influence of the charge away from the nanotube. In contrast, the presence of the semiconductor enhances the influence of the point charge (see Color Plates, p. 343)

fields is altered. Nevertheless, a difference between the metallic and semiconducting tubes remains, although it is attenuated compared to the electrically grounded case. The effective linear charge density of semiconducting/DNA hybrids is still enhanced over that of the DNA alone because of reduced polarizability of the embedded tube, without change in actual net linear charge density. The main point is that the difference in effective electric fields outside the DNA/CNT hybrid based on the type of nanotube is maintained regardless of whether one assumes the nanotube to be electrically isolated or grounded. These two extremes simply control the quantitative value of the difference in field.

Qualitatively, the phosphate charge near a metallic nanotube is converted into a dipole because of compensating screening positive charges on the nanotube. The field of the dipole is weaker and falls more rapidly with distance than that of a charge near a semiconducting nanotube. In Fig. 10.6 we show a numerically calculated distribution of potential around a point charge near an interface between two materials [32].

The charge-nanotube interaction changes the binding energy by only about 0.03 eV/nm, justifying our separate treatment of electrostatics and structure. However, the effective field of the semiconducting case is enhanced significantly – by about 40% at a distance of 3.4 Angstroms from the charge (and away from the nanotube), at a location where it might bind to fixed positive counter-ions on an ion-exchange column.

There is, however, a potentially serious objection to the image charge model. That arises from the fact that if charges on the surface of a cylinder

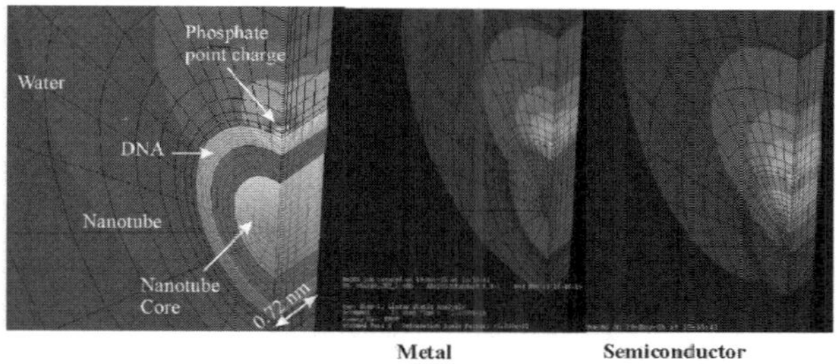

Fig. 10.7. Numerically computed electrostatic potential (a.u.) surrounding a point charge near a metallic or semiconducting nanotube. While the geometry and periodicity of charges reduces the difference due to metallic and semiconducting cores, it remains sufficient to explain differentiated interaction with external (e.g., fixed) charges (see Color Plates, p. 344)

are closely packed (separated by distances small compared to their distance from the cylinder), then they represent an effective cylindrical sheet of charge. The field inside a cylinder with uniformly distributed charge vanishes, and so there would be no image charges generated. In fact, this may offer the reason why other charged surfactants, such as SDS, while effective dispersants, have not yielded separability based on effective charge. The distribution of charge on DNA-CNT may well be close to optimal ($\approx$7 Angstroms separation, $\approx$ 4 Angstroms from nanotube) for simultaneous dispersability and separability.

Figure 10.7 shows numerically computed potential near a charge using the cylindrical geometry of a nanotube, and imposing periodic boundary conditions to match recurring charges along a polymeric backbone. The cylindrical geometry and periodicity of charges diminishes the difference between metal and semiconducting cases compared to the simple isolated charge near a flat bimaterial interface. Nevertheless the difference remains large enough to result in differentiated attachment to external charges – the basis of chromatographic separation.

10.5 Effects of Protein Adsorption on the Electronic Properties of Single Walled Carbon Nanotubes

A promising application for carbon nanotube-based electronic devices is as a sensor, for example, to detect biological molecules. Only recently has the detection of a small number of biomolecules via measurement of the electron transport of nanotubes been reported [33, 34]. In this section we report on the direct detection of a small number of Cytochrome c (Cytc) molecules adsorbed onto an individual SWNT transistor via changes in the electron transport properties [35].

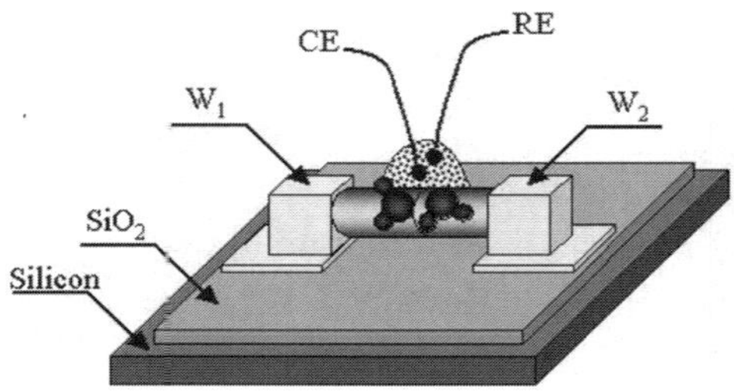

Fig. 10.8. Schematic diagram of a SWNT device. The nanotube is on an SiO_2 surface and connected to two gold electrodes, which serve as source and drain. Electron transport through the nanotube can be controlled with the electrochemical gate, which is achieved by controlling the potential between the nanotube and a reference electrode (Ag) using a bipotentiostat and a counter electrode (Pt)

The SWNT devices (Fig. 10.8) were prepared following a method similar to those reported in literature [36]. Before exposing the SWNTs to buffer and protein solutions, the nanotubes were characterized by measuring I-V curves at various gate voltages applied to the back of the chips. The characterization shows that while most of the SWNTs are p-type semiconducting, some are metallic. For semiconducting SWNTs, there is also a variation in the conductance, which allowed us to study the response of different SWNTs to the adsorption of the protein.

The SWNTs were brought into contact with phosphate buffer (10 mM, pH = 7.0) using a solution cell. We used Ag wire as a quasi-reference electrode, Pt wire as a counter electrode and a bipotentiostat to control the electrochemical potential "gate" of the nanotube. We introduced the protein solution into the cell after a stable conductance was obtained in the buffer solution. The protein Cytc (Fluka) was dissolved in phosphate buffer and the solution was freshly prepared for each experiment. The redox activity of Cytc was first verified electrochemically on graphite (HOPG) [37] whose surface properties are similar to those of the nanotube. The facile electron transfer between Cytc and the graphite electrode indicates that the adsorbed protein orients its redox center, heme group, close to the electrode.

Immediately after the protein solution was introduced the conductance of the nanotube began to decrease and eventually reached a stable value. The average time of the adsorption process lasts for a few minutes, which is in good agreement with our previous study [37] of the adsorption of Cytc on HOPG using real time AFM imaging. The decrease in conductance for the p-type SWNT is expected for the adsorption of positively charged protein because it decreases the concentration of p-type carriers.

The adsorption-induced conductance decrease is observed for all semiconducting SWNTs, but the amount of decrease varies from one nanotube to another. For example, while conductance decrease can vary by two-fold, the difference in the relative change ($\Delta G/G$) between the nanotubes is negligible. We believe that this variation is mainly related to the electronic characteristics of the nanotubes, which determines the sensitivity of the conductance on the surface charge. This is consistent with the argument presented later that induced charge is dominated by the quantum capacitance. In that case,

$$\frac{\Delta G}{G} \propto \frac{\Delta N}{N} \tag{10.8}$$

where N is the number of charge carriers. Because this quantity is defined by Fermi statistics, and so is approximately given by

$$N \approx exp(-V/k_b T) \rightarrow \frac{\Delta N}{n} \approx \Delta V \quad \text{(assumed to be fixed)} \tag{10.9}$$

where V is the gap between Fermi energy and the valence band, showing that the relative change in conductance is expected to be a constant if one assumes that the effective change in potential due to the adsorbed species is the same from tube to tube.

We have found a strong correlation between the amount of protein-induced conductance decrease and the dependence of the I-V characteristics on V_g measured before the nanotubes were exposed to the solution. For metallic nanotubes, the I-V characteristic curves are completely insensitive to V_g and the adsorption of Cytc induces no detectable change in the conductance. For semiconducting nanotubes, the greater the sensitivity of the I-V curves to V_g, the greater is the protein-induced conductance decrease.

If we assume that the protein adsorption-induced conductance change is caused purely by its charge, we can estimate the number of adsorbed proteins independent of the AFM data [35, 38]. This task requires measurements of the conductance vs. the electrochemical gate potential. Figure 10.9 shows the typical result of such measurements in phosphate buffer (10 mM). The data were recorded by measuring the current through the nanotube at small fixed bias voltage (10 mV) applied between the ends of the nanotube while the potential of the nanotube was swept with respect to the reference electrode. At positive potential, the conductance is small, but increases quickly as the potential decreases towards negative values. The dependence shown in Fig. 10.9 is expected for a p-type SWNT and is in agreement with recent results by other groups [39, 40]. From the conductance (G) vs. potential (V) plot, we have estimated ($\approx 10^{-6}$ S/V over the linear range of the data), which allows us to relate charge change (ΔQ) to the conductance change (ΔG) according to $\Delta Q = C(\Delta G/\Delta V)^{-1}\Delta G$, where C is the capacitance, which is dominated by the quantum capacitance ($\approx 4 \times 10^{-10}$ F/m) of the SWNT [40]. For a 1 μm SWNT, the quantum capacitance is $\approx 4 \times 10^{-16}$ F. Using the above relation, because the change in charge on the nanotube equals the change in charge because of Cytc molecules, the number of Cytc molecules for a given conductance decrease can be determined if the charge per protein is known.

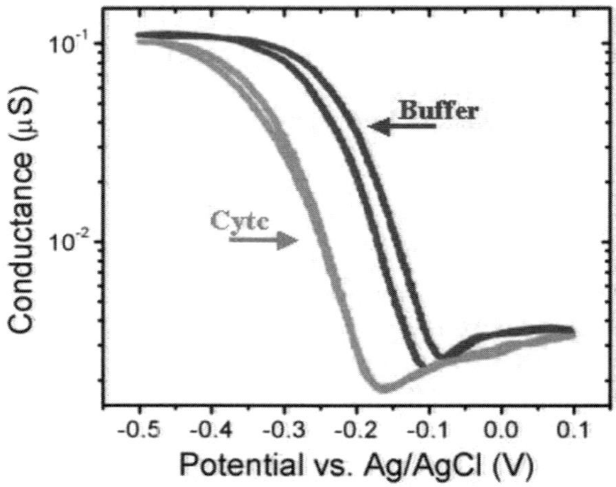

Fig. 10.9. SWNT conductance as function of the electrochemical potential in phosphate buffer and 200 μM Cytc in the phosphate buffer solution

The surface charge of Cytc is well characterized and it is $\approx +10$e per protein at pH $= 7$ [41]. For the adsorption-induced conductance, we estimated that the number of Cytc molecules is $\approx$25, which is consistent with the AFM data [35, 38].

As expected, the adsorption of Cytc causes a shift in the G vs. V plot toward negative potentials. Note that the protein adsorption does not change the slope nor the shape of the plot, which is consistent with the finding that the capacitance of the SWNTs is dominated by the quantum capacitance. The average shift is approximately 0.09 V. Using the above-mentioned capacitance, the corresponding charge change brought by the adsorbed molecules of Cytc is $\approx +200$e. Therefore, the number of adsorbed Cytc molecules on the SWNT is $\approx$20. This is in agreement with the number of molecules estimated from the AFM images [38] and the adsorption plots. Our results show that a small number of proteins can be detected by measuring the transport properties of the SWNTs.

References

1. N.C. Seeman: Trends Biotechnol **17**, 437–43 (1999)
2. A.P. Alivisatos, et al.: Nature **382**, 609–11 (1996)
3. C.A. Mirkin, R.L. Letsinger, R.C. Mucic, J.J. Storhoff: Nature **382**, 607–9 (1996)
4. M.R. Arkin: Science **273**, 475–80 (1996)
5. S. Wang, et al.: Nature Materials **2**, 196–200 (2003)
6. S.R. Whaley, D.S. English, E.L. Hu, P.F. Barbara, A.M. Belcher: Nature **405**, 665–8 (2000)

7. G.R. Dieckmann, et al.: J Am Chem Soc **125**, 1770–7 (2003)
8. D.S. Wilson, and J.W. Szostak: Annu Rev Biochem **68**, 611–47 (1999)
9. M.J. O'Connell, et al.: Science **297**, 593–6 (2002)
10. M.J. O'Connell, et al.: Chem. Phys. Lett. 342, 265–271 (2001)
11. D. Chattopadhyay, I. Galeska, and F. Papadimitrakopoulos: J Am Chem Soc **125**, 3370–5 (2003)
12. R. Krupke, F. Hennrich, H. Lohneysen, and M.M. Kappes: Science **301**, 344–7 (2003)
13. R.B. Weisman: Nat Mater **2**, 569–70 (2003)
14. M. Zheng, et al.: Nat Mater **2**, 338–42 (2003)
15. M. Zheng, et al.: Science **302**, 1545–8 (2003)
16. S.M. Bachilo, et al.: J. Am. Chem. Soc. **125**, 11186–11187 (2003)
17. J.R. Williamson: Annu Rev Biophys Biomol Struct **23**, 703–30 (1994)
18. T.C. Marsh, J. Vesenka, and E.A. Henderson: Nucleic Acids Res **23**, 696–700 (1995)
19. M.S. Strano, et al.: Journal of Physical Chemistry B **107**, 6979–6985 (2003)
20. P.G. Collins, K. Bradley, M. Ishigami, and A. Zettl: Science **287**, 1801–4 (2000)
21. V. Derycke, R. Martel, J. Appenzeller, and P. Avouris: Applied Physics Letters **80**, 2773–2775 (2002)
22. S.H. Jhi, S.G. Louie, and M.L. Cohen: Phys Rev Lett **85**, 1710–3 (2000)
23. D.C. Sorescu, K.D. Jordon, and P. Avouris: Journal of Physical Chemistry B **105**, 11227–11232 (2001)
24. W. Saenger, *Principles of Nucleic Acid Structure* (Springer-Verlag, New York, 1984)
25. *Discover* with *pcff* force-field, Accelrys Inc.
26. G.S. Manning: Q. Rev Biophys **11**, 179–246 (1978)
27. G.S. Manning: Biophys Chem **101-102**, 461–73 (2002)
28. L.X. Benedict, S.G. Louie, and M.L. Cohen: Physical Review. B. Condensed Matter. **52**, 8541–8549 (1995)
29. J.F. Nye, *Physical Properties of Crystals, Their Representation by Tensors and Matrices* (Clarendon Press, Oxford, 1985)
30. A.J.M. Spencer, *Continuum Mechanics* (Longman, London, 1980)
31. J.D. Jackson, *Classical Electrodynamics* (John Wiley & Sons, 1999)
32. Computed using a commercial finite element program, ABAQUS, version 6.4, HKS (Rhode Island, USA).
33. R.J. Chen, et al.: Proc Natl Acad Sci U S A **100**, 4984–9 (2003)
34. A. Star, J.-C.P. Gabriel, K. Bradley, and G. Gruner: Nano Letters **3**, 459–463 (2003)
35. S. Boussaad, N.J. Tao, R. Zhang, T.J. Hopson, and L.A. Nagahara: Chem. Commun. **13**, 1502–1503 (2003)
36. C. Zhou, J. Kong, E. Yenilmez, and H. Dai, H.: Science **290**, 1552–5 (2000)
37. S. Boussaad, N.J. Tao, and R. Arechabaleta: Chemical Physics Letters **280**, 397–403 (1997)
38. B.R. Azamian, J.J. Davis, K.S. Coleman, C.B. Bagshaw, and M.L. Green: J Am Chem Soc **124**, 12664–5 (2002)
39. S. Kazaoui, N. Minami, N. Matsuda, H. Kataura, and Y. Achiba: Applied Physics Letters **78**, 3433–3435 (2001)
40. S. Rosenblatt, et al.: Nano Letters **2**, 869–872 (2002)
41. G.R. Moore, and G.W. Pettigrew, in Cytochrome c: evolutionary, structural and physicochemical aspects (Springer-Verlag, 1990)

11 Electrical and Mechanical Properties of Nanotubes Determined Using In-situ TEM Probes

J. Cumings and A. Zettl

In this chapter, we describe experiments performed on carbon and boron nitride nanotubes inside a transmission electron microscope (TEM). We demonstrate that it is possible to peel open the outer layers of a multiwall carbon nanotube to expose the inner shells. In this configuration, it is then possible to extract the inner shells in a telescopic extension mode. We present investigations of the forces associated with this extension and the electrical resistance of the nanotube during the telescoping process. We also described a series of studies of field emission and electrical conductivity using boron nitride nanotubes.

11.1 Introduction

11.1.1 Carbon and BN Nanotubes

Carbon nanotubes can be conceptually considered as originating from single-layer sheets of graphite, rolled up into perfectly seamless tubes with nanometer dimensions [1]. The sp^2 carbon-carbon covalent bond in graphite is among the strongest known bonds from chemistry, and carbon nanotubes therefore are mechanically robust. The bonds between layers of graphite, are relatively weaker Van der Waals bonds that give graphite its natural lubricating properties. Carbon nanotubes also have unique electronic properties, different from graphite, and present an exciting potential for nanoscale electronic devices.

The predominant synthesis methods for carbon nanotubes produce two distinct kinds of nanotube materials: single-wall nanotubes and multi-wall nanotubes. In a typical single-wall synthesis, the nanotubes vary in diameter from 1.0 nm to 1.4 nm, and can bond to one another in a regular parallel array on a triangular lattice. Such composite structures are sometimes referred to as ropes or bundles. A multi-wall nanotube, on the other hand, is a composite structure comprising many single-layer nanotubes, all nested concentrically. The spacing between the layers is close to the natural Van der Waals layer-layer spacing for graphite, 3.4 Å. Figure 11.1 shows transmission electron microscopy (TEM) images of single-wall and multi-wall carbon nanotubes.

Nanotubes can also be formed from other layered materials, notably boron nitride (BN) [3]. Boron and nitrogen lie to either side of carbon on the periodic table, and are the only other elements that can readily participate in

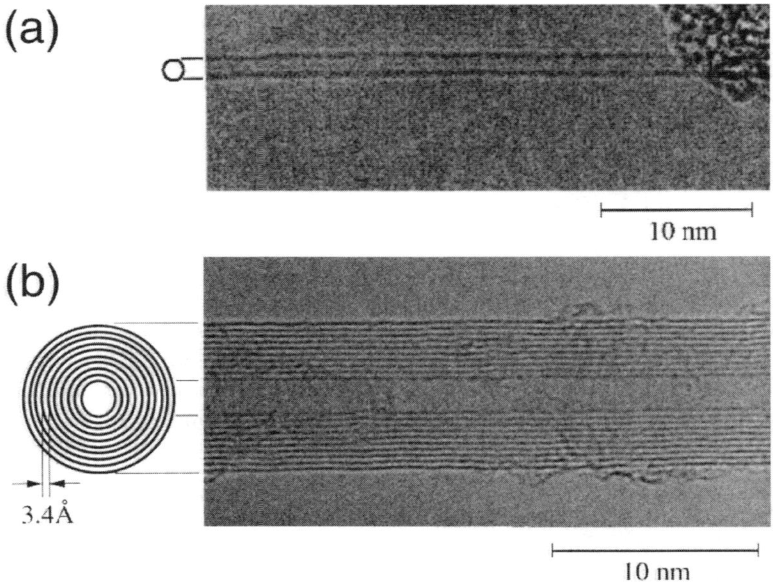

Fig. 11.1. Carbon nanotubes: (**a**) a typical single-wall nanotube; (**b**) a multi-wall nanotube [2]

sp^2 bonding. Bulk hexagonal boron nitride is a layered material analogous to graphite with similar structural properties [4]. The electronic properties, however, are markedly different. Due to the ionicity of the B–N bond, boron nitride has a bandgap of 5.8 eV in the graphitic phase [5]. Because of this, boron nitride nanotubes all have approximately the same bandgap ($\sim$ 5 eV) independent of the diameter or chirality of the nanotubes [6,7]. This could be useful for technological applications as opposed to carbon nanotubes which can be either metallic or semiconducting, depending on the structure. Figure 11.2 shows examples of single and multi-wall BN nanotubes, as examined by TEM. BN nanotubes are predicted to be more energetically stable structures that carbon nanotubes [8] and are expected to be more resistant to oxidation.

The virtually defect-free graphitic honeycomb network of carbon atoms of both single-wall carbon nanotubes and multi-wall carbon nanotubes suggests that various elastic moduli, such as the axial Young's modulus, might experimentally approach "ideal" graphite values. An early, but comprehensive overview of relevant theory is presented by Ruoff and Lorents [9]. Experiments bear out many of the theoretical predictions with axial Young's moduli of order 1 TPa being reported (see Table 11.1). A review of experimental results for carbon nanotubes is presented by Salvetat et al. [10].

Fig. 11.2. TEM images of BN nanotubes (courtesy W.Q. Han)

There are several theoretical studies of the mechanical properties of non-carbon nanotubes. However, due to a paucity of good samples, there has only been one experimental investigation to our knowledge. The thermal vibrations of a boron nitride MWNT have been analyzed, and yield a Young's modulus of 1.22 ± 0.24 TPa [11].

The extremely high moduli and strengths for nanotubes, summarized in Table 11.1, suggest that they might make good mechanical elements on the nanoscale.

Table 11.1. Mechanical Properties of Nanotubes

	Material	SWNT	SWNT Rope	MWNT Arc-Grown	MWNT Chemically Grown	$B_xC_yN_z$ Nanotubes
Young's Modulus	Theoretical	$\sim$ 500 GPa [9]	0.4–0.8 TPa [12, 13]	$\sim$ 1 TPa [12–14]		BN: $\sim$ 0.9 TPa [15, 16]
		$\sim$ 1 TPa [12, 14, 17, 18]				BC_3: $\sim$ 0.9 TPa [15, 16]
		$\sim$ 350 GPa [19]				BC_2N: $\sim$ 1.0 TPa [15, 16]
		$\sim$ 1.2 TPa [15, 16]				C_3N_4: $\sim$ 0.6 TPa [16]
	Experimental	1.25 TPa [20]		1.8 TPa [21]	30 GPa [22]	BN: 1.22 TPa [11]
		$\sim$ 1 TPa [23]		1.2 TPa [24]	450 GPa [25]	
		2.8–3.6 TPa [26]		0.8 TPa [22]		
				0.1–1.2 TPa [27]		
				0.27–0.95 TPa [28]		
				1.7–2.4 TPa [26]		
Shear Modulus	Theoretical	$\sim$ 0.45 TPa [12, 14, 29]		$\sim$ 0.45 TPa [12, 14]		
	Experimental		$\sim$ 1 GPa [23]			
Bending Strength	Theoretical	0.155 nm/d^2 Crit. Curvature [30]				
	Experimental			14 GPa [24]		
Tensile Strength	Theoretical	5–10% strain [31–34]				
	Experimental	> 45 GPa [35]	> 45 GPa [35]	11–63 GPa [28]	1.72 GPa [25]	

11.1.2 TEM Nanomanipulation

Nanomanipulationis a powerful tool. Atomic manipulation experiments using an STM (scanning tunneling microscope) have already demonstrated that nanomanipulation can facilitate the creation of atomic-scale structures that would be difficult to form or study by any other means [36–40]. The main problem with STM is that it requires a flat, conducting surface to work on. Furthermore, the process of acquiring an image of the surface requires scanning the STM tip back and forth many times, and can it can take several minutes just to acquire a single image. Atomic force microscopy (AFM) [41] and high-speed STM [42] lift some of these limitations, but the full benefits of nanomanipulation have yet to be realized. An alternative approach is to combine the manipulation capabilities of an STM, AFM, or related scanned probe with the imaging capabilities of high-resolution TEM. This allows continuous, real-time imaging during nanomanipulation on both conducting and insulating materials.

A nanomanipulation probe, in its simplest form, could simply be thought of as an STM tip. Instead of the location of the tip being controlled by a computer which maps out the surface, control of the position and voltage of the tip is given directly to the operator. It is possible, in principle, to modify surfaces, manipulate atoms, probe the mechanical and electronic properties of nanostructures, and build nanoscale mechanical and electronic systems. A number of such nanomanipulation stages have been developed for TEM applications [43–47]. The stages allow measurements to be carried out on delicate nanosystem geometries which are inherently unamenable to scanned probe techniques, all with real-time TEM imaging. This presents exciting opportunities for the study of physical properties of nanoscale systems, including nanotubes.

The biggest limitation to performing nanomanipulation experiments inside a TEM are the space constraints. The quality of the TEM magnetic lenses dictate the resolution of the microscope, and the highest quality lenses necessarily have the least room for samples inside. The objective lens comprises two conical anvils of magnet iron. The gap between the pole faces varies from microscope to microscope, but is typically about 2 mm. The specimen holder which loads the sample into the microscope can therefore only be about 1 mm in height. Mechanical vibrations can also present a problem for performing experiments inside a TEM, as small vibrations from room noise can completely blur out the atomic-scale features of a TEM image. Working within these constraint makes the design and construction of any custom TEM specimen stage a challenging task. The experiments described below were performed using a custom-built nanomanipulation stages, adapted to high-resolution TEM's.

11.2 Studies of Carbon Nanotubes

In order to perform measurements inside the TEM on individual nanotubes, it is useful to have specimens from which nanotubes protrude in great numbers. In the case of carbon nanotubes, an easy source of such material is the soft inner boule material from a multiwall carbon nanotube arc synthesis [48]. This material contains many small fibers, and when one of these fibers is broken, the broken end has many pristine multiwall nanotubes protruding from it. It is also possible to process nanotubes into tangled mats, sometimes referred to as "bucky paper" [49], and a torn edge of this bucky paper can also be used for manipulation experiments.

Either the boule fibers or the bucky paper can be broken into pieces approximately 50 µm by 500 µm. These pieces can be easily handled with tweezers and attached to the manipulation setup. In a typical setup, the nanotubes are glued in place with conductive adhesive. Sometimes a clean metal counter-electrode is used, and in this case a 50 µm-diameter gold wire serves nicely.

11.2.1 Electrically-Induced Mechanical Failure of Multiwall Carbon Nanotubes

We describe electrically-induced mechanical failure experiments performed at high bias on multiwall carbon nanotubes (MWNT's) inside the TEM. For these experiments, MWNT were protruding from boule fibers glued to the TEM manipulation stage. The stationary electrode was a 50 µm-diameter gold wire. The voltage was applied with a battery-powered voltage source, and the current was monitored with a current amplifier. Both current and voltage were recorded on a computer-based 12-bit analog-to-digital acquisition system.

Typical current-voltage characteristics for carbon MWNT's are shown in Fig. 11.3. The top image shows the first voltage ramp after making contact to the nanotube. The initial resistance was 57 kΩ, which is a typical value. Several general properties of MWNT I-V curves are visible here. First, there is a linear low-field region that is from 0.1–0.3 V wide. at bias values higher than this, the conductance typically increases, as can be seen from the nonlinearity in the I-V curve. At higher values still (above approximately 1.5–2.0 V), the conductance *decreases*, as can also be seen from the nonlinear I-V curve (negative curvature). In the first plot in Fig. 11.3, the current voltage curve shows a non-reversible behavior; the conductance was decreased by increasing the voltage. This is a generic effect, and is apparently caused by current- and field-induced annealing of the contact between the nanotube and the gold metal (or perhaps between the nanotube and its anchor in the boule fiber). Sometimes the annealing effect leads to a gradual charge in the conductance (as in the upper plot of Fig. 11.3) or to discrete jumps in the conductance (as

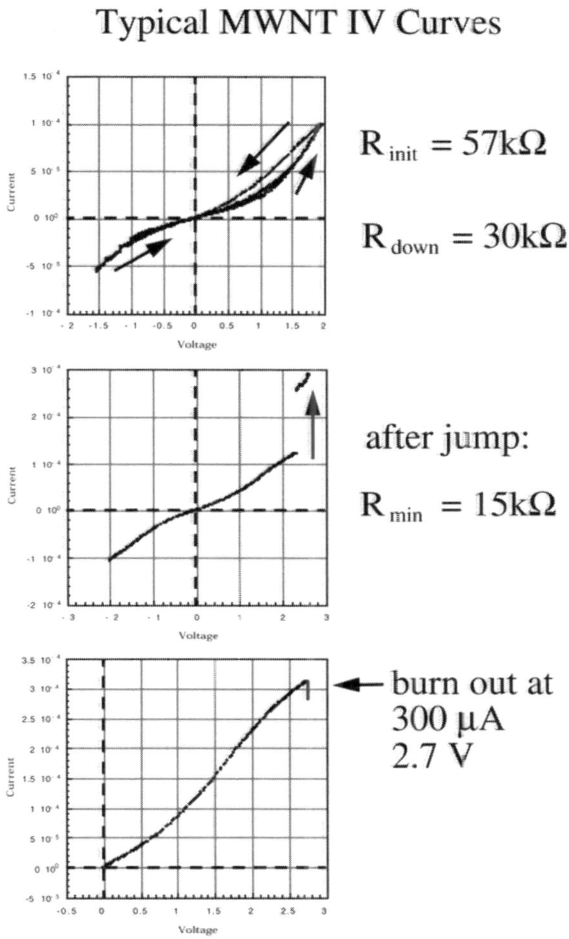

Fig. 11.3. Current-voltage curves taken on a carbon MWNT leading up to failure inside the TEM

in the middle plot of Fig. 11.3). After all annealing is completed, the nanotube from Fig. 11.3 has a low-bias resistance of 15 kΩ, which is typical.

This non-reversible annealing effect is *not* the cause of the nonlinear increase in current between 0.2 V and 1.5 V, as repeated curves through this region track this nonlinear behavior reversibly. This nonlinear effect could be due to conduction through more sub-bands in the nanotubes, or through a field-activated tunnelling phenomenon, such as tunnelling to inner nanotube shells [50], or hopping between localized states. The high-field nonlinearity above approximately 1.5 V (negative curvature) is probably associated with current saturation mechanisms in nanotube transport, and may be indicative

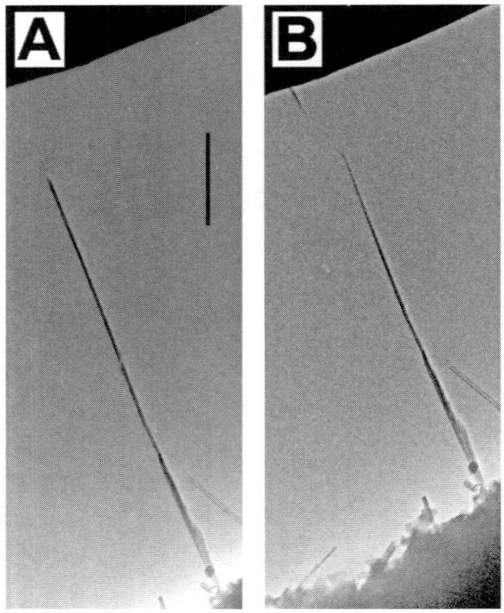

Fig. 11.4. TEM images of a nanotube before (**a**) and after (**b**) electrical failure inside a TEM. The scale bar (vertical in (**a**)) is 500 nm

of the onset of dissipative transport. In this region, we have observed effects which suggest that the nanotubes are undergoing current-induced heating. These are an increased motion (random walk) of carbon onions Van der Waals bonded to the sides of the nanotubes, as well as the evaporation of metal beads deposited on the sidewalls of the nanotubes. This annealing effect has also led to a method for decreasing the contact resistance of nanotubes. Some nanotubes have considerable ($\sim$10 MΩ or greater) contact resistance when initially contacted in the TEM by the gold electrode. If the nanotube is connected in series with a current-limiting resistor of 1 MΩ or 100 kΩ, and the voltage is gradually increased to $\sim$5 V, the two-terminal resistance of the nanotube decreases greatly, often coming into the range of 10–100 kΩ, or less.

At high biases, the nanotubes typically fail by forming a completely open circuit, as shown in Fig. 11.4. Typical failure voltages are between 1.5 V and 4 V, and typical failure currents are between 100 and 300 μA. During the electrical failure, there is very little loss of carbon, unlike what is typically observed for a macroscopic fuse. The details of the failure mechanism have been studied carefully elsewhere [50–52] in oxidizing environments, where multiwall carbon nanotubes are observed to fail layer-by-layer. This failure mode is not observed in the present studies in the TEM, however, as the vacuum prevents the controlled oxidation of nanotube layers. Instead, the nanotube typically fails by a complete cleavage at some region near the middle of the nanotube. The cleaved ends of the nanotubes after failure often exhibit a stepped, sharpened structure, suggesting that there is slightly more loss of carbon from the outer layers of the nanotube than from the inner layers. This is distinct from the peeling and sharpening effect, where only the outer layers are damaged, and the inner layers are left intact (described below).

11.2.2 Peeling and Sharpening Multiwall Carbon Nanotubes

Carbon nanotubes, due to their unique mechanical and electrical properties, are attractive candidates for a host of applications including catalysts [53], biological cell electrodes [54], nanoscale electronics [55, 56], and scanned probe microscope and electron field emission tips [57, 58]. For many such applications it would be desirable to control or shape nanotube geometry. For example, the "ideal" scanned probe, field emission, or biological electrode tip would be long, stiff and tapered for optimal mechanical response, and have an electrically conducting tip. Although recent progress has been made in the growth of nanotubes at pre-selected sites [59] and the modification of nanotube ends through chemical etching [60], finely-controlled nanotube shaping has not been possible.

A simple and reliable method has been developed that allows highly controlled engineering or shaping of multiwall carbon nanotubes (MWNT's) [61]. With this method average MWNT's are easily converted into "ideal-geometry" tips for scanned probe, field emission, or biological insertion applications. The shaping process involves electrically-driven vaporization of successive layers (i.e. tube walls) of the MWNT. Outer nanotube layers are successively removed near the end of the nanotube, leaving the core nanotube walls intact and protruding from the bulk of the MWNT. This peeling and sharpening process can be repeatedly applied to the same MWNT until the very innermost small-diameter tube or tubes protrude, often with a tip radius of curvature comparable to that of a single, single-walled nanotube.

The method has been demonstrated in a transmission electron microscope (TEM) configured with a custom-built mechanical/piezo manipulation stage with electrical feedthroughs to the sample. Figure 11.5 shows high resolution TEM images of a conventional arc-grown MWNT at different evolutionary stages in the peeling and sharpening process. The left end of the nanotube (not seen in the image) is attached to a stationary zero-potential gold electrode. To the right (also not shown) is a larger nanotube which serves as the "shaping electrode"; it is attached to the manipulator whose potential can be externally controlled. Figure 11.5a shows the MWNT in its pristine, as-grown state. For Fig. 11.5b, the shaping electrode has been momentarily brought into contact with the MWNT and a carbon onion has been inadvertently transferred from the shaping electrode to the MWNT, but the applied voltage (2.4 V) and current (170 μA) are below the shaping threshold and no peeling or sharpening has taken place. For Fig. 11.5c, the shaping electrode has been brought into contact with the tip of the MWNT at 2.9 V and 200 μA; numerous layers of the MWNT have been peeled away near its end and the

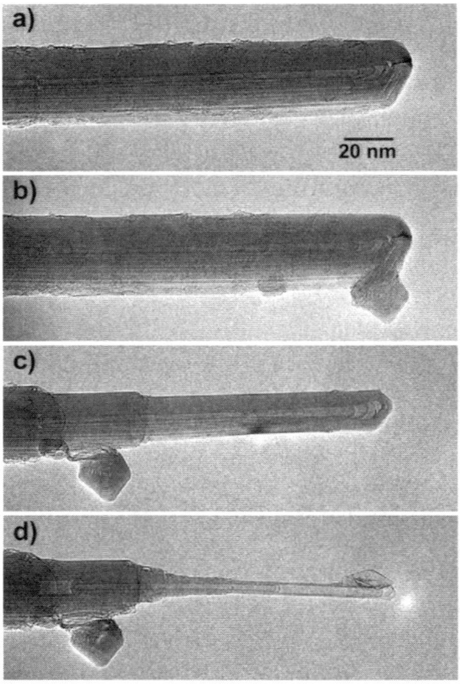

Fig. 11.5. Transmission electron microscope images of a multiwalled carbon nanotube being shaped: (**a**) the nanotube in its pristine form. It contains approximately 37 walls and has outer radius 12.6 nm; (**b**) a carbon onion has been inadvertently transferred to the nanotube end from the shaping electrode, but no attempt has been made to shape the nanotube; (**c**) & (**d**) the results of subsequent peeling and sharpening processes; the onion has simultaneously been displaced to a benign position down the tube axis. The shaped or "engineered" nanotube in (**d**) is thick and mechanically rigid along most of its length (not seen in the image) but tapers stepwise to a fine sharp (and electrically conducting) tip, ideal for scanned probe microscopy or electron field emission applications. The final long nanotube segment contains three walls and has outer radius 2.1 nm

MWNT now has a stepped diameter and is significantly sharpened. The carbon onion has been displaced to a benign position further down the tube. We note that the newly exposed tip of the MWNT appears undamaged. For Fig. 11.5d, the peeling and sharpening process has been repeated, resulting in a MWNT with highly desirable characteristics for many nanotube applications. The dominant protruding segment now consists of a three-walled electrically conducting nanotube with a radius of just 2.5 nm. We note that although we have used an in-situ TEM configuration to document the sharpening and peeling process, this is not a requirement. The process can be performed "blind" and monitored from the electrical characteristics of the nanotube alone. In addition, the "shaping" electrode is easily replaced by a conventional conducting substrate.

This peeling and sharpening effect is distinct from the cleavage effect typically observed in electrical failure of nanotubes (see above). The peeling and sharpening is effected by using current-limiting on the voltage source (a 1 MΩ resistor), and bringing the manipulation tip briefly in contact with the nanotube at gradually larger and larger bias voltages. The peeling and sharpening is sometimes observed without current limiting only on extremely large diameter nanotubes (larger than ~30 nm).

The physics behind the novel peeling and shaping process is intriguing. It appears improbable that uniform Joule heating of the nanotube would result in the observed behavior. It is more likely that MWNT's conduct ballistically [62], and the energy to break the carbon bonds and remove the MWNT layers originates from highly localized dissipation at defect scattering sites, located primarily at the ends of the tube. The fact that only the outer layers are affected suggests that electrical current in MWNT's flows primarily in just the outer carbon layers of the tube, consistent with the conclusions of recent MWNT magnetotransport experiments [63].

11.2.3 Telescoping Nanotubes: Linear Bearings and Variable Resistors

There is great current interest in shrinking the size scale of various mechanical systems, and dramatic recent advances have been achieved in the fabrication of microelectromechanical systems (MEMS) [64]. Ultimately, such systems are expected to reach well into the nanometer domain, and hence considerations of the mechanical behavior of materials at the nano- or atomic-scale, including those related to atomic perfection and friction, become critically important. Carbon nanotubes have unique mechanical [27, 65, 66] and electronic [67,68] properties that make them outstanding candidates for nanomechanical or nanoelectromechanical applications.

In analogy to the well-known lubricating properties of Van der Waals bonded graphite, the individual cylinders of MWNT's might be expected to easily slide or rotate with respect to one another, forming near-ideal linear and rotational nanobearings. Recent theoretical calculations [69–71] indicate that the MWNT interlayer corrugation energy is indeed small, favoring such motion. For a MWNT, one could envision an extension mode much like the "telescoping" of a mariner's traditional spyglass. Some prior evidence for inadvertent MWNT telescopic extension can be found in severe mechanical stress failure mode studies, including MWNT's embedded in a stressed polymer composite [66] and for MWNT's torn apart in quasi-static tensile stress measurements performed in a scanning electron microscope [28]. Controlled and reversible telescoping of MWNT's has been more difficult to achieve.

A major difficulty in initiating controlled telescoping in MWNT's is the commonly capped ends which seal in all inner core nanotube cylinders. Even

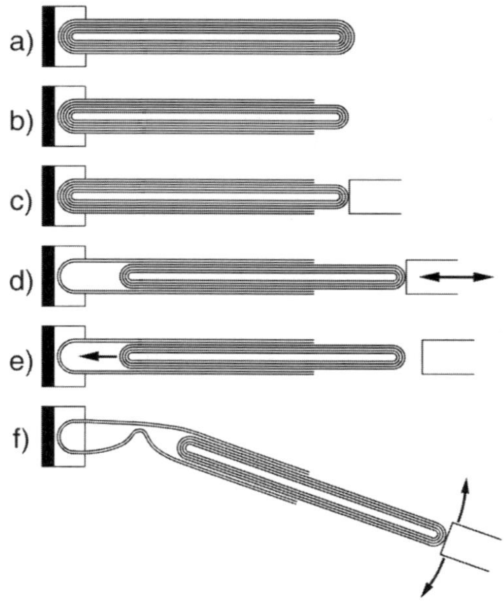

Fig. 11.6. A schematic representation of the experiments performed inside the TEM. Parts (**a**), (**b**), and (**c**) show the process of opening the end of a multiwall nanotube, exposing the core tubes, and attaching the nanomanipulator to the core tubes. Parts (**d**), (**e**) and (**f**) depict three different classes of subsequent experiments performed. In d) the nanotube is repeatedly telescoped, while observations for wear are performed. In (**e**), the core is released and pulled into the outer shell housing by the attractive Van der Waals force. In (**f**) lateral deflections of the nanomanipulator are performed to observe for kinking in the nanotube for different telescoped configurations

if the MWNT ends are opened by methods such as acid etching, it is difficult to selectively contact just the core tubes. One can exploit the peeling and sharpening technique described above to attach a moveable nanomanipulator [27] to only the core nanotubes within a MWNT. In-situ manipulation of the nanotube core allows controlled reversible telescoping to be achieved, and associated forces to be quantified. Robust ultra-low friction linear nanobearings and (constant-force) nanosprings are demonstrated. Mechanical nanotube failure modes, including reversible buckling and reversible collapse, are also investigated.

Figure 11.6 shows schematically the configurations used inside the TEM for the different mechanical experiments. An as-grown MWNT produced by conventional arc-plasma methods is first rigidly mounted (a), and the free end of the MWNT is then engineered to expose the core tubes (b). In (c) the nanomanipulator is brought into contact with the core tubes and, using electrical current, is spot-welded to the core. Figure 11.6c is the common

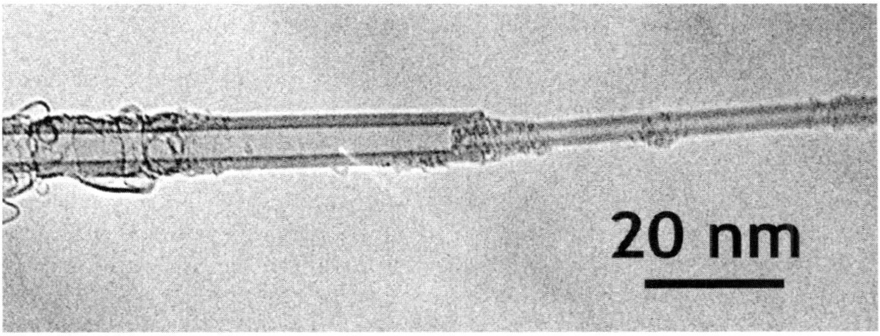

Fig. 11.7. A TEM image of a telescoped nanotube. This particular nanotube originally had 9 shells, but on telescoping a 4 shell core has been nearly completely extracted

starting point for all of the experiments to be described here. Sub-figures (d), (e), and (f) show three different classes of experiments. In d) the manipulator is moved right and left, thus telescoping the core out from, or reinserting it into, the outer housing of nanotube shells. The extration/reinsersion process can be repeated numerous times, all the while viewing the MWNT at high TEM resolution to test for atomic-scale nanotube surface wear and fatigue. In (e) the manipulator first telescopes out the inner core, then fully disengages, allowing the core to be drawn back into the outer shells by the intertube Van der Waals energy-lowering force. A real-time video recording of the core bundle dynamics gives information pertaining to Van der Waals and frictional forces. In (f), a partially-telescoped nanotube is subjected to additional transverse displacements, and reversible mechanical failure modes such as buckling and complete collapse are induced.

Figure 11.7 shows a TEM image of a MWNT in a fully telescoped position. Using higher resolution imaging than that used for Fig. 11.7, we determined that this MWNT originally had 9 walls, with an outer diameter of 8 nm and an inner diameter of 1.3 nm. After extension, a 4 nm diameter core segment (consisting of 4 concentric walls) has been almost completely extracted from the outer shell structure. The telescoping process was found to be fully reversible, in that the core could be completely pushed back into the outer shells, restoring the MWNT to its original "retracted" condition. The process of extending and retracting the core was repeated many times for several different MWNTs, and in all cases no apparent damage to the "sliding" surfaces, i.e. the outer tube of the core or the inner tube of the shell structure, was observed, even under the highest TEM resolution conditions (~2.5 Å). The apparent lack of induced defects or other structural changes in the nanotube contact surfaces at the atomic level suggests strongly that these near atomically-perfect nanotube structures may be wear-free and will not fatigue even after a very large number of cycles.

Telescoping Forces: A Linear Bearing

In the engineering of macroscopic bearings, the moving parts are typically cycled 10^3 to 10^9 times before definitive conclusions about wear can be drawn. This is because the damage from a single cycle is microscopic and cannot be readily observed by eye or even conventional microscopy. Here, atomic-scale imaging of the structure is possible and shows that after all cycles, from the first cycle onward, the atomic structure of the nanotubes is unaffected by the motion. Hence it is possible to conclude that the nanotube sections are near-perfect sliding surfaces, apparently free from wear for all cycles. Interestingly, in all MWNT's that we have examined, all repeated sliding motion for a given MWNT was observed to take place between the same two nanotube shells (and similarly no "multiple" telescoping was observed where the total length of the nanotube might become more than double the length of the original MWNT, as would occur from "sticking" of the segments at some point in their extension). We interpret this repeatability as a self-selection process where the most perfect surfaces offer the least resistance to motion. Importantly, even after repeated motions, the same surfaces remained the "most favored" ones, again providing evidence for no sliding-induced wear on the active surfaces. (Of course, in a many-walled MWNT, even the catastrophic failure (i.e. fusing) of one surface pair would not render the MWNT bearing unusable, as another (nearly equally perfect) surface pair would simply become the active elements). In some cases, there is a small amount of residual amorphous carbon produced during the initial core exposure process. Interestingly, because of the extremely small interwall clearance in MWNT's, such contamination appears to have no effect on the bearing action as it is simply brushed away upon reinsertion of the core section into the nanotube housing. Hence, MWNT-based linear bearings are self-cleaning and immune from typical contaminant-induced wear.

Several internal forces are associated with telescoping MWNT's. To first order these consist of the Van der Waals-derived force and possible static and dynamic frictional forces. The Van der Waals force is given by

$$F_{\mathrm{vdW}} = -\nabla U(x) \tag{11.1}$$

where the Van der Waals energy [72] is given by $U(x) = 0.16Cx$ joules with C the circumference of the "active" nanotube bearing cylinders and x the length of the overlap between the core section and the outer walls, both measured in meters. The Van der Waals energy lowering gained by increasing the tube-tube contact area tends to retract the extended core of a telescoped MWNT. Interestingly, since the active intertube contact area decreases linearly with core tube extension, this restoring force is independent of contact area, or equivalently, independent of core extension. Hence, a telescoped nanotube with only one active (sliding) surface pair is expected to act as a *constant force* spring.

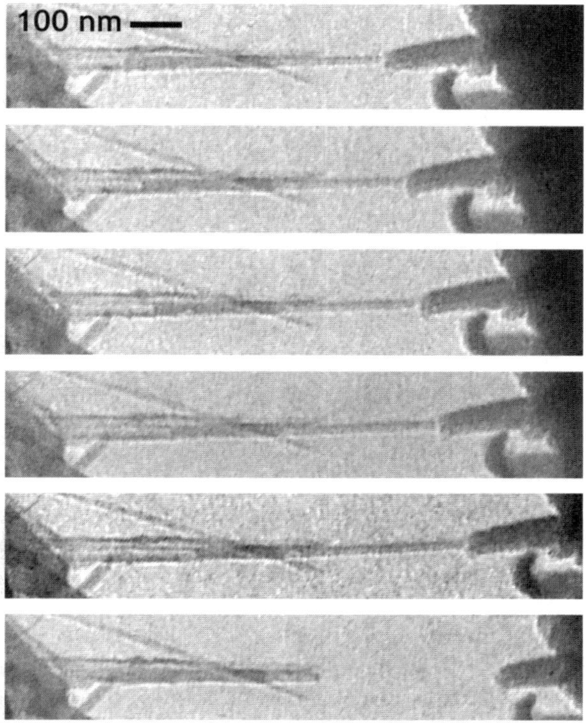

Fig. 11.8. Selected frames of a video recording of the in-situ telescoping of a multiwall nanotube. In the first five frames, the core nanotubes are slowly withdrawn to the right. In the sixth image, which occurred one video frame after the core was released, the core has fully retracted into the outer nanotube housing due to the attractive Van der Waals force

To determine experimentally if F_{vdW} dominates nanotube linear bearing dynamics we have used the configuration described in Fig. 11.6. The core tubes of a MWNT were first telescoped using the manipulator. Lateral deflections of the manipulator were used to fatigue and eventually break the spot weld, thus releasing the core segment. The resulting accelerated motion of the released core segment was recorded using a continuous video system tied to the TEM imaging electronics.

Figure 11.8 shows several selected frames from one such video recording. The upper five frames show the core segment being slowly and successively telescoped to the right (the structure in the left third of the image seen crossing the MWNT at about a 30° angle is another nanotube unrelated to the experiment and it is not in physical contact with the subject MWNT). Just after the fifth frame the manipulator has released the core segment. The sixth and final frame, which occurred one video frame after the release of the nanotube, shows the core after it rapidly and fully retracted inside the outer shells of the MWNT. The dimensions for the core segment of the MWNT of Fig. 11.8 yield a core segment mass 2.9×10^{-16} g. Combining this with $C = 57$ nm and the initial extension of 330 nm, (11.1) leads to complete retraction of the core tubes in 4.6 nsec. This is at least consistent with the experimental observation that the complete contraction occurred in less than one video frame (33 msec).

From TEM observations, we may also draw conclusions about the static and dynamic friction between concentric shells of a multiwall nanotube. While macroscopic models of friction between solids dictate that friction is proportional to normal force, independent of contact area, modern microscopic models of friction predict that friction is in fact proportional to contact area [73]. In macroscopically rough samples, the actual contact occurs at point asperities, and the microscopic contact area is proportional to the total normal force. Nanotube shells, however, are atomically smooth, so any interlocking between the shells (due, for example, to the atomic corrugations) is best estimated by using the entire surface area of contact. The F_{vdW} retraction force for the nanotube in Fig. 11.8 is calculated to be a mere 9 nN. This indicates that the static friction force is small, with $f_s < 2.3 \times 10^{-14}$ Newtons-per-atom (6.6×10^{-15} Newtons-per-Å^2). Furthermore, from the fact that the tube fully retracts, we conclude that the dynamic friction $f_k < 1.5 \times 10^{-14}$ Newtons-per-atom (4.3×10^{-15} Newtons-per-Å^2). Friction is an important concern in small-scale systems, such as microelectromechanical systems, or MEMS [64], and recent atomic-scale frictional force measurements [74, 75] using conventional materials yield values approximately three orders of magnitude greater than the upper limit frictional forces found here for MWNT surfaces.

It is also possible to directly demonstrate through the images in Fig. 11.8 that the retraction force is, in fact, constant throughout the telescoping of the nanotube. The nanotube which was used as the anchor to spot weld to the tip of the core nanotube section was actually poorly anchored to the manipulator tip. It could be flexed in and out of the manipulator tip with a reasonable Hooke's law spring constant. By pushing this anchor up against a long nanotube, and using the known Young's modulus for MWNT [21], we estimated the spring constant for this nanotube deflection to be ~0.1 Newton-per-meter. In Fig. 11.9, we show the same images offset to that the manipulator tip (as noted by the kinked nanotube in the bottom right corner of each image) is at the same horizontal position in each frame. In these images it is possible to notice that the tip is at the same extension in each frame, implying that the retraction force is constant throughout the extension. We can also estimate the force from the spring constant to be ~5nN, and we estimate that it is constant throughout the telescoping to within ±0.3 nano-Newton. Of course, the absolute accuracy of this order-of magnitude estimate is questionable, and is based on assumptions of linearity of the manipulator tip force response. The calculation is also subject to errors in the Young's modulus of the nanotube

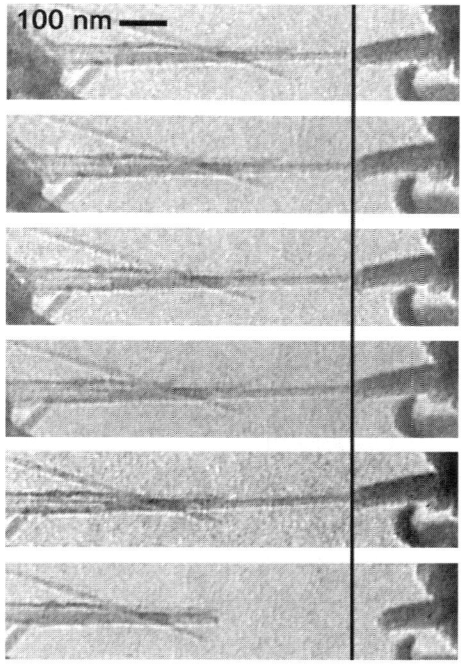

Fig. 11.9. The same images as Fig. 11.8. Here, the images have been offset to demonstrate that the anchoring nanotube, immediately to the right of the vertical line in the upper 5 images, feels a constant retraction force throughout the telescoping

used to calibrate it, but it is a good order-of-magnitude estimate of the retraction force. It also demonstrates the constancy of the retraction force, and the value agrees well with the calculations in the previous paragraph, which yield 9 nano-Newtons.

A note of interest is that if the (already generally small) friction between nanotube shells is indeed proportional to the length of the overlapping sections, then minimum bearing friction would be obtained for the shortest possible nanotube core/housing overlap. Several possibilities exist to achieve this configuration. First, a very short MWNT might be used from the outset. Second, the core of a long MWNT could be telescoped nearly all the way out, yielding a short contact area. The third possibility takes advantage of "bamboo" configurations [76] which sometimes occur in MWNTs, where the inner nanotube sections do not extend along the entire length of the MWNT, but rather represent an end-to-end series of shorter, fully capped, nanotube segments residing inside the continuous housing of outer nanotube shells. Figure 11.10a shows a close-up TEM image of such a bamboo joint inside a larger MWNT. In Fig. 11.10b, the core segment on the right has been telescoped

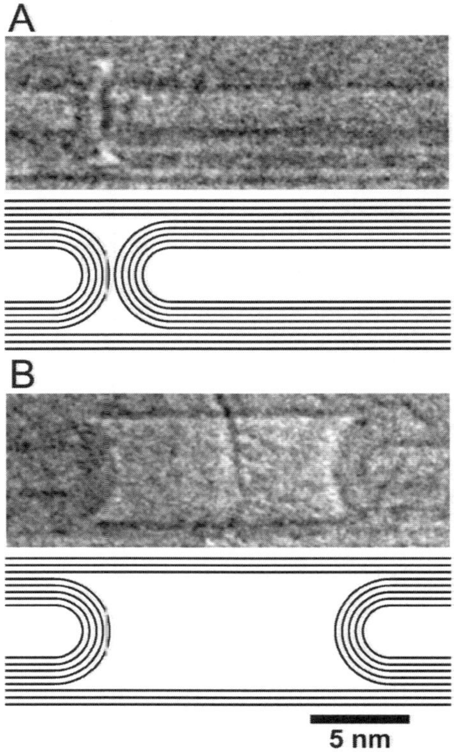

Fig. 11.10. TEM images of a bamboo section of a multiwall nanotube: (**a**) an as-grown bamboo section; (**b**) the same area after the core tubes on the right have been telescoped outward. Bamboo sections are independent and are free to slide inside their nanotube housing. The line drawings beneath the images are schematic representations to guide the eye

out, cleanly separating the bamboo joint. This is the first demonstration that bamboo nanotube core segments are truly independent sections with weak end-to-end binding. This observation has important practical implications for common-axis bearings: short independent bamboo sections might serve as ultra-low friction linear or rotational bearings that are firmly embedded in a common long, stiff cylindrical housing.

We now briefly consider lateral deformations of partially telescoped nanotubes. It has been predicted and observed that nanotubes can, upon lateral deformation, form kinks [65] or even fully collapse [77]. MWNT's with large inner diameters and few concentric shells are particularly susceptible to kinking and collapse. Although kinked or fully collapsed nanotubes have been observed experimentally using static TEM [65, 77] or atomic force microscopy methods [78], in-situ high-resolution controlled and reversible deformation studies have been difficult. As outlined in Fig. 11.6f, using partially telescoped nanotubes we can study the kinking and collapse of a controlled nanotube system. We can directly alter the inner diameter and aspect ratio of the hollow nanotube, apply lateral forces, and in real time observe resulting failure modes. As expected, we find that a MWNT will kink and collapse much more readily after the inner core has been removed. One particular nanotube had 60 original layers with an outer diameter of 43 nm. A 40 layer core was telescoped out to a maximum extension of 150 nm, leaving an outer shell housing of just 20 layers with an inner diaster of 29 nm. The housing was supported at the base and the inner core section of the tube was still engaged in the housing for a length of 200 nm. When the manipulator was driven laterally to approximately $\sim 5°$ angular displacement, the housing shells developed a kink in the middle of the large inner diameter section. At $\sim 26°$ displacement the kink was severe and resembled the schematic in Fig. 11.6f. At any displacement angle, the telescoped core section was still mobile, and could be moved back and forth inside the unkinked portion of the outer shell housing. At small kink angles less than $\sim 10°$, the core could be inserted past the kink position, forcing the kink to disappear and re-inflating the outer shells to their original circular cross section. At more severe bending angles, in excess of $\sim 20°$, the kink blocked the inner core section from being fully inserted. Hence, suitable kinking of the outer shell housing provides an effective motion stop for nanotube core insertion.

The above results demonstrate that MWNT's hold great promise for nanomechanical or nanoelectromechanical systems (NEMS) applications. Low-friction low-wear nanobearings and nanosprings are essential ingredients in general NEMS technologies. The expected order 1–10 nsec transit time for complete nanotube core retraction implies the possibility of exceptionally fast electromechanical switches.

Telescoping Resistance: A Variable Resistor

The study of carbon nanotube electronic transport is an active experimental and theoretical research area. It has been demonstrated that single-walled carbon nanotubes (SWNT's) are physical realizations of ideal quantum wires with ballistic conduction [79–81]. Electronic conduction through multiwall carbon nanotubes (MWNT's), on the other hand, is complicated by inter-shell interactions and possible wavefunction interference between electronic states on different shells of the composite tube. For charge transport in such MWNT's the experimental understanding is complicated, with different studies variously supporting quantized electrical conductance, conductance dominated by transport along the outermost shell, diffusive scattering, and current-induced damage [52, 61–63, 80, 82–84]. Part of the difficulty is that most transport experiments are performed without detailed knowledge of the MWNT geometrical characteristics as might be obtained via high-resolution

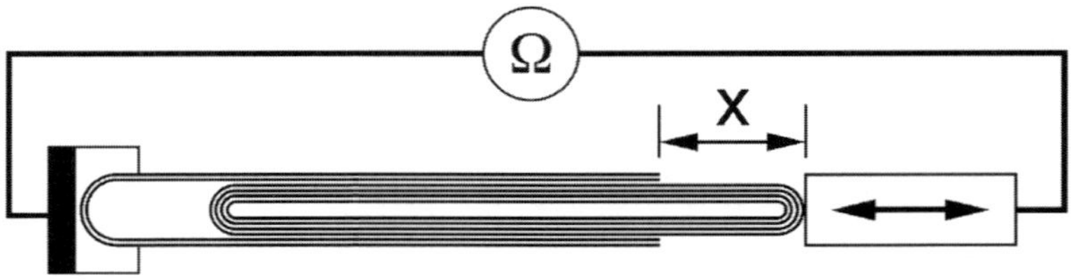

Fig. 11.11. Schematic diagram of telescoped MWNT resistance measurements. On the left is a nanotube anchored inside the TEM, and on the right is the manipulation tip. The nanotube is reversibly telescoped while the dc electrical resistance is measured with a low noise current amplifier

transmission electron microscopy (TEM). In addition, substrate or nanotube surface contamination effects may play a role.

Recently, it was demonstrated that the core shells of MWNT's could be controllably telescoped from their outer shell housings, thus constituting low-friction linear bearings [61, 85–87]. Such experiments were performed in-situ inside a high-resolution TEM, allowing simultaneous imaging of the MWNT system. An interesting measurement enabled by this geometry is the determination of the electrical resistance, R, between the ends of a telescoped MWNT as the telescoping distance, x, is varied. In this Letter, we report on the results of such measurements. For all nanotubes studied, R is found to increase nonlinearly with increasing x. The functional form of $R(x)$ and the electric field dependence of R are consistent with predictions for a one-dimensional localized system, with a room-temperature carrier localization length 1000–1500 nm. $R(x)$ is highly reproducible, suggesting that telescoped MWNT's constitute reliable high-resolution nanoscale rheostats.

The experimental starting point was MWNT's whose core tubes had been partially exposed and mechanically contacted using methods previously described [61, 85]. Briefly, one end of a bundle of surfactant-free (i.e. unprocessed) arc-grown MWNT's was embedded in silver paint and attached to a stationary electrode. The furthest protruding nanotube was peeled open and the protruding core tubes were contacted via spot welding to a piezo-driven nanomanipulator electrode. During the experiment, the dc electrical resistance between the two electrodes was monitored as the nanomanipulator (reversibly) telescoped the inner core tubes out from, or collapsed them back into, the outer nanotube housing.

Figure 11.11 shows schematically the experimental configuration. The telescoping or extraction distance x was monitored both via the nanomanipulator piezo drive voltages and analysis of direct TEM video recordings of the telescoping process. To aid in the analysis, each video frame contained critical data overlays. Throughout the telescoping process only one pair of adjacent concentric nanotube shells displayed relative motion, i.e. there was no multi-segment telescoping behavior. Typical MWNT initial (i.e. fully col-

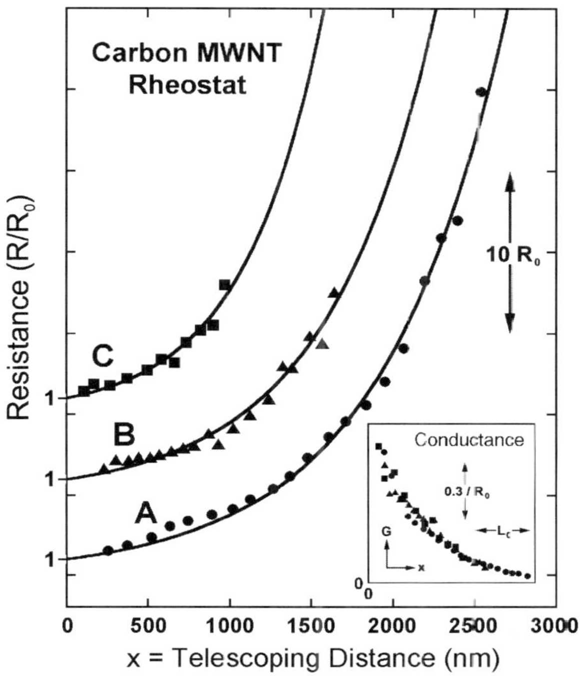

Fig. 11.12. Resistance R vs. telescoping distance, x, for three different nanotubes. The data are normalized to $R_0 = R(x \to 0)$. The inset shows the same data plotted as conductance ($1/R$) as a function of normalized telescoping distance (x/L_0). Note that the conductance data are also nonlinear and collapse onto a universal curve

lapsed) lengths were from 1 to several microns. The nanotube system was always suspended in high vacuum inside the TEM (Topcon 002B, operating at 100 keV), thus eliminating potentially spurious substrate effects.

Figure 11.12 shows, for three independent MWNT's labeled A, B, and C, the resistance R between the ends of the telescoping nanotube system as a function of telescoping or core extraction distance x. The solid symbol data points have been normalized to R_0, the resistance of the $x \to 0$ unextended MWNT. The values of R_0 ranged from 3.5 kΩ to 110 kΩ. No extrinsic contact resistance is subtracted from the data we report here. Data such as those shown in Fig. 11.12 were recorded for successive extension and contraction runs and were found to be reproducible without evidence of significant hysteresis or temporal drift. The resistance data of Fig. 11.12 are clearly not linear in telescopic extension. To fit the data to a nonlinear functional form, we tentatively choose

$$R(x) = R_0 \exp\left(\frac{x}{L_0}\right), \tag{11.2}$$

with L_0 a characteristic length to which we attach a physical meaning below. The solid lines in Fig. 11.12 are fits of (11.2) to the three experimental $R(x)$ data sets. Fitting parameters for the three curves are given in Table 11.2, along with other relevant nanotube parameters obtained via TEM imaging, such as nanotube length and core and housing outer diameter. The inset of Fig. 11.12 shows the same data plotted as conductance, $G(x) = 1/R(x)$, as a function of the normalized telescoping distance, x/L_0; the conductance also exhibits nonlinearity with positive second derivative. From this observation and simple calculus manipulations we can rule out various functional forms for $R(x)$: any series combination of linearly changing resistance or conductance, any logarithmic dependence, or any power law, $R(x) \propto x^{\alpha}$, where $\alpha \leq 1$. Functional forms that are consistent with this observation include an exponential functional form, such as (11.2), or a power law with $\alpha > 1$. We also note that in our telescoping experiments we have never observed conductance resonances or plateaus as might be expected from mercury-dipping experiments using non-telescoped MWNT's [62, 82, 83].

We now discuss these experimental findings. We first examine two candidate models for the resistance behavior: diffusive transport and junction tunneling. In the (semi-classical) diffusive transport model the increase in resistance with telescoping results from strictly geometric considerations. Two hollow and concentric, tightly fitting metal cylinders can show an increasing R that is not strictly linear with increasing telescoping distance x. Even assuming an isotropic conductivity tensor, this model is difficult to solve exactly, but in the limit of very thin cylinder walls or very long telescoping distances (as we have for the MWNT system), the model predicts a linear increase in R with increasing x. Under no circumstances does the generalized model yield the observed unique nonlinear length dependence.

In the junction tunneling model, the dominant contribution to the resistance arises from charge carriers tunneling from the telescoped nanotube core to the nanotube housing, while along the core and housing regions themselves the charge transport is assumed ballistic with no inherent length de-

Table 11.2. The sizes of the nanotubes telescoped in this study, and the parameters extracted from the fits to the data shown in Fig. 11.12. The nanotube core section inner diameters are not given as they could not be unambiguously resolved from the microscopy images, but from observations of other nanotubes in our sample, the inner diameters are typically 1–3 nm. The R_0 and L_0 values are obtained by fitting the resistance data to (11.2).

Nanotube	Housing Outer Diameter	Core Outer Diameter	Core Length	R_0	L_0
A	25 nm	18 nm	2.6 μm	3.5 kΩ	760 nm
B	18 nm	7 nm	1.7 μm	54 kΩ	670 nm
C	20 nm	5 nm	1.1 μm	110 kΩ	490 nm

pendence to the resistance. A tunneling probability directly proportional to the core/housing overlap junction area then leads to a linear decrease in conductance, $1/R(x)$, with increasing x. This inset of Fig. 11.12 clearly shows that the conductance also decreases nonlinearly, ruling out this possible explanation. Indeed, our data suggest that the conductance of the sliding contact between the nanotube core and its housing is actually quite good, in agreement with independent experimental [50] and theoretical [88, 89] studies. The addition of a contact resistance to the junction tunneling model does not improve the fit to the experimental conductance data (this would cause a functional form with negative second derivative).

We propose that the correct interpretation of the observed nonlinear $R(x)$ is on-tube quantum interference and localization of electron wavefunctions similar to that first proposed by Landauer and Thouless for low-dimensional systems with defect scattering [90, 91]. In the original localization model, a conduction channel has a distribution of scattering sites, and the effect of increasing the length of the channel is to cause a linear increase in the number of scatters in the path of the transport current. Due to quantum interference, however, each scatter introduces a multiplicative factor in the averaged total system resistance, resulting in a resistance that increases exponentially with the length x of the system. The model also predicts an exponential dependence of resistance on temperature. While the signature of localization in the temperature dependence of resistance has been experimentally verified for a host of low-dimensional systems, including carbon nanotubes and graphitic materials [92–95], the unambiguous observation of an exponential length dependence of the resistance of a one-dimensional system has remained experimentally elusive[1].

The critical parameter in the localization model is the localization length l_c, representing the probability amplitude decay length of the (non-extended) electronic wavefunctions. For a two-terminal resistance measurement as employed for our telescoped MWNT system, the total resistance can be expressed within the localization model as [97][2]

$$R(x) = R_0 \exp\left(\frac{2x}{l_c} - 1\right) + R_c, \tag{11.3}$$

with R_c the resistance of the fully collapsed nanotube. With the substitutions $l_c = 2L_0$ and $R_0 = R_c$, (11.3) is in fact identical to (11.2). From our previous fits in Fig. 11.12 we find that the localization length l_c ranges approximately from 1000 nm to 1500 nm in the telescoped MWNT systems here investigated (see Table 11.2). The correspondence between the scaling factor

[1] This theoretical model has also been implicated in recent measurements of nonlinear resistance in single-walled nanotubes [96].

[2] Our experiments are performed at room temperature where conventional localization theory is applicable. At very low temperature, where electron correlation effects play a major role, (11.3) would require modification.

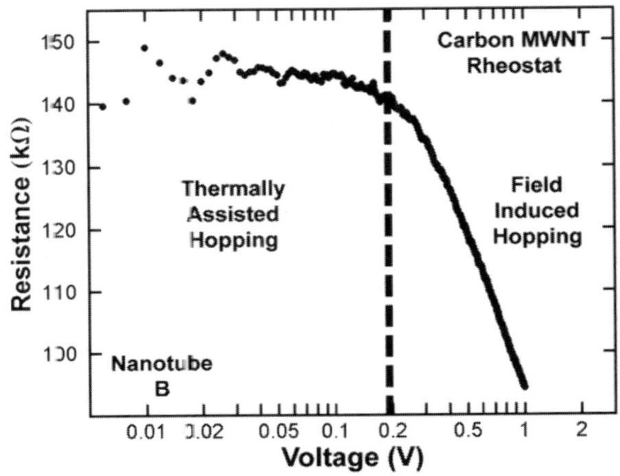

Fig. 11.13. Resistance vs. applied voltage of nanotube B when it is telescoped at a distance of $x = 580\,\mathrm{nm}$. The vertical *dashed line* is the calculated delineation between thermally-assisted carrier hopping and electric field assisted carrier hopping (see (11.4) and text)

R_0 and the collapsed tube "contact resistance" R_c suggests that R_c enters into the MWNT localization problem in the same way as do the intrinsic scattering sites, namely as a multiplicative factor.

Localization theory makes additional, experimentally testable predictions concerning the electric field E dependence of the electrical resistance [98]. For low applied E, ohmic conduction should be observed, while for high applied E the range of accessible states is increased leading to nonlinear conductance. The different field regimes can be characterized by the parameter [98]

$$\beta = \frac{eEl_c}{k_B T}, \tag{11.4}$$

where e is the electron charge, k_B is Boltzmann's constant, and T is the absolute temperature. For $\beta \ll 1$ transport is dominated by thermally-assisted hopping, while for $\beta \gg 1$ transport is dominated by field-induced hopping. We have measured the electric field dependence of the telescoped MWNT system, as shown in Fig. 11.13 for nanotube B at a fixed telescopic extension. At low applied voltages the resistance is approximately constant with some fluctuations due to experimental noise. At high applied voltages, however, the resistance decreases precipitously with increasing voltage. The vertical dashed line in Fig. 11.13 corresponds to $\beta = 1$, calculated using $l_c = 1340\,\mathrm{nm}$ as determined from the $R(x)$ fit for nanotube B. This demarcation agrees well with the observed transition from ohmic to strongly nonlinear resistance, providing additional support for carrier localization.

What is the origin of the "disorder" leading to carrier scattering and localization in the telescoped nanotube system? Some experiments have suggested that current through MWNT's flows almost exclusively on the outermost nanotube shell [63]. In a possible model, structural defects in this shell could then be a source of system disorder. In the present experiments, the outermost shell is not connected to both electrodes, but it is possible that the current flows in the outermost shell of the core segment until reaching the core/housing overlap, from which point current flows in the outermost shell of the housing segment. Thus, the effective length of the "outer" nanotube shell increases as the MWNT is telescoped. Alternatively, there is some evidence that even in non-overlapped regions of the nanotube, the transport current may probe more than one nanotube shell [50]. For typical nanotubes at room temperature, it has been suggested that approximately 3–7 shells contribute significantly to the current. Interlayer chirality mismatches can yield an effective disorder potential (analogous to that originally considered by Anderson) for the charges traversing the otherwise defect-free nanotube. Indeed, it has been demonstrated in theoretical studies that atomic mismatch between otherwise defect-free shells can lead to diffusive transport along the axis of MWNT's [99].

Our transport results have important implications for the degree of phase coherence for charge carriers in telescoped MWNT's. The nonlinear $R(x)$ behavior observed here implies that the carriers maintain a degree of phase coherence between scattering events, and that the phase coherence length $l_\phi \gtrsim l_c$. The exact relationship implied between the phase coherence length and the localization length is not clear, mainly because the regime of $l_\phi \sim l_c$ is not theoretically well understood [100]. An unambiguous exponential length dependence would be a clear sign of $l_\phi \gg l_c$. However, the data presented here do not unambiguously prove an exponential relationship. Such a proof typically demands data that vary over many decades. A general functional form of $R(x)$ is not known, to our knowledge, in the regime of $l_\phi \sim l_c$. The exponential form probably persists as l_ϕ is reduced, giving rise to other nonlinear forms for $l_\phi \sim l_c$ (such as a power law as suggested above for our data), then eventually to a linear form for $l_\phi \ll l_c$ (weak localization). However, even with conservative estimates, the phase coherence length is surprisingly long (perhaps as long as the nanotube itself), given that the experiments were performed at room temperature. The unique properties of nanotubes (including low concentrations of defects, weak electron–phonon coupling, and unusual Debye temperature and screening length [101–103]) are most likely responsible for the implied long l_ϕ values. This long phase coherence length observed here for MWNT's may have relevance to phase-coherent scanning tunneling microscopy, dual-tipped phase-coherent electron sources [104], or quantum computing systems.

From a nanoscale rheostat applications viewpoint, telescoped MWNT's have attractive mechanical and electrical characteristics. They may find util-

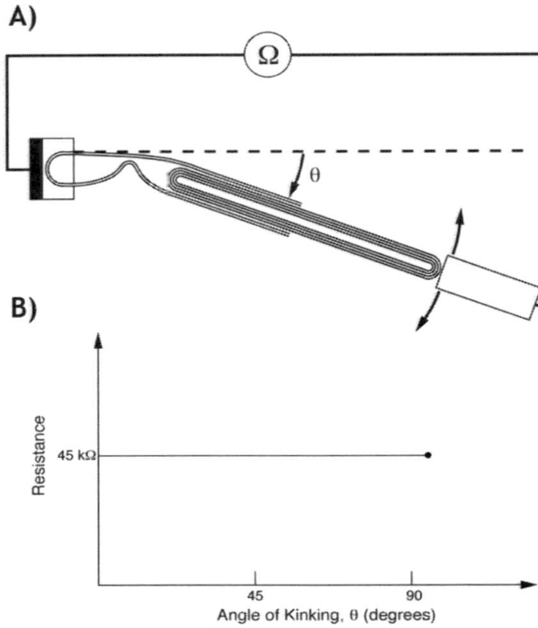

Fig. 11.14. Schematic of an experiment to measure resistance versus kinking for a MWNT

ity as tuning resistors, local strain gauges, or position sensors. Highly sensitive position encoders with nm-scale resolution have been previously developed [41, 105, 106], but in general those devices require large-scale detection systems making them impractical for efficient integration in self-contained nano-scale systems.

We now present a small set of preliminary studies on the resistance of MWNT studied as a function of bending and kinking. As shown schematically in Fig. 11.14A, a MWNT is first partially telescoped out, then lateral forces are applied to the ends of the core structure to induce controlled "kinking" in nanotube fabric. The kinking occurs in the large diameter "hollowed out" region of the tube, were the housing tubes are no longer supported by the inner core tubes. The kinking position can be controlled by varying the extension of the core tubes. All the while, the resistance between the ends of the tube is monitored. Figure 11.14B schematically shows the measured electrical resistance as the tube is kinked from its unperturbed straight configuration (θ=0°) to a severe right-angle kink (θ=90°). Surprisingly, no resistance changes are observed even for such extreme kinking. All kinking here performed was fully reversible, with no permanent tears in the nanotube fabric. Our findings suggest that even severely mechanically deformed nanotubes are good electrical conductors (somewhat akin to flexible electromagnetic waveguides). It should be noted here that the bias on the nanotube and the gain on the current amplifier have been optimized to detect small changes in the resistance. Resistance changes as large as ~25% may not have been detected during the experiments reported here, and a more careful study of the effects of kinking on nanotube transport would be beneficial.

11.3 Studies of Boron Nitride Nanotubes

We describe field emission and contact current-voltage measurements on individual BN nanotubes. Figure 11.15 shows the setup for the measurements. The nanotubes were manipulated with a piezo-driven manipulation stage inside the TEM. The BN nanotubes are double-wall nanotubes synthesized by arc discharge [107], and purified by ultrasonic assisted filtration [108]. The as-purified BN nanotube samples are electrically insulating in bulk, but for the TEM in-situ measurements, the nanotubes were mixed with a conductive epoxy (Epo-Tek H20E) in a volume ratio of approximately 1:1. The resulting composite was then cleaved before loading into the TEM, and TEM imaging verified that clean nanotubes were protruding from the surface of the composite. The second electrode for the electrical measurements was a 50 μm gold wire. For field emission experiments, the tips of the protruding nanotubes were positioned 6 micron from the surface of the wire, and for contact IV measurements, the tips of individual nanotubes were gently brought into contact with the gold wire by the piezo manipulator. Control experiments were also performed to verify that the conductive epoxy itself does not field emit at the voltages applied in these experiments. The current was monitored in all cases with a high gain preamplifier.

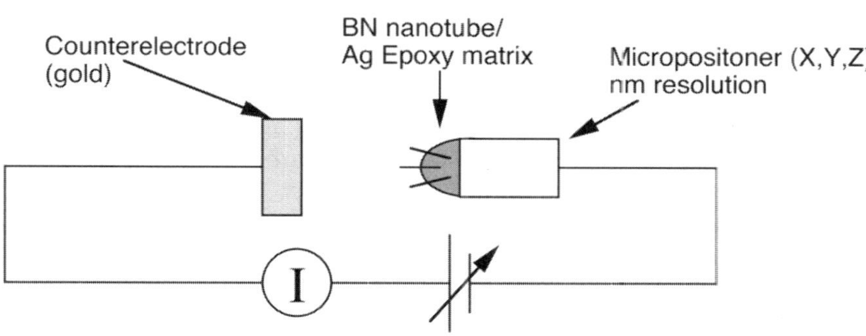

Fig. 11.15. The experimental setup for BN nanotube electrical measurements inside the TEM

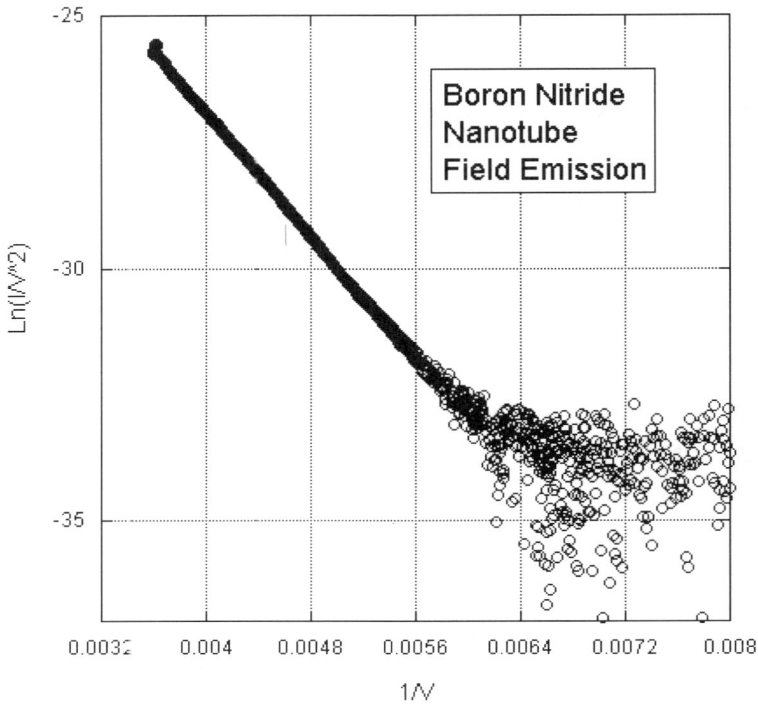

Fig. 11.16. Fowler–Nordheim plot of the current-voltage characteristic of field emitting BN nanotubes

11.4 Electron Field Emission from BN Nanotubes

Figure 11.16 shows the results of field emission experiments. The turn-on voltages are approximately 150 volts. This is twice the turn-on voltage measured in control experiments performed using carbon nanotube samples in similar geometries. The current densities, however, are similar in both cases. One notable difference between the field-emission behavior of BN and carbon nanotubes is the current-voltage characteristic. Generally, field emission is characterized using the theory of Fowler and Nordheim [109], which predicts a linear relationship between $\ln(\frac{I}{V^2})$ and V^{-1}. For carbon nanotubes, however, the theory often breaks down, frequently with the current showing saturation at high fields, or two distinct Fowler–Nordheim slopes [110]. For the field emission for BN nanotubes we have observed a single slope in the Fowler–Nordheim plot. Additionally, carbon nanotubes usually have a noisy, switching behavior most likely associated with gas molecules adsorbed to the tips of the nanotubes [111]. Boron nitride nanotubes, however, show stable field emission with less noise than for typical carbon nanotube samples. This may have implications for the use of BN nanotubes as stable field emission sources for lighting and flat panel displays.

From the slope of the Fowler–Nordheim characteristic, it is possible to extract the work-function of the field emitting material. Following the analysis in [110], it is possible to model the field at the tip of the nanotube by modelling the nanotube tip as a hemisphere. In this model, the field at the tip of the nanotubes can be expressed as

$$E_{\mathrm{loc}} = \frac{V_{\mathrm{applied}}}{(\alpha R_{\mathrm{tip}})}, \tag{11.5}$$

where R_{tip} is the radius of the nanotube tip (in nm) and α is a parameter which captures the details of the geometry of the field emission environment. In this model, the work function (in eV) can then be extracted from the Fowler–Nordheim slope (S_{FN}) as

$$\Phi = \left(\frac{-S_{\mathrm{FN}}}{6.8\alpha R_{\mathrm{tip}}} \right)^{2/3}. \tag{11.6}$$

As in [110], we take the enhancement factor to be $\alpha \sim 10$. Over repeated field-emission experiments performed using BN nanotubes inside the TEM, the slope of the Fowler–Nordheim characteristic varies from -2600 to -3100. From independent TEM observations, the radius of the tips of the BN nanotubes has been measured to be approximately 1 nm. From these quantities, it is possible to make a rough estimate of the work function in the range of 11–13 eV. This large work function is obviously unphysical, and may indicate that field emission from BN nanotubes does not truly obey the Fowler–Nordheim theory. If instead we use the expected work function for hexagonal BN (approximately ~6 eV), (11.6) can be used to calculate the radius of the tip of the field emitting nanotube. From this analysis, $R_{\mathrm{tip}} = 3$ nm.

It is surprising that boron nitride nanotubes can pass any current at all, given that they are wide bandgap semiconductors. Such a counter-intuitive observation deserves careful inspection. We note that the nanotubes in these experiments were observed in the TEM where the are subjected to possible electron beam-induced damage. It may be possible that this beam damage affects the intrinsic properties of the nanotubes, leading to the conductivity. Control experiments, however were performed outside the TEM in a vacuum field-emission setup. In this setup, the nanotubes exhibit field emission with characteristics similar to those observed in the TEM. Therefore, electron beam damage is not essential for field emission from boron nitride nanotubes. The nanotubes are also observed in the TEM to be partly covered with amorphous material from the binder of the conducting epoxy used in the preparation of the nanotube composites. It may be that this material plays some role in the field emission, and future studies using mixtures with different types of binder material, or little to no binder may help address this issue.

An additional useful future study would be to look carefully at the current noise during field emission from boron nitride nanotubes. Here, it was observed that BN nanotubes seem to have less noise during field emission, but it would be useful to put this observation on a more solid empirical footing. For instance, the noise frequency spectrum could be carefully measured [112] during field emission from both BN nanotubes and from carbon nanotubes (both SWNT and MWNT). Comparison of the noise spectra would show whether BN nanotubes are indeed less noisy than carbon nanotubes, and may even help elucidate the source of the noise.

11.5 Electrical Breakdown and Conduction of BN Nanotubes

The field emission properties of BN nanotubes are surprising given that they are predicted to be wide bandgap semiconductors [6, 8], and therefore electrically insulating. To test whether or not individual nanotubes are in fact insulating, the nanotube tips were brought into contact with the gold counter-electrode. For all BN nanotubes observed, there was no conduction at low bias. It is known that often carbon nanotubes can have high contact resistance, but we rule out contact resistance for the case of BN nanotubes, because the voltage can be increased through 10 V with no current or damage to the tube. When this is done with carbon nanotubes, the nanotubes will mechanically fail or burn-out between 2 V and 4 V, even in cases when there is no conductance at low bias. BN nanotubes, therefore, are a good dielectric material up to approximately 10 V. At higher bias voltages, however, some of the BN nanotubes observed pass current, but do so in a reversible, non-destructive manner. Figure 11.17 shows a typical current-voltage relation for individual BN nanotubes. The breakdown generally occurs between 12 V and 25 V, and has a characteristic reminiscent of gas-discharge tubes. In the lower plot of Fig. 11.17, a sharp, reversible turn-on can be seen at 22 V. Such behavior might be characteristic of an avalanche effect leading to breakdown conductivity.

Acknowledgements

We are indebted to M.L. Cohen, P.G. Collins, U. Dahmen, T. Freeman, M.S. Fuhrer, D. Goldhaber-Gordon, W.Q. Han, S.G. Louie, D.K. Owen and E.A. Stach for valuable contributions to the work described here. This research was supported in part by the Director, Office of Energy Research, Office of Basic Energy Sciences, Division of Materials Sciences, of the U.S. Department of Energy under contract No. DE-AC03-76SF00098.

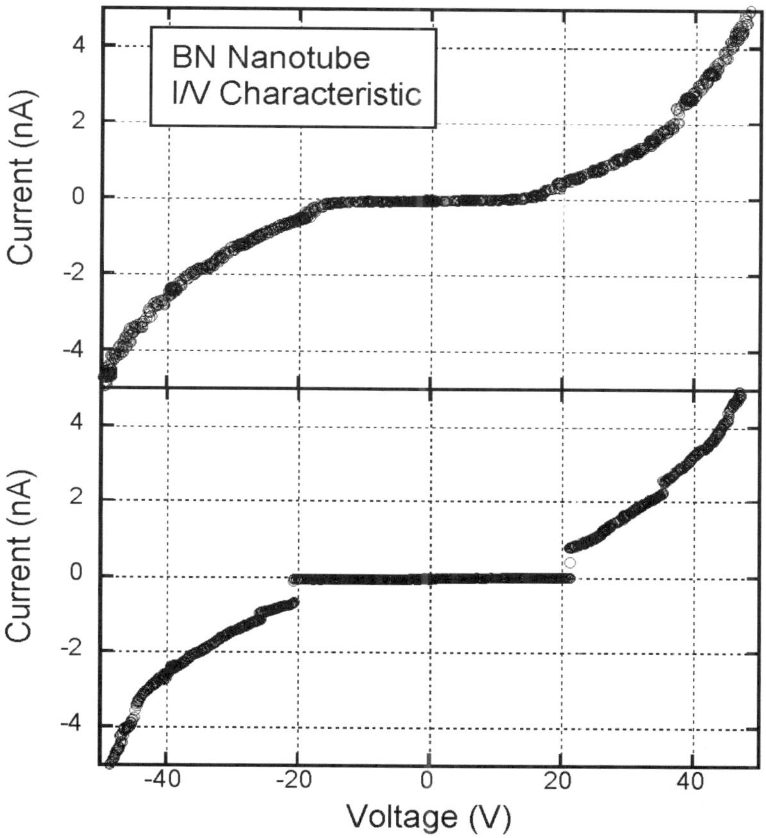

Fig. 11.17. The current-voltage characteristic of typical BN nanotubes under contact conditions. In both cases, the nanotubes protrude from the conducting matrix approximately 400 nm

References

1. S. Iijima: Nature **354**, 56 (1991)
2. N.G. Chopra: Synthesis and Characterization of $B_xC_yN_z$ Nanotubes. PhD Disseratation, University of California, Berkeley (1996)
3. N.G. Chopra, R.J. Luyken, K Cherrey et al: Science **269**, 966 (1995)
4. R.S. Pease: Acta Crystallogr. **5**, 356 (1952)
5. A. Zunger, A. Katzir, A. Halperin: Phys. Rev. B **13**, 5560 (1976)
6. A. Rubio, J.L. Corkill, M.L. Cohen: Phys. Rev. B **49**, 5081 (1994)
7. M. Terauchi, M. Tanaka, T. Matsumoto et al: Journal of Electron Microsc. **47**, 319 (1998)
8. X. Blase, A. Rubio, S.G. Louie et al: Europhys. Lett. **28**, 335 (1994)
9. R.S. Ruoff, D.C. Lorents: Carbon **33**, 925 (1995)

10. J.P. Salvetat, J.M. Bonard, N.H. Thomson et al: Appl. Phys. A **69**, 255 (1999)
11. N.G. Chopra, A. Zettl: Solid State Commun. **105**, 297 (1999)
12. J.P. Lu: Phys. Rev. Lett. **79**, 1297 (1997)
13. D. Sanchez-Portal, E. Artacho, J.M. Soler et al: Phys. Rev. B **59**, 12678 (1999)
14. J.P. Lu: Elastic properties of single and multilayered nanotubes. In: *Fullerenes '96* (Elsevier, Oxford, UK 1997) pp 1649–52
15. E. Hernandez, C. Goze, P. Bernier et al: Phys. Rev. Lett. **80**, 4502 (1998)
16. E. Hernandez, C. Goze, P. Bernier et al: Appl. Phys. A **68**, 287 (1999)
17. N. Yao, V. Lordi: J. Appl. Phys. **84**, 1939 (1998)
18. V.N. Popov, V.E. Van Doren, M. Balkanski: Phys. Rev. B **61**, 3078 (2000)
19. C.F Cornwell, L.T. Wille: Solid State Commun. **101**, 555 (1997)
20. A. Krishnan , E. Dujardin, T.W. Ebbesen TW et al: Phys. Rev. B **58**, 14013 (1998)
21. M.M.J. Treacy, T.W. Ebbesen, J.M. Gibson: Nature **381**, 678 (1996)
22. J.P. Salvetat, A.J. Kulik, J.M. Bonard et al: Adv. Mater. **11**, 161 (1999)
23. J.P. Salvetat, G.A.D. Briggs, J.M. Bonard et al: Phys. Rev. Lett. **82**, 944 (1999)
24. E.W. Wong, P.E. Sheehan, C.M. Lieber: Science **277**, 1971 (1997)
25. Z.W. Pan, S.S. Xie, L. Lu et al: Appl. Phys. Lett. **74**, 3152 (1999)
26. O. Lourie, H.D. Wagner: J. Mater. Res. **13**, 2418 (1998)
27. P. Poncharal, Z.L. Wang, D. Ugarte et al: Science **283**, 1513 (1999)
28. M.F. Yu, O. Lourie, M.J. Dyer et al: Science **287**, 637 (2000)
29. C. Downs, J. Nugent, P.M. Ajayan et al: Adv. Mater. **11**, 1028 (1999)
30. B.I. Yakobson, C.J. Brabec, J. Bernholc: Phys. Rev. Lett. **76**, 2511 (1996)
31. B.I. Yakobson: Appl. Phys. Lett. **72**, 918 (1998)
32. P. Zhang, P.E. Lammert, V.H. Crespi: Phys. Rev. Lett. **81**, 5346 (1998)
33. M.B. Nardelli, B.I. Yakobson, J. Bernholc: Phys. Rev. Lett. **81**, 4656 (1998)
34. M.B. Nardelli, B.I. Yakobson, J. Bernholc: Phys. Rev. B **57**, R4277 (1998)
35. D.A. Walters, L.M. Ericson, M.J. Casavant et al: Appl. Phys. Lett. **74**, 3803 (1999)
36. G. Binnig, H. Rohrer, C. Gerber et al: Phys. Rev. Lett. **50**, 120 (1983)
37. G. Binnig, H. Rohrer: Scanning tunnelling microscopy. Helv. Phys. Acta **55**, 726 (1982)
38. D.M. Eigler, E.K. Schweizer: Nature **344**, 524 (1990)
39. M.F. Crommie, C.P. Lutz, D.M. Eigler: Science **262**, 218 (1993)
40. M.T. Cuberes, R.R. Schlittler, J.K. Gimzewski: Appl. Phys. Lett. **69**, 3016 (1996)
41. G. Binnig, C.F. Quate, C. Gerber: Phys. Rev. Lett. **56**, 930 (1986)
42. R. Curtis, T. Mitsui, E. Ganz: Rev. Sci. Instrum. **68**, 2790 (1997)
43. Z.L. Wang, P. Poncharal, W.A. de Heer: J. Phys. Chem. Solids **61**, 1025 (2000)
44. Z.L. Wang, P. Poncharal, W.A. de Heer: Microsc. Microanal. **6**, 224 (2000)
45. W.K. Lo, J.C.H. Spence: Ultramicroscopy **48**, 433 (1993)
46. A.M. Minor, J.W. Morris, E.A. Stach: Appl. Phys. Lett. **79**, 1625 (2001)
47. E.A. Stach, T. Freeman, A.M. Minor et al: Microsc. Microanal. **7**, 507 (2001)
48. T.W. Ebbesen, P.M. Ajayan: Nature **358**, 220 (1992)
49. A.G. Rinzler, J. Liu, H. Dai et al: Appl. Phys. A **67**, 29 (1998)

50. P.G. Collins, P. Avouris: Appl. Phys. A **74**, 329 (2002)
51. P.G. Collins, M.S. Arnold, P. Avouris: Science **292**, 706 (2001)
52. P.G. Collins, M. Hersam, M. Arnold et al: Phys. Rev. Lett. **86**, 3128 (2001)
53. M. Freemantle: Chem. Eng. News **74**, 62 (1996)
54. P.J. Britto, K.S.V. Santhanam, P.M. Ajayan: Bioelectroch. Bioener. **41**, 121 (1996)
55. P.G. Collins, A. Zettl, H. Bando et al: Science **278**, 100 (1997)
56. S.J. Tans, R.M. Verschueren, C. Dekker: Nature **393**, 49 (1998)
57. H.Dai H, J.H Hafner, A.G. Rinzler et al: Nature **384**, 147 (1996)
58. W.A. de Heer, A. Chatelain, D. Ugarte: Science **270**, 1179 (1995)
59. J. Kong, H.T. Soh, A.M. Cassell: Nature **395**, 878 (1998)
60. S.C. Tsang, Y.K. Chen, P.J.F. Harris et al: Nature **372**, 159 (1994)
61. J. Cumings, P.G. Collins, A. Zettl: Nature **406**, 586 (2000)
62. S. Frank, P. Poncharal, Z.L. Wang et al: Science **280**, 1744 (1998)
63. A. Bachtold, C. Strunk, J.P. Salvetat et al: Nature **397**, 673 (1999)
64. W. Trimmer: *Micromechanics and MEMS: classic and seminal papers to 1990.* (IEEE Press, New York 1997)
65. S. Iijima, C. Brabec, A. Maiti et al: J Chem Phys **104**, 2089 (1996)
66. H.D. Wagner, O. Lourie, Y. Feldman et al: Appl. Phys. Lett. **72**, 188 (1998)
67. N. Hamada, S. Sawada, A. Oshiyama: Phys. Rev. Lett. **68**, 1579 (1992)
68. R. Saito, M. Fujita, G. Dresselhaus et al: Appl. Phys. Lett. **60**, 2204 (1992)
69. J.C. Charlier, J.P. Michenaud: Phys. Rev. Lett. **70**, 1858 (1993)
70. A. Kolmogorov, V. Crespi: B. Am. Phys. Soc. **45**, 254 (2000)
71. V.H. Crespi, P. Zhang, P.E. Lammert: Sliding, stretching, and tapering: recent structural results for carbon nanotubes. In: *Electronic Properties of Novel Materials–Science and Technology of Molecular Nanostructures*, ed by H. Kuzmany, J. Fink, M. Mehring, S. Roth (American Institute of Physics, College Park, MD 1999) pp 364–368
72. L.X. Benedict, N.G. Chopra, M.L. Cohen et al: Chem. Phys. Lett. **286**, 490 (1998)
73. B.N.J. Persson: Surf. Sci. Rep. **33**, 83 (1999)
74. R.W. Carpick, D.F. Ogletree, M. Salmeron: Appl. Phys. Lett. **70**, 1548 (1997)
75. M. Enachescu, R.J.A. Van Den Oetelaar, R.W. Carpick et al: Phys. Rev. Lett. **81**, 1877 (1998)
76. S. Iijima, P.M. Ajayan, T. Ichihashi: Phys. Rev. Lett. **69**, 3100 (1992)
77. N.G. Chopra, L.X. Benedict, V.H. Crespi et al: Nature **377**, 135 (1995)
78. M.R. Falvo, G.J. Clary, R.M. Taylor et al: Nature **389**, 582 (1997)
79. C.T. White, T.N. Todorov: Nature **393**, 240 (1998)
80. A. Bachtold, M.S. Fuhrer, S. Plyasunov et al: Phys. Rev. Lett. **84**, 6082 (2000)
81. A. Javey, J. Guo, Q. Wang et al: Nature **424**, 654 (2003)
82. C. Berger, Y. Yi, Z.L. Wang et al: Appl. Phys. A **74**, 363 (2002)
83. S. Sanvito, Y.K. Kwon, D. Tomanek et al: Phys. Rev. Lett. **84**, 1974 (2000)
84. H. Dai, E.W. Wong, C.M. Lieber: Science **272**, 523 (1996)
85. J. Cumings, A. Zettl: Science **289**, 602 (2000)
86. J.C. Charlier, J.P. Michenaud: Phys. Rev. Lett. **70**, 1858 (1993)
87. M.F. Yu, B.I. Yakobson, R.S. Ruoff: J. Phys. Chem. B **104**, 8764 (2000)
88. D.H. Kim, K.J. Chang: Phys. Rev. B **66**, 155402 (2002)
89. C. Buia, A. Buldum, J.P. Lu: Phys. Rev. B **67**, 113409 (2003)

90. R. Landauer: Philos. Mag. **21**, 863 (1970)
91. D.J. Thouless: Phys. Rev. Lett. **39**, 1167 (1977)
92. G.J. Dolan, D.D. Osheroff: Phys. Rev. Lett. **43**, 721 (1979)
93. D.J. Bishop, D.C. Tsui, R.C. Dynes: Phys. Rev. Lett. **44**, 1153 (1980)
94. M.S. Fuhrer, M.L. Cohen, A. Zettl et al: Solid State Commun. **109**, 105 (1999)
95. Y. Koike, S. Morita, T. Nakanomyo et al: J. Phys. Soc. Jpn. **54**, 713 (1985)
96. P.J. de Pablo, C. Gomez-Navarro, J. Colchero et al: Phys. Rev. Lett. **88**, 036804 (2002)
97. S. Datta: *Electronic transport in mesoscopic systems* (Cambridge University Press, Cambridge New York 1995)
98. N. Apsley, H.P. Hughes: Philos. Mag. **31**, 1327 (1975)
99. S. Roche, F. Triozon, A. Rubio et al: Phys. Lett. A **285**, 94 (2001)
100. P.A. Lee, T.V. Ramakrishnan: Rev. Mod. Phys. **57**, 287 (1985)
101. Z. Yao, C.L. Kane, C. Dekker: Phys. Rev. Lett. **84**, 2941 (2000)
102. J. Hone: Phonons and thermal properties of carbon nanotubes. In: *Carbon Nanotubes: Synthesis, Structure, Properties, and Applications*, ed by M.S. Dresselhaus, G. Dresselhaus, P. Avouris (Springer Verlag, Berlin 2001) pp 273–286
103. F. Leonard, J. Tersoff: Phys. Rev. Lett. **83**, 5174 (1999)
104. N. de Jonge, Y. Lamy, K. Schoots et al: Nature **420**, 393 (2002)
105. J. Clarke: Physica B & C **126**, 441 (1984)
106. R.V. Jones: P.I. Electr. Eng. **117**, 1185 (1970)
107. J. Cumings, A. Zettl: Chem. Phys. Lett. **316**, 211 (2000)
108. K.B. Shelimov, R.O. Esenaliev, A.G. Rinzler et al: Chem. Phys. Lett. **282**, 429 (1998)
109. R.H. Fowler, L. Nordheim: Proc. R. Soc. Lon. Ser.–A **119**, 173 (1928)
110. P.G. Collins, A. Zettl: Phys. Rev. B **55**, 9391 (1997)
111. K.A. Dean, B.R. Chalamala: Appl. Phys. Lett. **76**, 375 (2000)
112. P.G. Collins, M.S. Fuhrer, A. Zettl: Appl. Phys. Lett. **76**, 894 (2000)

12 Nanomanipulator Measurements of the Mechanics of Nanostructures and Nanocomposites

F.T. Fisher, D.A. Dikin, X. Chen, and R.S. Ruoff

While a great deal of progress has been made in understanding the theoretical fundamentals of nanodevices and nanomaterials, to date the complexities of performing mechanical experiments at the nanoscale have limited the ability to compare theoretical and computational predictions with experimental data. In this chapter we will discuss some of the critical challenges that hamper mechanical experiments at the nanoscale, and present how these challenges have been, and are currently being, addressed by next-generation tools designed to perform nanoscale mechanical experiments. In particular, we will focus on how our work and understanding have evolved through our experience in this area. We will then present a number of examples of how our current nanomanipulator has been used to perform nanoscale mechanical testing of a variety of nanostructures, including: tensile loading of individual nanostructures, nanoscale pullout tests of nanostructures embedded in a matrix material, and forced vibrational resonance to measure the modulus of nanorods and nanotubes. We will conclude by discussing relevant near- and long-term goals and challenges of nanoscale mechanical experimentation.

12.1 Introduction

As discussed throughout this book, the recent explosion in nanotechnology research throughout the world is driven by the common theme of exploiting the potential of designing materials, structures, and devices at the nanoscale. The possibility and potential of working at this scale. which today perhaps seems apparent, was a revolutionary idea when first presented by Feynman in his seminal lecture at Caltech in 1959 [1]. However, the discovery of multiwalled carbon nanotubes (MWNT) by Iijima in 1991 [2], combined with advances in electron microscopy and other nanoscale characterization techniques, has led to the current explosion in interdisciplinary research efforts in nanoscale science and technology. The common theme that exists throughout these ongoing research efforts is the necessity for strong interdisciplinary collaborations between different scientific fields, requiring coordination between theory, computational modeling, and experimental measurements.

One important material driving the current nanotechnology revolution is the carbon nanotube. Due to their outstanding physical, electrical, and thermal properties, carbon nanotubes have been proposed for use in a number of potential applications, including: materials reinforcement, field emission displays, electronic nanowires and devices, chemical sensors, and nanoscale fluidic systems [3]. In the area of field emission displays, research has progressed to the point where companies such as Samsung and NEC have demonstrated product-quality devices based on carbon nanotubes [4, 5]. Due to space constraints the reader is referred to a number of excellent general texts on carbon nanotubes [6–8] and recent literature reviews discussing their mechanical properties [9–11].

For the potential applications of carbon nanotubes listed above, the ability to perform accurate, quantitative, and repeatable experiments to determine the mechanical properties of nanostructures is desirable. These mechanical tests, and related experiments, are complicated by the nanometer-scale dimensions of these samples. Particular challenges in the design of such experiments include: (i) sample placement within the appropriate test configuration; (ii) controlled and repeatable attachment of the samples to the experimental apparatus; (iii) controlled application of the desired loading conditions at the necessary resolution; and (iv) measurement of the mechanical deformation at sufficient resolution.

The goal of this chapter is to highlight the current "state-of-the-art" with regards to experimental techniques designed to quantitatively measure the mechanical properties of materials at the nanoscale. Excellent work has been reported using various types of scanning probe microscopy (SPMs), and in particular atomic force microscopy (AFM), to probe the mechanical properties of nanostructures; however, such manipulations are limited in that they are two-dimensional and rely on a large substrate support [12–20]. While such approaches are of interest for certain types of problems and continue to be developed, they are not of universal use for fully nanoscale three-dimensional manipulation. Thus for purposes of this chapter discussion will be primarily limited to those techniques in which the nano-sized sample can be *lifted* from a surface such that underlying substrate effects can be eliminated. Such nanomanipulation techniques, while at the moment more difficult and time-consuming than scanning probe-based methods, are more flexible and offer the greatest potential for future quantitative nanoscale experimental methods development. In addition, while a number of other research groups are currently doing inspiring and exciting work in this area, rather than attempt to provide an overview of the entire field, we will focus on our ongoing research efforts at Northwestern. It is our hope that this approach will provide a glimpse of the understanding and experience that we have developed over the years working in this area. Finally, we note that while a majority of the work described below is related to the mechanical properties of carbon nanotubes, the extension of these experimental efforts to other nanoscale materials and structures is ongoing.

The organization of the rest of the chapter is as follows. In Sect. 12.2 we discuss the design of our first generation nanomanipulator device as well as our recent advances in nanoscale testing stage development. In Sect. 12.3 we describe the application of these nanomanipulators to perform mechanics measurements on the nanoscale. Here two general types of nanomechanics tests are described: the tensile loading of individual carbon nanotubes and nanocoils, including individual nanotube pullout tests from a polymer matrix, and the induced vibrational resonance of nanostructures. We end the chapter by summarizing this work and identifying future directions of research in this area.

12.2 Nanomanipulators

12.2.1 Initial Nanomanipulator Development

The motivation for this work is the desire to develop the experimental capability to adapt standard macroscale mechanical test methods for nanoscale materials and structures. However, because at least one of the characteristic dimensions is smaller than the wavelength of light, techniques requiring visual feedback must be operated within a transmission electron microscope (TEM) or a scanning electron microscope (SEM) in order to resolve the image of the nanostructure. Of these two techniques, TEM offers the highest resolving power, down to the sub-nanometer scale, due to the high energy of the incident electrons used in the imaging. However, TEMs are typically characterized by very confined internal sample chambers which severely limit the size of the nanomanipulation tool that can be inserted within the TEM. While SEMs do not provide the resolution quality of TEMs, recent technological developments have led to significant enhancements in their resolving power, which for high-end models is quickly approaching the sub-nanometer scale. Such resolution is more than sufficient for all but the smallest nanoparticles.

The first nanomanipulation tool developed by our group, which was designed and built for use in an SEM, is highlighted in Figs. 12.1 and 12.2. Described in more detail elsewhere, this nanomanipulator can probe, select, and handle nanometer-sized objects via nanometer positional resolution along one rotational and three linear degrees of freedom [21]. Note that this device has been constructed using commercially available stages and actuators, and the entire manipulator occupies roughly 50 cm^3. Actuator electrical control signals are fed through isolated vacuum feedthroughs passing through the SEM main chamber wall, allowing each articulator element to be addressed individually from outside the vacuum chamber. The device allows two modes of motion:

– *coarse motion*: up to 10 mm linear motion along the X- and Y- degrees of freedom, up to 6 mm linear motion in the Z-direction, and 360° along the

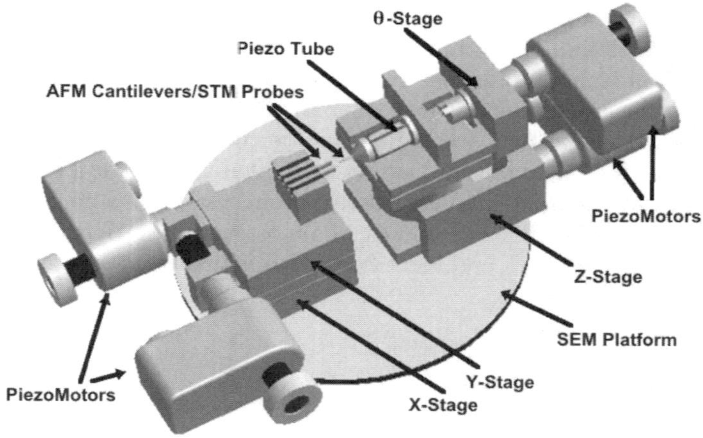

Fig. 12.1. Schematic of the first-generation nanomanipulator for use within an SEM. (Reproduced with permission from M.F. Yu, M.J. Dyer, G.D. Skidmore, H.W. Rohrs, X. Lu, K.D. Ausman, J.R. von Ehr, and R.S. Ruoff, *Nanotechnology*, 10, p. 244, 1999) (see Color Plates, p. 344)

Fig. 12.2. Photograph of the first-generation nanomanipulator device (see Color Plates, p. 344)

single-axis rotational probe (θ-stage) with an average rotation step size of 1 mrad, and

- *fine motion*: piezo bimorph (one direction) or piezotube (three direction) continuous motion with nanoscale resolution, dependent only on the stability of the power supply and the vibration isolation of the device.

This initial nanomanipulator has been used for a number of experimental studies related to the mechanical behavior of nanostructures. As described later in this chapter, these mechanics tests can generally be classified as being of two types: direct tensile loading of nanostructures and induced mechanical resonance.

12.2.2 Recent Nanoscale Testing Stage Development

The nanomanipulation stage described above has contributed to our understanding of the mechanics of carbon nanotubes. Based on the knowledge and experience obtained from using this device, we have actively sought to develop next-generation nanomechanical testing stages to enhance our experimental capability in this area. Below we highlight three recent advances in this area.

Next-Generation 10-DOF Nanomanipulator

Within the last several years there has been a push to develop 3D nanomanipulation systems which can operate within an electron microscope [21–24]. Several companies now specialize in micro and nanoscale actuators and positioning components for use within an SEM, including Kleindiek Nanotecknik and Klocke Nanotechnik (Germany) and Zyvex LLC (USA).

Based on the success of the four degree of freedom (DOF) nanomanipulator described above, we have recently designed and built a next-generation 10-DOF nanomanipulator using an open platform strategy, such that it is compatible with a number of SEM and optical microscope systems. Primary system requirements considered were fine nanoscale precision, macroscale (coarse) range of motion, compatibility with high-vacuum systems, low profile to avoid interference within the SEM chamber, and ease of rearrangement of the manipulator components to provide maximum versatility. To minimize costs the system was designed using only readily available commercial components. A modular approach allows a variety of front-end tools, such as AFM and STM tips, piezoelectric-based actuators (piezo- tubes and plates), and MEMS and NEMS devices to be easily integrated onto the nanomanipulator.

This next-generation nanomanipulator consists of 4 independent stages as shown in Fig. 12.3. Each stage has the same coarse alignment ($\sim$ 20 nm in all directions) and is based on macro-sized translating stages driven by Picomotors (New Focus, San Jose, California). The total lateral dimensions of the nanomanipulator are approximately 170 mm by 170 mm. The height of the nanomanipulator is $\sim$ 30 mm, so as not to interfere with the imaging capabilities of the SEM. The advantages of this manipulator are that it is easy to assemble and is relatively inexpensive, particularly when compared to commercially available versions. Other characteristics are high positional accuracy, large available workspace, easy re-configuration compatible with multiple front-end tools, open-loop stability, high linearity, compact size, and high load capacity. These advantages make it useful for a wide range of applications including high precision manipulation, probing, and nanoscale materials testing.

Microdelivery System for Nanoscale Sample Clamping

A critical element in performing accurate and repeatable mechanical measurements using the nanomanipulator is the attachment of nanostructures

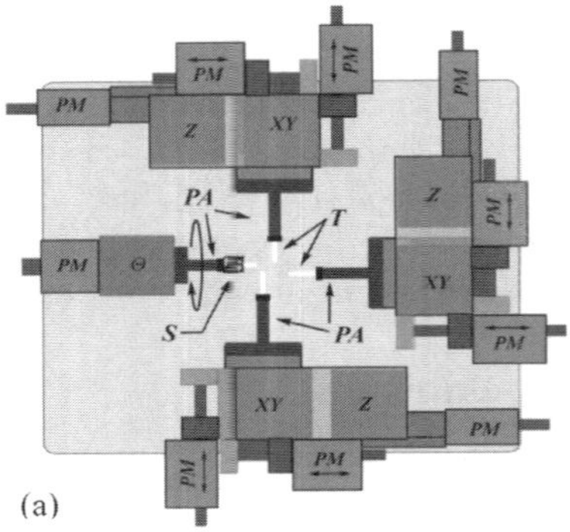

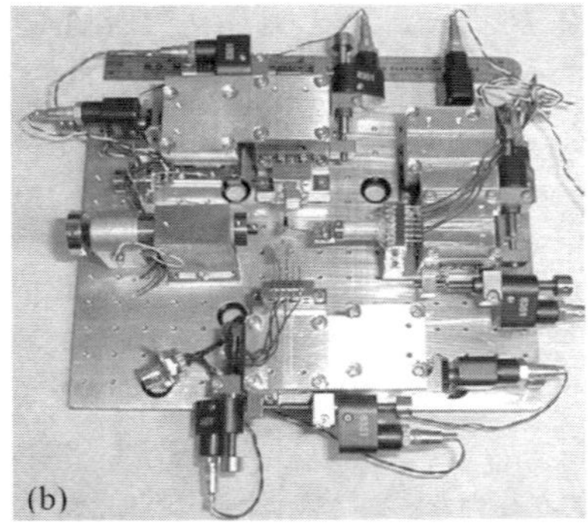

Fig. 12.3. (**a**) Schematic of the 10-DOF nanomanipulator. (**b**) Photograph of a particular configuration. XY, Z are translation stages, θ is the rotation stage, PM are Picomotors, PA are piezoactuators, T are front-end tools, and S is the sample source. The ruler in (**b**) is 6 inches long.

to the nanomanipulator front-end tool (typically an AFM cantilever). While researchers in the past have relied on Van der Waals forces or other relatively weak adhesive interactions between the object and the front-end tool surface, in many cases this is insufficient for picking up or manipulating a sample. Thus, clamps formed by the electron beam induced decomposition (EBID) of residual hydrocarbons in the SEM system have played an important role in many of our earlier nanomechanics tests performed using the nanomanipulator.[1] [38–41] Because the clamp is the intermediate medium for load transfer from the AFM cantilever to the nanostructure, knowledge of the structure and strength of the clamp is critical. For example, we have modeled the impact of the clamp boundary conditions on the measured resonance response of microscale quartz fibers [26].

While we have found EBID to be a practical way to fabricate clamps for our nanoscale experiments, it is somewhat limited in that there is a lack of control over critical experimental conditions that influence clamp strength. Currently, we can only regulate the electron beam dose, deposition time, and exposure region for the EBID process, but complete control of EBID clamp deposition is beyond our control since it depends on the residual gas in the vacuum chamber. In addition, EBID is tedious because residual hydrocarbons are sparse in a moderately clean SEM. For example, a clamp that is only 300 nm long by 200 nm wide, of sufficient strength to hold a carbon nanotube for tensile loading to fracture, can take 1 to 2 hours to deposit. Also, the

[1] We have found EBID to be a convenient and non-destructive method to make clamps; another advantage is that testing in the SEM can be done during the same session that the clamps are made. While deposits (clamps) can also be made using a system such as dual beam SEM-FIB, the focused ion beam can severely damage the nanostructures.

contact resistance of the carbonaceous nanoclamp deposit is high, on the order of a few megaohms, which imposes an undesirable limit for electrical measurements on nanostructures.

Due to these restrictions, we have recently designed and built a new microdelivery system for the accelerated fabrication of our nanoclamps. The primary goal of this system is to develop a reliable and repeatable method to attach nanostructures to conducting or insulating surfaces for manipulation and mechanical loading, and to do so in a manner which results in a low interface resistance between the nanostructure and the metal electrode (if applicable) to permit conductivity measurements. Supplementary goals that were considered included: one step *in situ* nanolithography for surface nanoelectrode, cantilevered probe, or nanojunction fabrication; controllable mass attachment on the nanoscale; and nanogap sealing. The primary design challenge was to achieve these goals in a way that was compatible with the high vacuum environment within an SEM.

The basis of the system is the controlled deposition of an appropriate precursor compound. The requirements for this material are that it is volatile, yet stable enough to undergo vapor phase transport; has a weak metal-to-ligand bond (low dissociation energy); should not react with the surface, or promote any surface processes; and decomposition should occur at the absorbed surface but not in the gas phase. The resulting ligand should be very volatile, stable against dissociation at the surface, and not leave any residual contamination. Given these requirements, a number of organometallic compounds were considered, including metal carbonyls such as $W(CO)_6$, $Cr(CO)_6$, and $Co_2(CO)_8$. Also considered were inorganic compounds such as $(AuCl_3)_2$, $NbCl_5$, and TiF_4. We ultimately decided on $W(CO)_6$, which is also typically used in conjunction with Focused Ion Beam (FIB) microscopes. Particularly useful is that the first bond dissociation energy is $\sim 2\,\text{eV}$, and that the vapor pressure increases by an order of magnitude from room temperature ($\sim 2.5x10^{-2}$ Torr at 25° C) to 50° C ($\sim 4x10^{-1}$ Torr). Thus as described below it is only necessary to heat this precursor to around 50 to 60° C for the deposition process to occur.

The first generation microdelivery system, shown in Fig. 12.4, is composed of: (1) a sealed container/crucible to hold the precursor material, (2) a high power resistor to act as a heater, (3) thermometer, (4) a copper sealing gasket, (5) a crucible cap, (6) a metal capillary extender, (7) a long flexible capillary (Teflon tubing) with a fine nozzle at the end which fits onto the extension, and (8) a delivery nozzle. The end portion of the capillary (i.e. the delivery nozzle) is attached to one of the nanomanipulator positioning stages, which allows exact positioning of the nozzle at the desired area or surface. This is critical for achieving optimal performance during deposition, as proximity is important to obtain maximum efficiency of the deposition rate [27]. In addition, it is also important to maintain high vacuum conditions in the SEM chamber by minimizing the gas flow out of the nozzle. As shown in the

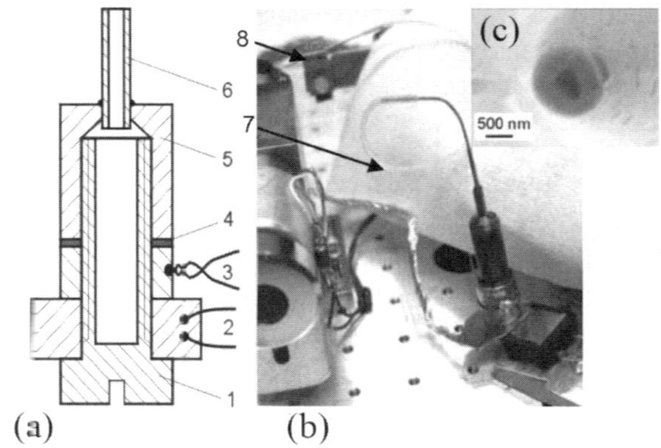

Fig. 12.4. Microdelivery system. (**a**) System schematic. (**b**) Photograph of the microdelivery unit seated on the base plate of the manipulator. The inset (**c**) shows as SEM image of the glass capillary nozzle

Fig. 12.4c, we have fabricated a glass capillary nozzle with an inner diameter of about 0.3 μm and a tapered shape that will provide a smaller beam spread for the microdelivery system. While such small capillaries are more difficult to handle, they open many new possibilities for controlled nanoscale deposition. In Fig. 12.5 we demonstrate one of our initial fabrication attempts, where the nozzle was a stainless steel dispensing needle used in syringes with an inner diameter of 250 μm and $W(CO)_6$ was used as the precursor compound. The temperature was maintained at approximately 60° C during the deposition, and the vacuum in the SEM chamber did not rise above $\sim 8x10^{-4}$ Pa during the entire deposition process.

While to date we have successfully used the clamp in a number of applications within our group, future modifications of the system to both enhance the quality of the produced clamps as well as reduce the contamination of the SEM chamber are anticipated. For example, in order to minimize contamination an extremely local deposition (at best $\sim 10 \times 10$ nm) is desired. Future versions of the microdelivery system will allow rapid on-off pulsing of the precursor to provide exquisite delivery of the precursor and minimize the introduction of contaminants within the SEM, as well as incorporate a small heater/cooler, in the form of a Peltier plate, to allow for more rapid temperature variation of the precursor material. In addition, we plan to pursue the use of a second gas phase (for example, water vapor) to increase the metal residue from the organometallic [28]. Future applications of this microdelivery system include both nanoclamp formation as well as its use in creating a suitable micro-environment for probing the chemical reactivity of the nanostructure or thin film to the gas phase precursor material.

Fig. 12.5. Microdelivery system in operation. (**a**) Si wafer attached to the θ-stage and tilted 30° with respect to the electron beam axis. (**b**) Image of a ~ 60 nm wide and ~ 7μm long wire deposited between two gold electrodes. (**c**) A free standing structure deposited on the sharp corner of the Si wafer

MEMS-Based Thermally Actuated Stage

We have recently demonstrated the design, fabrication, and characterization of a MEMS-based stress-strain device micromachined using deep reactive ion etching (DRIE) for nanomechanics experiments [29]. As shown in Fig. 12.6, the device is based on thermal actuation with integrated motion amplification structures, permitting a working displacement range from nanometers up to 10 μm. Potential uses of this stage are both as a front-end tool for the nanomanipulators described earlier, as well as a stand-alone tool for use in a TEM. A distinguishing characteristic of this design is the large thickness (130 μm) of the free-standing stage components. This results in a rigid out-of-plane structure, such that the offset in height between the fixed and movable stages is less than 40 nm over the working range of the device. Based on modeling of the current device, an applied force of 0.27 μN is predicted for a minimum displacement of 30 nm; the primary limitation on minimum displacement resolution is the stability of the power supply. This force increases linearly up to 88 μN at a maximum displacement of 9.7 μm, such that

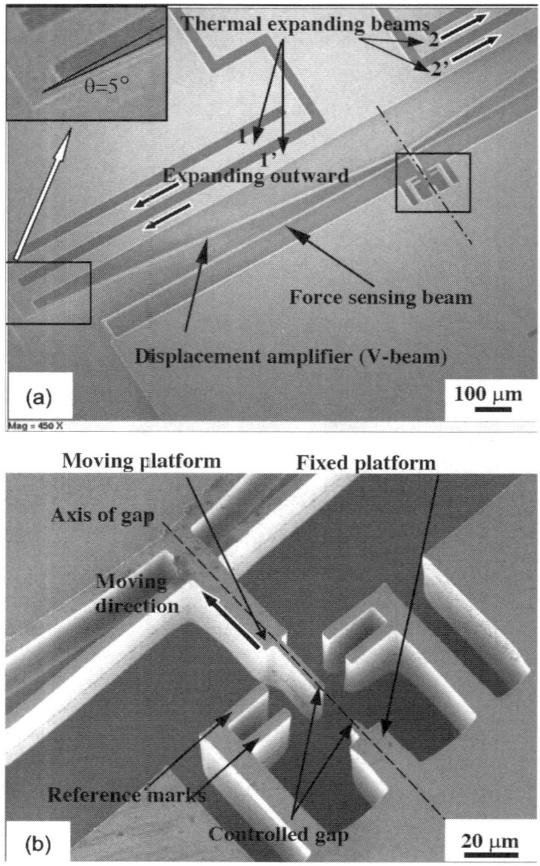

Fig. 12.6. (**a**) SEM image of the MEMS-fabricated loading stage. Thermal expansion beams 1-1' and 2-2' are symmetric relative to the center-line (*dashed line*) of the device. Beams 1 and 2 are 970 μm long, while beams 1' and 2' are 1000 μm long, and the width of the beams is 50 μm. The upper left inset shows the connection between the thermal expansion and V-shaped beams. (**b**) Higher magnification image of the central gap region. (Reproduced with permission from S. Lu, D.A. Dikin, S. Zhang, F.T. Fisher, J. Lee, and R.S. Ruoff, *Rev. Sci. Instrum.* **75**, p. 2154, 2004)

for this particular stage geometry the force constant of the stage is 9 N/m. One benefit of this design is that future stages based on this methodology can be custom-designed to provide a desired force constant appropriate for the structure to be tested. Additional benefits of this new stage are its potential for mass fabrication for standardized testing and the fact that it is TEM-compatible due to its small size. Finally, the highly co-planar opposing stages are well-suited for the tensile loading of thin films.

The design of the present device overcomes a number of limitations that are inherent to other nano- and micro-positioners fabricated using MEMS techniques. For example, while actuation on the order of a few hundred nanometers has been demonstrated using electrostatically actuated x-y translators combined with e-beam lithography [30, 31], the relatively large electrostatic forces required to load nanostructures such as carbon nanotubes would require very large devices or very high driving voltages [32]. Micromachined structures have also been used to mount and test nanotubes; [33, 34] however, because they rely on separate piezo drives/motors as actuators, the total size of the system is much larger than could be achieved with a fully integrated self-actuated MEMS device. While a number of MEMS testing stages devices based on thermal actuation have been designed, they have not been able to achieve displacement resolutions on the order to tens of nanometers [35–37]. Additional benefits of the current design are that nanoscale displacements are based on small temperature changes ($\sim$ 1°C) and current flow through the sample is eliminated, preventing thermal effects from influencing the mechanical measurements. Details of the device fabrication and characterization are available in the literature [29].

In application, the test specimen is prepared separately and placed across the opposing platforms shown in Fig. 12.6. The ends of the nanostructure can be clamped to the top surface of each platform using a number of techniques, such as the EBID method described previously or by simply taking advantage of Van der Waals or other adhesive interactions. After careful characterization of the stand-alone device (with no test specimen present), the change in the device characteristics due to the presence of the nanostructure specimen will allow one to determine the mechanical behavior of the sample. For example, by obtaining the difference in displacements of the movable platform with and without the nanostructure for a particular input power, the spring constant of the nanostructure can be obtained as shown in Fig. 12.7. With a given input power (corresponding to a particular force), the displacement is measured before (y) and after (y') a specimen is mounted between the stages. Because the sample stiffness is in parallel with the overall stiffness of the device, the applied force F is given as $F = Ky = (K + k)y'$. Thus the spring constant of the sample, k, is $k = K(y - y')/y'$, where K is the spring constant of the force sensing beam.

We estimate that for this current device a minimum displacement resolution of 5 nm and a corresponding force resolution of 45 nN are feasible [29]. Thus for a particular nanostructure geometry, the range of moduli that can be measured can be estimated; for a 50 nm diameter nanorod, moduli between 13 GPa and 6.4 TPa, and tensile strengths between 23 MPa and 11 GPa, are obtained. However, for extremely stiff materials and structures with small cross-sectional areas, the resolution of the elongation measurement will limit the utility of a given loading device for a particular material and sample geometry. As an example, we consider a (10,10) single walled carbon nanotube

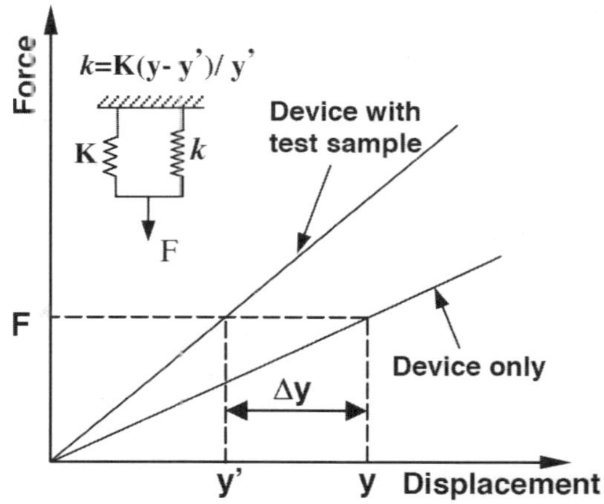

Fig. 12.7. Sample stiffness measurement using the MEMS thermally-actuated testing stage. (Reproduced with permission from S. Lu, D.A. Dikin, S. Zhang, F.T. Fisher, J. Lee, and R.S. Ruoff, *Rev. Sci. Instrum.* **75**, p. 2154, 2004)

with a gage length of 20 μm that breaks at 10% strain at an applied force of ∼ 150 nN. The thermal actuator can provide the necessary tensile force which would result in a ∼ 2 μm displacement of the moving platform. However, because the force is distributed between both the specimen and the force-sensing beam, a larger overall force (∼ 18.2 μN) is required. In this case, a testing stage custom-designed for a particular set of loading conditions (force and displacement resolutions) for a particular sample is envisioned.

The success of this initial device design suggests that this methodology will be useful in producing a variety of MEMS stress-strain stages to yield the force and displacement resolution necessary to test a variety of nanostructures. Projects related to both the application of this stage for various nanocomposites and nanomaterials, as well as future design iterations based on this initial design to improve the performance of the testing stage, are underway.

12.3 Nanomanipulator-Based Mechanics Measurements

12.3.1 Tensile Loading of Nanostructures

Tensile Loading of Carbon Nanotubes

We have primarily conducted two types of mechanical experiments using the nanomanipulators described earlier in this chapter: tensile loading of nanostructures attached to two opposing AFM cantilever tips, and resonance studies of the mechanically- and electrically-induced vibration of nanostructures

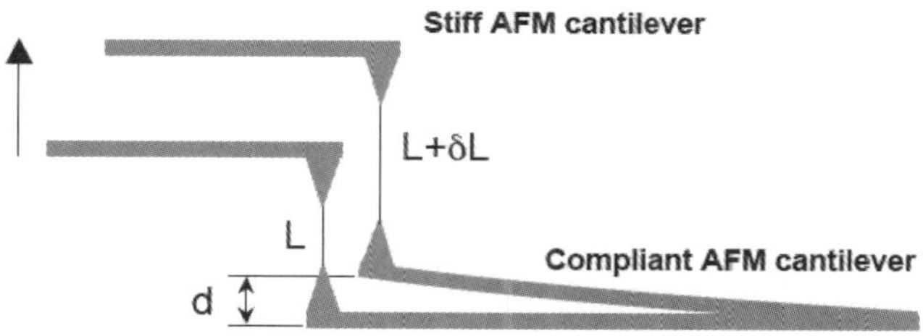

Fig. 12.8. Schematic of tensile loading experiments conducted using the nanomanipulator

attached on one end to an AFM tip. For tensile loading experiments, a general schematic of the experiment is shown in Fig. 12.8. The nanostructure to be tested is connected to two opposing AFM cantilevers, one "compliant" and one "stiff", such that the ratio of the cantilever spring constants is typically greater than 10. It is critical to ensure that the structures are attached to the AFM tips such that relative slipping between the sample and the tips is prevented. This is typically done through electron beam-induced decomposition (EBID)[2] of residual carbonaceous material within the SEM [38–41]. Once the structure to be tested has been properly attached to the AFM tip(s), the stiff (top) cantilever is driven upward, resulting in the lower cantilever being bent upward by a distance d, while the nanostructure is stretched from its initial length of L to $L + \delta L$ due to the force exerted by the AFM tips. The force is calculated as kd, where k is the force constant of the lower cantilever. The strain of the nanostructure is $\delta L/L$ and is calculated based on SEM imaging of the experiment. In those cases where the sample is only attached to the AFM tip at one end and is assumed to be fixed at the other, the applied force and resultant elongation of the nanostructure are both calculated based on the deflection of the sole cantilever.

Using the nanomanipulator in this manner, we have conducted a number of experimental studies to determine the mechanical properties of various forms of carbon nanotubes. In particular, using this nanomanipulator we were able to subject individual MWNTs [22] (see Fig 12.9) and single-walled carbon nanotube (SWNT) ropes [23] (see Fig. 12.10) to tensile loading, as well as realize the sliding between nested shells of a MWNT [24] as shown in Fig. 12.11. Noteworthy from these results were identification of the "sword-

[2] Even for SEM chambers under high vacuum, such residual materials are present and can be used to make these sorts of clamping deposits. However, recent experience has shown that the EBID process is not well-suited for ultra-clean or ultra-high vacuum SEM chambers, which motivated the microdelivery system development described earlier in this chapter.

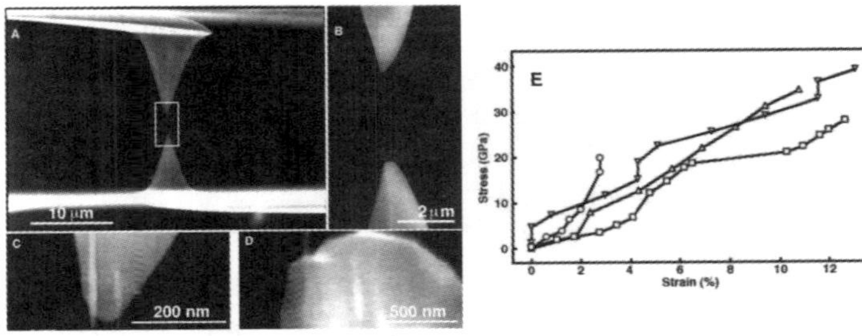

Fig. 12.9. Tensile loading of an individual MWNT. (**a**, **b**) SEM images of two AFM tips holding a MWNT. (**c**, **d**) High magnification SEM images showing the EBID attachments on both AFM tips after fracture. (**e**) Plot of stress versus strain curves for individual MWNTs. (Reprinted with permission from M.F. Yu, O. Lourie, M. Dyer, K. Moloni, T.F. Kelly, and R.S. Ruoff, *Science* **287**, p. 637, 2000. Copyright 2000 AAAS)

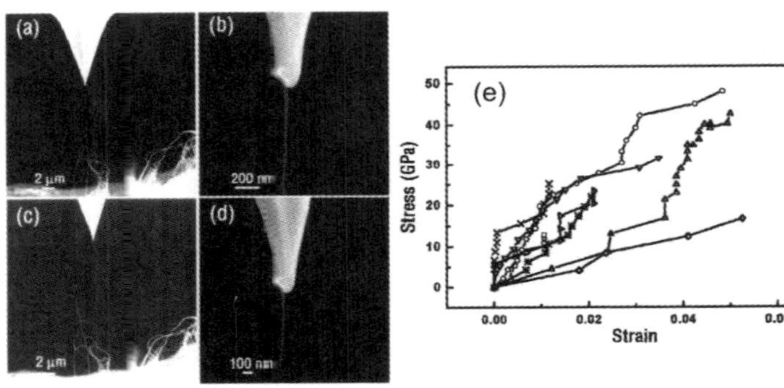

Fig. 12.10. SWNT rope tensile loading experiment. (**a**, **c**) A tensile loaded SWNT rope between an AFM tip and a SWNT "paper" sample before (**a**) and after (**c**) failure of the rope. (**b**, **d**) EBID attachment of the SWNT rope to the AFM tip before (**b**) and after (**d**) rope failure. (**e**) Stress-strain curves obtained for eight SWNT ropes. Stress is calculated using the cross-sectional area of the perimeter SWNTs. (Reprinted with permission from M.F. Yu, B.S. Files, S. Arepalli, and R.S. Ruoff, *Phys. Rev. Lett.* **84**, p. 5552, 2000. Copyright 2000 by the American Physical Society)

in-sheath" failure mechanism of MWNTs, as well as the fact that the loading of the SWNT ropes was only carried by those SWNTs on the perimeter. The reader is referred to the references for a more complete description of this work.

For these tensile loading experiments, AFM cantilevers represent the most reliable and size-compatible force sensor on the nanoscale. Accurate calibration of the AFM cantilevers used as the "front end" tools of the nanomanipulator is critical, particularly given the typical range of manufacturer specifi-

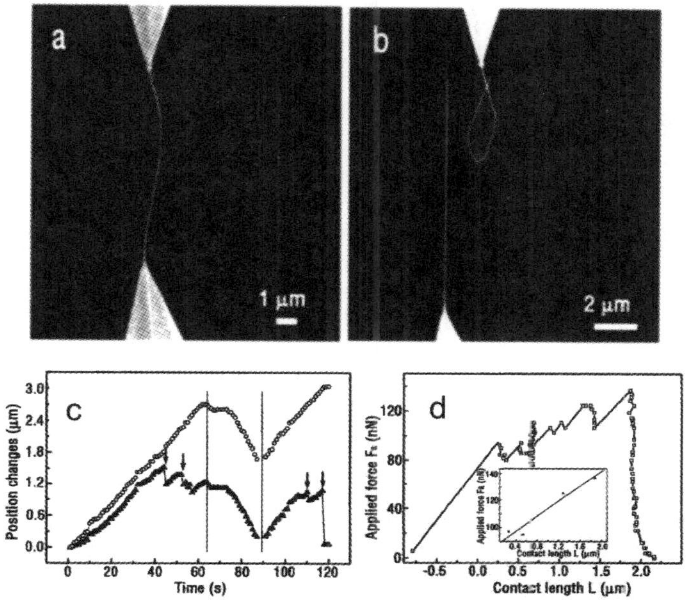

Fig. 12.11. Sliding and pulling out of nested MWNT shells. (**a**, **b**) SEM images showing the "sword-in-sheath" breaking mechanism of MWNTs. (**c**) Displacement of the stiff AFM tip (*top curve, open circles*) and compliant AFM tip (*lower curve, open triangles*) of as a function of time. (**d**) Applied tensile force F_a versus $L(t)$. Inset in (**d**): linear fit to F_a at "stick" events (marked by arrows in (**c**)) as a function of $L(t)$. (Reproduced with permission from *J. Phys. Chem. B.* **104**, 8764-8767, 2000. Copyright 2000 Am. Chem. Soc.)

cations as shown in Table 12.1. From our experience, we have found that the most reliable cantilever calibration method is the resonance technique [25]. First, the planar dimensions of the AFM cantilever are measured with an optical microscope, after which the cantilever is loaded onto the nanomanipulator and the cantilever thickness measured within the SEM. Finally, an *ac* signal from a function generator is applied and the cantilever is excited at resonance when the input signal frequency is equal to the cantilever natural frequency. As shown in Table 12.2, our calibration results on MikroMasch NSC 12 cantilevers fit within the manufacturer specifications; however, the large range in the specifications requires that individual calibration of each AFM cantilever be performed for accurate data interpretation. An additional benefit of this calibration method is that it can be done immediately before mechanical tests in SEM without any modification to the experimental setup, which makes the calibration process compatible with subsequent mechanical measurements involving each individual cantilever.

Table 12.1. Manufacturer specifications for NSC12 AFM cantilevers (MikroMasch)

AFM Cantilever length (μm)	Resonance Frequency (kHz)			Force constant (N/m)		
	min	ave	max	min	ave	max
300	23	28	34	0.20	0.35	0.70
350	17	21	24	0.10	0.30	0.40

Table 12.2. Experimental force constant calibration results for NSC12 AFM cantilevers (MikroMasch)

AFM Cantilever	Length (μm)	Experimental resonance frequency (kHz)	Experimental force constant (N/m)
1	300	30.1	0.35
	350	23.0	0.30
2	300	34.7	0.56
	350	26.2	0.38
3	300	37.0	0.64
	350	27.8	0.42
4	300	35.0	0.57
	350	26.2	0.37

Mechanics of a Carbon Nanocoil

A recent development in carbon nanostructures is the synthesis of carbon nanocoils [42–44]. Because of their unique three-dimensional helical structure, carbon nanocoils have been proposed for a variety of applications, including resonating elements, nanoscale springs, and novel reinforcement in high-strain composite materials. Using the nanomanipulator and tensile-loading test described above, the direct measurement of the mechanical properties of these carbon nanostructures was achieved as shown in Fig. 12.12 [45]. The shear modulus G of the nanocoil was determined based on the spring constant of the nanocoil K, defined by the total applied load divided by the total elongation, and the geometric parameters of the nanocoil. At low-strain levels, we found that the nanocoil behaves as an elastic spring with a spring constant $K = 0.12$ N/m, with a characteristic upturn in the spring constant at higher strain levels as the nanocoil straightens. While high-resolution TEM images show that the nanocoil is actually comprised of two (or perhaps more) individual "wires" that are tightly coupled, after repeated loading of the nanocoil we could find no evidence of relative slippage between these individual "wires". In addition, we found that the nanocoil could be extended up to 42% strain without evidence of plastic deformation, such that the nanocoil returned to its relaxed geometry after loading. Our experimental results indicate that the effective shear modulus of the coils was approximately 2 GPa, which is much lower than the shear modulus estimated for high-quality carbon nanotubes.

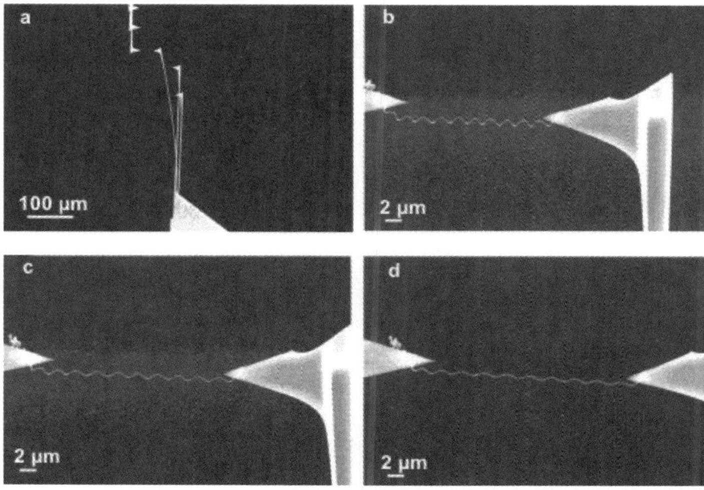

Fig. 12.12. Tensile loading of a carbon nanocoil. (**a**) Carbon nanocoil clamped between two AFM cantilevers. The left cantilever is stiffer than the right cantilever. (**b**) Relaxed nanocoil prior to loading. (**c, d**) Nanocoil at a relative elongation of (**c**) 20% and (**d**) 33%. (Reproduced with permission from *Nano Lett.* **3**, 1299-1304, 2003. Copyright 2003 Am. Chem. Soc.)

Single Nanotube Pullout Tests from a Polymer Matrix

One application of carbon nanotubes that has received considerable attention in the literature is the use of CNTs as the reinforcing phase in a polymer composite [46,47]. Similar to traditional polymer matrix composites, the nanotube-polymer interface plays a crucial role in determining the effectiveness of load-transfer between the polymer matrix and the embedded nanotubes. While a variety of experimental techniques have been developed for traditional micron-sized fiber composites to quantify the fiber-matrix interface [48–54], the extension of these techniques to nanotube-polymer systems is exceedingly difficult due to the small size of the nanotubes.

In standard pullout tests, the interfacial shear strength (IFSS) τ_s is a commonly used interfacial parameter. Here the IFSS of the fiber-polymer interface at failure can be calculated by dividing the peak debonding force F_d by the interfacial area, $\tau_s = F_d/(\pi DL)$, where L is the embedded fiber length and D is the fiber diameter. While the IFSS only represents an average value at the time of failure, it can still be useful for qualitative understanding of the quality of the interface region; to further understand the debonding process, researchers have developed more complicated theories to predict the stress distribution in composites during inclusion pull-out [53,54].

Preliminary work investigating the interfacial strength of nanotube-polymer composites suggests that the interface may be particularly strong, and in many cases much stronger then the interface in traditional micron-

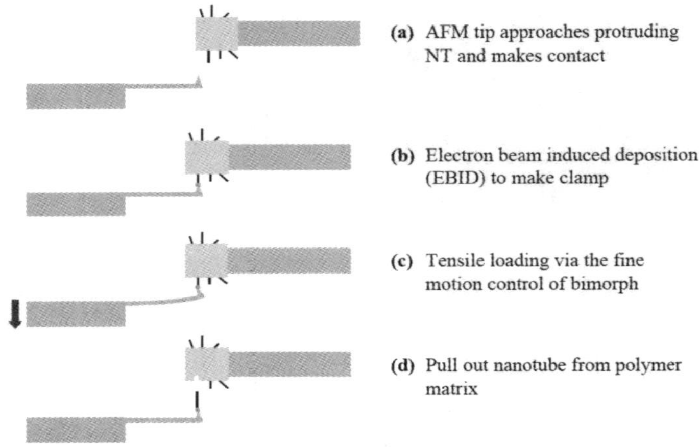

Fig. 12.13. Schematic of single nanotube pullout from a polymer matrix

sized composites. Using molecular mechanics simulations and elastic calculations, the IFSS of a carbon nanotube-polystyrene composite system has been estimated to be approximately 160 MPa [55]. This can be compared to an experimentally determined IFSS of 500 MPa estimated for arc-grown MWNTs in an epoxy matrix, which was noted to be much higher than typically reported for conventional fibers within a polymer matrix [56]. However, more recent experimental work reporting a nanoscale pullout test using atomic force microscopy on a MWNT polyethylene-butene matrix suggests an interfacial shear stress of 47 MPa for that system [57]. Continued development and refinement of nanoscale pullout tests will provide a valuable tool with which to quantify the critical load-transfer properties of nanotube-polymer systems.

Our initial efforts to develop a nanoscale analogy to the single-fiber pullout test are presented schematically in Fig. 12.13. Similar to the test methods described above, we use the nanomanipulator within an SEM to position an AFM cantilever in contact with a nanotube protruding from the fracture surface of a nanotube-polymer composite. Once the nanotube is attached to the AFM cantilever tip, the fine motion of the piezoelectric bimorph is used to apply the tensile load to the embedded nanotube, which is measured via the deflection of the AFM cantilever with a pre-calibrated stiffness constant. An SEM image of an unsuccessful nanotube pullout test, where "sword-in-sheath" failure of the arc-grown MWNT rather than failure of the MWNT-polymer interface is exhibited, is shown in Fig. 12.14. Such results are qualitatively consistent with the strong nanotube-polymer interface suggested for these materials.

During the process of conducting nanoscale pullout tests on MWNTs embedded in polycarbonate, an interesting phenomena was observed which

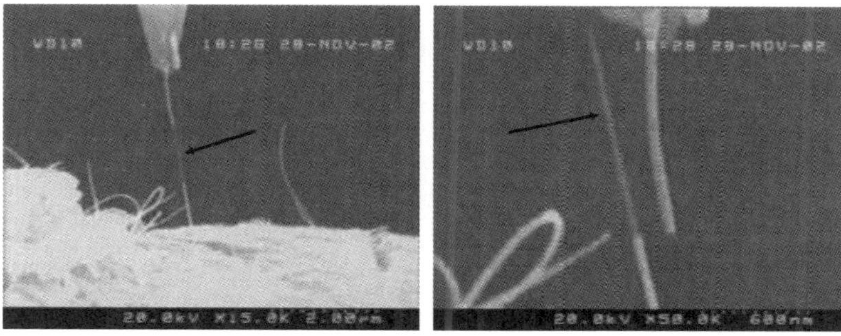

Fig. 12.14. Typical "sword-in-sheath" failure of the embedded MWNT observed during an unsuccessful nanotube pullout test. Arrows denote inner MWNT layers

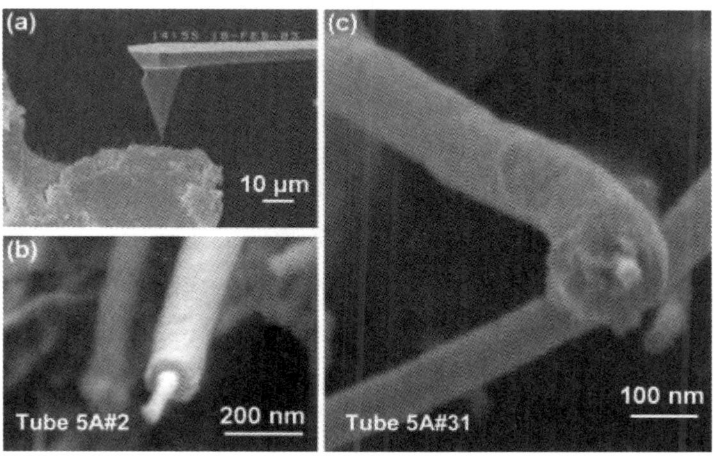

Fig. 12.15. (**a**) Far-field SEM image of the setup of the nanopullout experiment. (**b, c**) High-resolution images of nanotube structures coated with a polymer sheath protruding from the MWNT-polycarbonate fracture surface. (Reproduced with permission from *Nano Lett.* **3**, 1593-1597, 2003. Copyright 2003 Am. Chem. Soc.)

is highlighted below and described in detail elsewhere [58]. As shown in Fig. 12.15, high-resolution images of the fracture surface indicate the presence of what appears to be a polymer coating (or sheathing) on the surface of nanotubes protruding from the fracture surface. The presence of this polymer sheathing is consistent with models of a non-bulk polymer interphase region which has been identified in nanotube-polymer composites [59–61].

In the process of contacting these structures with an AFM tip in order to conduct nanotube pullout testing, we discovered an unusual effect that further supports our hypothesis of polymer sheathing. As shown in Fig. 12.16, upon contact with the AFM tip the polymer sheathing suddenly contracted and balled up, exposing a much thinner structure that may be

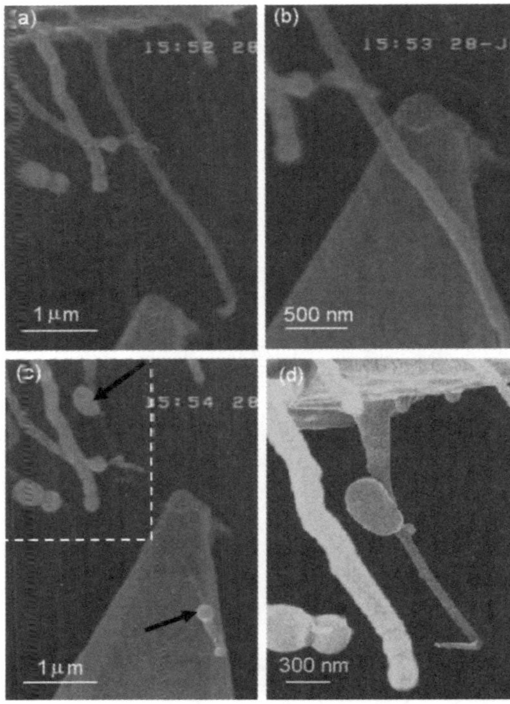

Fig. 12.16. SEM time-lapsed images of balling up of the polymer sheath after contact with the AFM tip. (**a**) The AFM tip approaches a coated MWNT. (**b**) The AFM tip is brought into contact with the coated MWNT. (**c**) After contact the MWNT fractures and the polymer coating balls up on each side of the contact point as highlighted by the arrows. (**d**) Higher magnification image of the upper polymer globule identified in (**c**). (Reproduced with permission from *Nano Lett.* **3**, 1593-1597, 2003. Copyright 2003 Am. Chem. Soc.)

either a bare nanotube or a MWNT coated with a much thinner layer of polymer. While the MWNT shown in Fig. 12.16 has fractured, either due to contact with the AFM tip or the subsequent balling of the polymer coating, in other cases the MWNT did not fracture during this experiment. Of 26 structures protruding from the same MWNT-PC fracture surface, 22 demonstrated a similar "balling up" response [58]. From video recording of these *in situ* SEM experiments, the apparent outer diameters of the coated MWNTs before and after this balling up phenomenon were measured; the volume of the newly formed globules matched the apparent decrease in volume of the coated MWNT (within experimental error). Estimates for the thickness of the polymer sheathing based on this balling up observation (42 nm) where similar to the sheathing thicknesses directly measured from SEM images similar to those shown in Fig. 12.15 (48 nm).

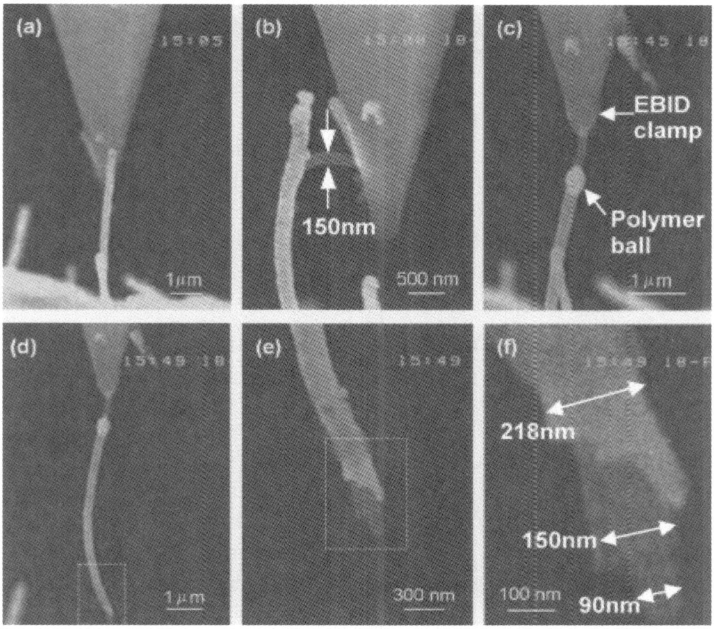

Fig. 12.17. Evidence of multiple polymer sheathing observed during nanomanipulation contact experiments. (**a**) Initial contact with a protruding (coated) MWNT by the AFM tip. (**b**) Inner structure partially detached from the outer sheath during bending. (**c**) The outer sheath balls up and the inner structure is clamped to the AFM tip. (**d, e, f**) Evidence of a double polymer sheath. (Reproduced with permission from *Nano Lett.* **3**, 1593-1597, 2003. Copyright 2003 Am. Chem. Soc.)

For those cases where the balling up of the polymer sheath did not occur, the coated tubes were clamped via EBID to the AFM tips and nanopullout tests from the polycarbonate matrix were conducted. As shown in Fig. 12.17, at the end where the coated MWNT was pulled from the fracture surface, high-resolution SEM images reveal what appear to be two distinct polymer layers coating the fractured MWNT. This is consistent with TEM results that indicate there are (in some cases) at least two distinct polycarbonate layers coating the MWNT. This suggests that here there may be two interfaces that will influence the effective mechanical properties of the system: the MWNT-inner polymer layer and inner polymer layer-outer polymer layer interfaces. Efforts to further elucidate this polymer sheathing phenomena and its impact on the mechanical behavior of nanotube-polymer systems are ongoing.

The long-term goal of this work is to develop a methodology to *quantify* the nanotube-polymer interface based on these pullout tests. Here the ability to obtain high-quality, quantifiable data is complicated by a number of factors, including: the ability to determine the initial embedded length of the nanotube within the polymer; the ability to align the embedded nanotubes perpendicular to the fracture surface; the "straightness" of the embedded nanotubes; and the application of a pure tensile load to the embedded nanotube. Due to these complications, we are currently investigating several potential methods to fabricate model nanotube-polymer composites to eliminate these obstacles from the interpretation of our pullout experiments. Continued development of this test methodology will in particular be useful in quantifying the influence of nanotube functionalization on the nanotube-polymer interface, where defects are intentionally introduced on the nanotube surface to provide sites for covalent bonding with the polymer matrix [62–65].

12.3.2 Induced Vibrational Resonance Methods

Background

Mechanical resonance of a nanofiber can be achieved by attaching the structure to a piezoelectric multilayer bender as shown in Fig. 12.18, where a supporting wire attached to the end of the actuator is used as a medium. By applying an *ac* electric field to the actuator and sweeping the driving frequency using a function generator, the oscillating bender plate can excite the harmonic resonances of the fiber as shown in Fig. 12.19.

An alternative method to drive resonance is via an *ac* electric field. Use of a driving electrical field to excite mechanical resonance in nanostructures is a simple, compact, and controllable method that can be utilized in sensors and actuators. The *ac* electric-field excitation of the mechanical vibration of a nanowire at its fundamental resonance frequency, in particular, has been used in the past few years in determining the elastic properties of synthesized quasi-one-dimensional nanoscale rods, tubes, and whiskers. Specifically, carbon nanotubes have been excited at their fundamental frequency and higher harmonics and their elastic moduli were determined based on the resonance frequency [66–68]. The method has also been used to determine the modulus

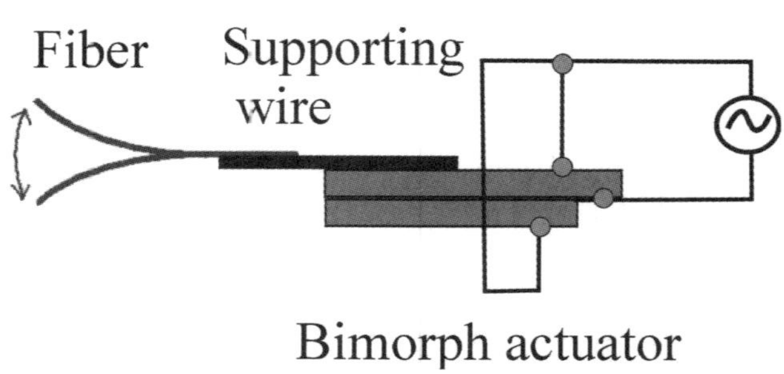

Fig. 12.18. Mechanical resonance via attachment to a piezoelectric bimorph actuator. (Reproduced with permission from X. Chen, S. Zhang, G.J. Wagner, W. Ding, and R.S. Ruoff, *J. Appl. Phys.* **95**, p. 4823, 2004)

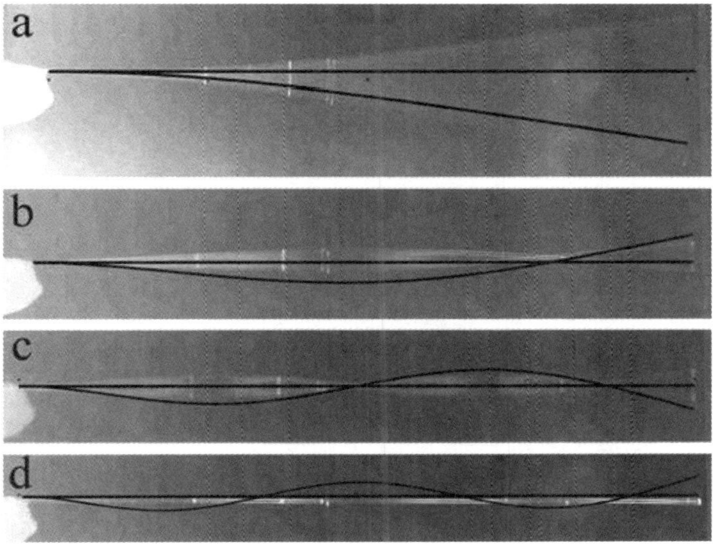

Fig. 12.19. Optical microscope images of the first four resonances of an oscillating quartz microfiber. Black lines are visual aides showing the theoretical mode shapes. (Reproduced with permission from X. Chen, S. Zhang, G.J. Wagner, W. Ding, and R.S. Ruoff, *J. Appl. Phys.* **95**, p. 4823, 2004)

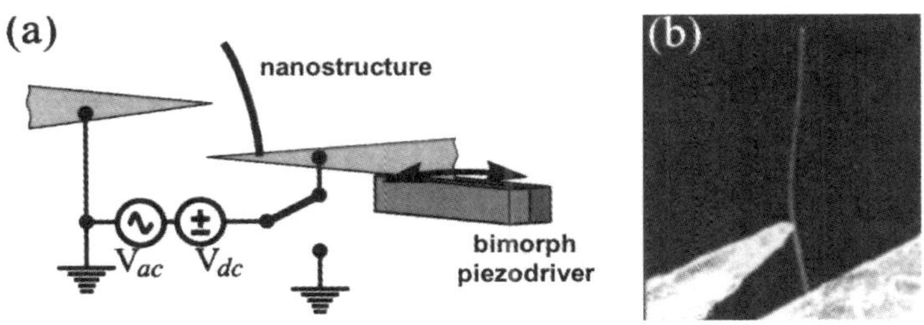

Fig. 12.20. *ac* electric field-induced resonance of a nanowire. (**a**) Schematic of the experimental setup. (**b**) SEM micrograph showing an SiC_2 nanowire clamped to one electrode and touched by another probe. (Reproduced with permission from D.A. Dikin, X. Chen, W. Ding, G. Wagner, and R.S. Ruoff, *J. Appl. Phys.* **93**, p. 226, 2003)

of other nanostructures [26, 69–72]. As shown in Fig. 12.20, two sharp scanning tunneling microscope (STM) tips are bonded onto two opposing position stages in a nanomanipulator and used as electrical leads for the experiment. The electrode on the right-hand side in Fig. 12.20a holds the nanowire and is fixed at the end of bimorph piezodriver. The periodic electric field established between the two STM tip electrodes can then excite nanostructure vibration based on polarization effects.

Once resonance of the structure has been determined, the modulus of the structure can be fit based on standard vibration analyzes. For example, for a fixed-free cantilevered beam with a uniform circular cross-section, simple beam theory [73] for flexural vibration gives the natural resonance frequency as

$$f_n = \frac{\beta_n^2}{2\,\pi}\,\frac{d}{L^2}\,\sqrt{\frac{E}{16\,\rho}};$$

where E is the modulus, ρ is the density, d and L are the diameter and length, respectively, and β_n is the eigenvalue obtained from the characteristic equation $\cosh(\beta_n)\,\cos(\beta_n) = -1$. While resonance vibration is a straightforward method for fitting the Young's modulus of nanostructures, we have found that tensile loading is a more powerful means with which to probe the mechanical behavior of these materials.

Resonance of Silica Nanowires

We have recently used our nanomanipulators in a set of experiments to measure the bending modulus of amorphous silica (SiO_2) nanowires (diameters $\sim$ 100 nm, lengths $>$ 10 µm) via both mechanical and *ac* electric field induced resonance [72]. Because the mechanical and electrical properties of amorphous inorganic materials are less predictable and more difficult to analyze than crystalline materials, it is critical that experiments to measure the nanoscale properties of these materials are undertaken. As explained in more detail elsewhere [72], several unexpected phenomena were observed in conducting resonance experiments on these materials. The flexibility of the experimental setup using the nanomanipulator opens up future research directions characterizing the electromechanical response of such nonmetallic nanowires.

We found that while the original shapes of the nanowires may not be perfectly straight and uniform (see Fig. 12.21), close agreement between the calculated mode shapes and the observed resonance behavior suggest that the stiffness of the nanowires can be treated as uniform along its length. Further, in one study a nanowire was driven in vibration for about 8 hours, corresponding to $\sim 10^9$ cycles, without any noticeable change in resonance behavior. Later, during the final testing of this nanowire we found that three nanosized particles had become attached to its perimeter, which resulted in a significant shift in the original resonance behavior (187.5 kHz) to multiple lower resonance frequencies (176.8, 165.4, and 151 kHz). While these three resonance peaks were observed for mechanically driven vibrations, only the larger two resonant frequencies were observed under *ac* electrical field excitation; this was attributed to the lowest resonant frequency being masked by blurring of the SEM image at that excitation frequency. As shown in Fig. 12.21, the three closely spaced mechanically induced resonances appear

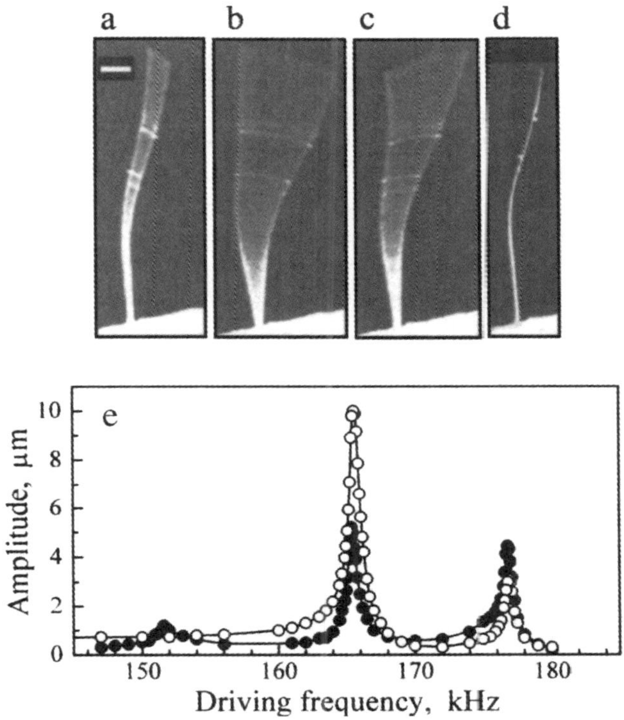

Fig. 12.21. (**a, b, c**) SiO_2 nanowire with three attached nanoparticles driven mechanically at three closely spaced resonant frequencies. (**d**) Original nanowire. (**e**) Corresponding amplitude *vs* frequency curves for both mechanically-driven (*solid circles*) and electrically-driven (*open circles*) vibration. Scale bar in (**a**) is 2 mm. (Reproduced with permission from D.A. Dikin, X. Chen, W. Ding, G. Wagner, and R.S. Ruoff, *J. Appl. Phys.* **93**, p. 226, 2003)

to have slightly different orientations, which apparently represent slightly different torsional motion. In Fig. 12.21(e), the oscillation amplitude for both mechanically and *ac* field driven vibration are shown as a function of driving frequency. The Q factor, defined as the resonance frequency divided by the half bandwidth, is indicative of the quality of the resonance and was about 80 in all cases. The relatively low value of Q for these experiments may be due to the amorphous structure of the SiO_2 nanowires, as well as the presence of a substantial axial component of the applied load, which would lead to frequency broadening of the resonance behavior. In addition, the relatively low value of Q may be due to additional damping effects and losses resulting from non-ideal nanowire clamping and imperfect boundary conditions.

We note that previous studies on amorphous SiO_x nanowires yielded anomalously low Young's modulus values relative to those of fused silica. For example, the mean value for Young's modulus for 3 SiO_x nanofibers was

47 GPa [72], while for 5 SiO_x nanofibers the mean value was 28 GPa [71]. These values should be contrasted with the value of bulk fused quartz of 72 GPa. There is, therefore, the question of whether the mechanical resonance method is providing the correct value for the modulus of the nanostructures. To address this issue, we have studied microscale quartz fibers as a way of qualifying the mechanical resonance method to obtain Young's modulus values for the ever-growing class of newly synthesized nanoscale fibers, wires, and tubes [26]. Specifically, if there are "problems" associated with microscale samples, it is likely that the same sorts of issues will arise with nanoscale samples. The Young's modulus of the quartz fibers has been determined by measuring the resonance frequency for a given geometry and then applying a model that treats the influence of the type of clamp. The mean value from measurements of the fundamental resonance on 14 different microfibers was 70 GPa, close to the standard value of 72 GPa for the Young's modulus of bulk fused quartz

In addition to the calculation of the nanowire modulus based on the resonance frequency, we have identified significant charge trapping in the silica nanowires as highlighted in Fig. 12.22 [72]. We observed that lowering the raster scan rate of the SEM influenced the nanowire vibrational response when resonance was induced via an *ac* electric field. As an example, in Fig. 12.22 all three SEM images were taken under identical experimental conditions except for the sweep rate, which results in a different time of irradiation and alters the subsequent interactions of the electron-beam and wire; at the lower scan rates, an unequal charge distribution alters the interaction between the SiO_2 nanowire and the external *ac* electric field. Based on this observation we developed a simple model, highlighted in the inset of Fig. 12.23, where the *ac* loading force is evenly distributed along the length of nanowire exposed to the electron flux. The close agreement between the

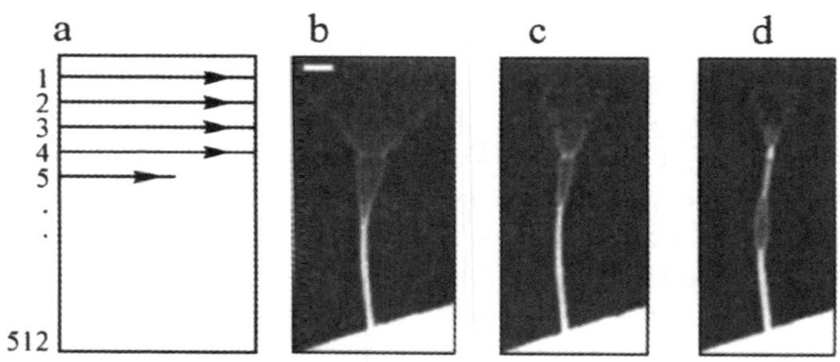

Fig. 12.22. Vibration modes obtained at three different scan rates of the SEM, together with a schematic representation of the scanning electron beam (**a**). The time interval between each closed scan line inside of one frame is: (**b**) 4 ms, (**c**) 17 ms, and (**d**) 50 ms. (Reproduced with permission from D.A. Dikin, X. Chen, W. Ding, G. Wagner, and R.S. Ruoff, *J. Appl. Phys.* **93**, p. 226, 2003)

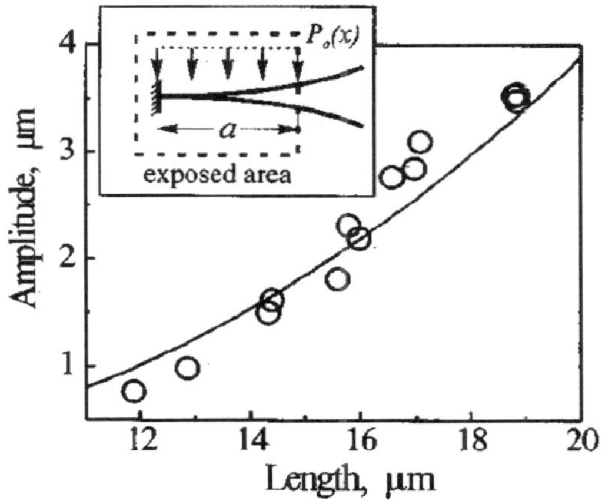

Fig. 12.23. Measured (*open circles*) and calculated (*solid line*) nanowire deflection versus the length of the nanowire exposed to the electron flux which results in the loading force $P_0(x)$. (Reproduced with permission from D.A. Dikin, X. Chen, W. Ding, G. Wagner, and R.S. Ruoff, *J. Appl. Phys.* **93**, p. 226, 2003)

measured and calculated deflection amplitudes given in Fig. 12.23 suggests that charging of the exposed section of the nanowire leads to the *ac* electric field applying an external force only along that section of the nanowire length. The charge trapping effects seen here are consistent with the local charging of conducting and insulating materials caused by electron bombardment [76] and can be used to characterize the intrinsic defects present in nonconductive nanowires. Charge trapping, dissipation, polarization, and electrical losses can be studied via future developments of this method.

12.4 Summary and Future Directions

While computational and theoretical efforts have been significant driving forces behind the recent thrust in nanotechnology research, experimental verification has in many cases been lacking due to the challenging task of performing nanoscale mechanical experiments. However, advances in the design and construction of nanomanipulators operating within an electron microscope are now providing the ability to perform accurate, quantitative, and reproducible experiments to determine the mechanical properties of nanostructures. Such nanomanipulation techniques, while at the moment more difficult and time-consuming than scanning probe-based methods, are perhaps more flexible and offer great potential, as they are further developed, for future quantitative nanoscale experiments.

While considerable progress in the area of nanoscale mechanical testing has been made within the last few years, a number of critical milestones have yet to be achieved. For example, tensile loading of an individual SWNT has yet to be achieved, and the application of a controlled, reversible twist of known torque along the axis of a nanotube has not been demonstrated. The influence of temperature, humidity, and chemical environment on the mechanics of nanostructures has not been explored, and the impact of loading rate and defect behavior (nucleation, propagation, and sensitivity) has yet to be addressed from an experimental perspective. In addition, as the capability to perform nanoscale mechanical tests such as those discussed in this chapter become more widespread, the development of more rigorous test methods and protocols will be desired.

We note that the test methods described in this chapter require a long and tedious process of iterative focusing-defocusing of the two-dimensional SEM image, as the depth information of the AFM tip and the nanostructure specimen is not directly available. In addition, there is an non-intuitive mapping between the motion control input device, a one degree of freedom slider, and the motion of the AFM tip, which requires sequential motion of nanomanipulator tools rather than the simultaneous motion of all degrees of freedom in a coordinated manner. To address these issues we foresee new visualization and control tools based on tele-operation and robotics to enable automated and efficient tool navigation. This would greatly reduce the time and expertise required to conduct these nanoscale experiments with no compromise of the quality of the experimental data. Such technology will also allow automated testing, facilitating data collection for a large number of repetitive experiments. These enhancements will allow the widespread adaptation of these techniques in the scientific and engineering communities.

Finally, while often supporting current theory, in some cases the use of these nanomanipulation tools has led to the discovery of unexpected phenomena, which in turn has led to a more complete understanding of nanoscale science. Thus, as nanotechnology research continues to progress, developments in experimentation using three-dimensional nanomanipulation strategies will most certainly continue to contribute to our understanding of science on the nanoscale.

Acknowledgements

R.S.R. would like to acknowledge support from NSF, NASA, ONR, and Zyvex for early portions of the work highlighted in this chapter. More recent support has been provided by the National Science Foundation grant: Nanorope Mechanics (NSF No. 0200797, Oscar Dillon and Ken Chong, Program Managers), the NASA Langley Research Center Computational Materials: Nanotechnology Modeling and Simulation Program, the NASA University Research, Engineering and Technology Institute on Bio Inspired Materials (BIMat) under award no. NCC-1-02037, and the Office of Naval Research

Mechanics of Nanostructures under grant award no. N000140210870. R.S.R. would also like to acknowledge the contribution of a number of students and colleagues who have contributed to this work over the years, including Min-Feng Yu, Henry Rohrs, Hui Huang, Oleg Lourie, Kevin Ausman, Tomek Kowalewski, Wing Kam Liu, Dong Qian, Greg Wagner, Richard Piner, Weiqiang Ding, Shaoning Lu, Zebin Huang, and Mark Dyer. We thank Will McBride and Kevin Kohlhaas for critically reading this manuscript.

References

1. R.P. Feynman: Engineering and Science (Caltech) **23**, 22 (1960)
2. S. Iijima: Nature **354**, 56 (1991)
3. P.M. Ajayan, O.Z. Zhou: *Applications of Carbon Nanotubes*. In: *Carbon Nanotubes*, ed by M.S. Dresselhaus, G. Dresselhaus, and P. Avouris (Springer-Verlag Berlin: Heidelberg 2001). Topics in Applied Physics, **80:** p. 391
4. D. Normile: Science **286**, 2056 (1999)
5. W.B. Choi, D.S. Chung, J.H. Kang, H.Y. Kim, Y.W. Jin, I.T. Han, Y.H. Lee, J.E. Jung, N.S. Lee, G.S. Park, J.M. Kim: Applied Physics Letters **75**, 3129 (1999)
6. M.S. Dresselhaus, G. Dresselhaus, P.C. Eklund: *Science of Fullerenes and Carbon Nanotubes* (Academic Press New York 1996)
7. P.J.F. Harris: *Carbon Nanotubes and Related Structures: New Materials for the 21st Century* (Cambridge University Press Cambridge 1999)
8. R. Saito, M.S. Dresselhaus, G. Dresselhaus: *Physical Properties of Carbon Nanotubes* (Imperial College Press London 1998)
9. D. Qian, G.J. Wagner, W.K. Liu, M.-F. Yu, R.S. Ruoff: Applied Mechanics Reviews **55**, 495 (2002)
10. R.S. Ruoff, D. Qian, W.K. Liu: Comptes Rendus Physique **4**, 993 (2003)
11. B.I. Yakobson, P. Avouris: *Mechanical Properties of Carbon Nanotubes*. In: *Carbon Nanotubes*, ed by M.S. Dresselhaus, G. Dresselhaus, and P. Avouris (Springer-Verlag Berlin: Heidelberg 2001). Topics in Applied Physics, **80:** p. 287
12. M.R. Falvo, G.J. Clary, R.M. Taylor II, V. Chi, F.P. Brooks Jr, S. Washburn, R. Superfine: Nature **389**, 582 (1997)
13. J.-P. Salvetat, A.J. Kulik, J.-M. Bonard, G.D.A. Briggs, T. Stöckli, K. Métérier, S. Bonnamy, F. Béguin, N.A. Burnham, L. Forró: Advanced Materials **11**, 161 (1999)
14. J.-P. Salvetat, G. Briggs, J.-M. Bonard, R. Basca, A. Kulik, T. Stöckli, N. Burnham, L. Forró: Physical Review Letters **82**, 944 (1999)
15. M.-F. Yu, T. Kowalewski, R.S. Ruoff: Physical Review Letters **85**, 1456 (2000)
16. D. Walters, L. Ericson, M. Casavant, J. Liu, D. Colbert, K. Smith, R. Smalley: Applied Physics Letters **74**, 3803 (1999)
17. E.W. Wong, P.E. Sheehan, C.M. Lieber: Science **277**, 1971 (1997)
18. W.D. Shen, B. Jiang, B.S. Han, S.S. Xie: Physical Review Letters **84**, 3634 (2000)
19. I. Chasiotis, W.G. Knauss: Experimental Mechanics **42**, 51 (2002)
20. S. Sundararajan, B. Bhushan: Sensors and Actuators A **101**, 338 (2002)

21. M.-F. Yu, M.J. Dyer, G.D. Skidmore, H.W. Rohrs, X. Lu, K.D. Ausman, J.R. von Ehr, R.S. Ruoff: Nanotechnology **10**, 244 (1999)
22. M.-F. Yu, O. Lourie, M. Dyer, K. Moloni, T.F. Kelly, R.S. Ruoff: Science **287**, 637 (2000)
23. M.-F. Yu, B.S. Files, S. Arepalli, R.S. Ruoff: Physical Review Letters **84**, 5552 (2000)
24. M.-F. Yu, B.I. Yakobson, R.S. Ruoff: Journal of Physical Chemistry B **104**, 8764 (2000)
25. J.E. Sader, I. Larson, P. Mulvaney, L.R. White: Review of Scientific Instruments **66**, 3789 (1995)
26. X. Chen, S. Zhang, G.J. Wagner, W. Ding, R.S. Ruoff: Journal of Applied Physics **95**, 4823 (2004)
27. K.T. Kohlmann, K.T., M. Thiemann, W.H. Brunger: Microelectronic Engineering **13**, 279 (1991)
28. K. Molhave, D.N. Madsen, A.M. Rasmussen, A. Carlsson, C.C. Appel, M. Brorson, C.J.H Jacobsen, P. Boggild: Nano Letters **3**, 1499 (2003)
29. S. Lu, D.A. Dikin, S. Zhang, F.T. Fisher, J. Lee, R.S. Ruoff: Review of Scientific Instruments, **75**, 2154 (2004)
30. J. Brugger, V.P Jaecklin, C. Linder, N. Blanc, P.F. Indermuhle, N.F. de Rooij: J. Micromech. Microeng. **3**, 161 (1993)
31. J.J. Yao, S.C. Arney, N.C. MacDonald: J. Microelectromech. Sys **1**, 14 (1992)
32. N.D. Mankame, G.K. Ananthasuresh: J. Micromech. Microeng **11**, 452 (2001)
33. G.G. Demczyk, Y.M. Wang, J. Cumings, M. Hetman, W. Han, A. Zettl, R.O. Ritchie: Mat. Sci. Eng. A **334**, 173 (2002)
34. P.A. Williams, S.J. Papadakis, M.R. Falvo, A.M. Patel, M. Sinclair, A. Seeger, A. Helser, R.M. Taylor, S. Washburn, R. Superfine: Applied Physics Letters **80**, 2574 (2002)
35. D.M. Burns, V.M. Bright: *SPIE Int. Soc. Opt. Eng.* (1997)
36. J.H. Comotois, V.M. Bright: Sensors and Actuators A **58**, 19 (1997)
37. T. Moulton, G.K. Ananthasuresh: Sensors and Actuators A **90**, 38 (2001)
38. A. Folch, A., J. Servat, J. Esteve, J. Tejada, M. Seco: Journal of Vacuum Science & Technology B **14**, 2609 (1996)
39. R.R. Kunz, T.M. Mayer: Journal of Vacuum Science & Technology B **6**, 1557 (1988)
40. K.T. Kohlmann, J. Chlebek, M. Weiss, K. Reimer, H. Oertel, W.H. Brunger: Journal of Vacuum Science & Technology B **11**, 2219 (1993)
41. H.W.P. Koops, R. Weiel, D.P. Kern, T.H. Baum: Journal of Vacuum Science & Technology B **6**, 477 (1988)
42. V.K. Varadan, J.N. Xie: Smart Materials & Structures **11**, 728 (2002)
43. C. Kuzuya, W. In-Hwang, S. Hirako, Y. Hishikawa, S. Motojima: Chemical Vapor Deposition **8**, 57 (2002)
44. M. Zhang, Y. Nakayama, L.J. Pan: Japanese Journal of Applied Physics Part 2-Letters **39**, L1242 (2000)
45. X. Chen, S.L. Zhang, D.A. Dikin, W.Q. Ding, R.S. Ruoff, L.J. Pan, Y. Nakayama: Nano Letters **3**, 1299 (2003)
46. K.T. Lau, D. Hui: Composites: Part B **33**, 263 (2002)
47. E.T. Thostenson Z.F. Ren, T.W. Chou: Composites Science and Technology **61**, 1899 (2001)
48. B.W. Kim, J.A. Nairn: Journal of Composite Materials **36**, 1825 (2002)

49. S. Zhandarov, E. Pisanova, E. Mader, J.A. Nairn: Journal of Adhesion Science and Technology **15**, 205 (2001)
50. E. Pisanova, S. Zhandarov, E. Mader, I. Ahmad, R. Young: Composites: Part A **32**, 435 (2001)
51. E. Pisanova, S. Zhandarov, E. Mader: Composites: Part A **32**, 425 (2001)
52. J.A. Nairn: Advanced Composites Materials **9**, 373 (2000)
53. C.K.Y. Leung, V.C. Li: Journal of Materials Science **26**, 5996 (1991)
54. R.J. Kerans, T.A. Parthasarathy: Journal of the American Ceramic Society **74**, 1585 (1991)
55. K. Liao, S. Li: Applied Physics Letters **79**, 4225 (2001)
56. H.D. Wagner, O. Lourie, Y. Feldman, R. Tenne: Applied Physics Letters **72**, 188 (1998)
57. A.H. Barber, S.R. Cohen, H.D. Wagner: Applied Physics Letters **82**, 4140 (2003)
58. W. Ding, A. Eitan, F.T. Fisher, X. Chen, D.A. Dikin, R. Andrews, L.C. Brinson, L.S. Schadler, R.S. Ruoff: Nano Letters **3**, 1593 (2003)
59. M.S.P. Shaffer, A.H. Windle: Advanced Materials **11**, 937 (1999)
60. F.T. Fisher: Nanomechanics and the Viscoelastic Behavior of Carbon Nanotube-reinforced Polymers. Ph.D. Dissertation. Northwestern University, Evanston, IL (2002)
61. F.T. Fisher, A. Eitan, R. Andrews, L.C. Brinson, and L.S. Schadler: Advanced Composites Letters **13** (2), 105–111 (2004)
62. A. Eitan, K. Jiang, R. Andrews, L.S. Schadler: Chemistry of Materials **15**, 3198 (2003)
63. H.W. Goh, S.H. Goh, G.Q. Xu, K.P. Pramoda, W.D. Zhang: Chemical Physics Letters **373**, 277 (2003)
64. J. Jang, J. Bae, S.H. Yoon: Journal of Materials Chemistry **13**, 676 (2003)
65. Y.P. Sun, K.F. Fu, Y. Lin, W.J. Huang: Accounts of Chemical Research **35**, 1096 (2002)
66. R. Gao, Z.L. Wang, Z. Bai, W.A. de Heer, L. Dai. M. Gao: Physical Review Letters **85**, 622 (2000)
67. P. Poncharal, Z.L. Wang, D. Ugarte, W.A. de Heer: Science **283**, 1513 (1999)
68. Z.L. Wang, R.P. Gao, P. Poncharal, W.A. de Heer, Z.R. Dai, Z.W. Pan: Materials Science & Engineering C **16**, 3 (2001)
69. J. Fujita, M. Ishida, T. Sakamoto, Y. Ochiai, T. Kaito, S. Matsui: Journal of Vacuum Science & Technology B **19**, 2834 (2001)
70. Z.L. Wang, Z.R. Dai, R.P. Gao, J.L. Gole: Journal of Electron Microscopy **51**, S79 (2002)
71. Z.L. Wang, R.P. Gao, Z.W. Pan, Z.R. Dai: Advanced Engineering Materials **3**, 657 (2001)
72. D.A. Dikin, X. Chen, W. Ding, G. Wagner, R.S. Ruoff: Journal of Applied Physics **93**, 226 (2003)
73. R.E.D. Bishop, D.C. Johnson, *Mechanics of Vibration* (Cambridge University Press Cambridge 1960)
74. K. Fukushima, S. Kawai, D. Saya, H. Kawakatsu: Review of Scientific Instruments **73**, 2647 (2002)
75. H.T. Miyazaki, Y. Tomizawa, K. Koyano, T. Sato, N. Shinya: Review of Scientific Instruments **71**, 3123 (2000)
76. W.S.M. Werner: Surface and Interface Analysis **31**, 141 (2001)

Color Plates

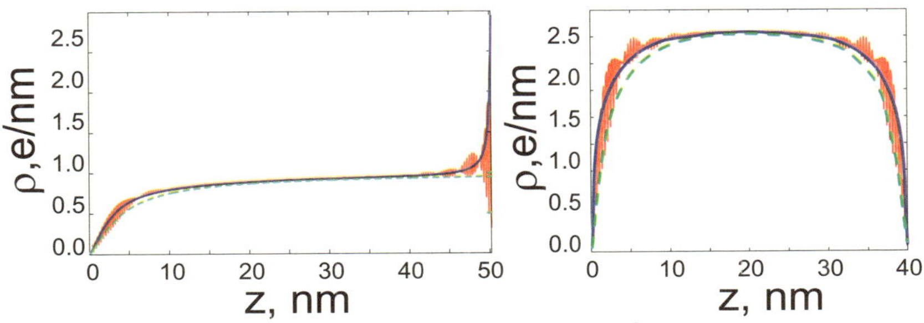

Fig. 1.5. Specific charge density for two devices: (right) string and (left) cantilever NEMS. The *solid oscillating (red) curve* is a result of the quantum mechanical calculation. The *solid (blue) line* is a solution of joint Poisson and Boltzmann equations. The *dashed (green) line* is the result of the analytical approximation

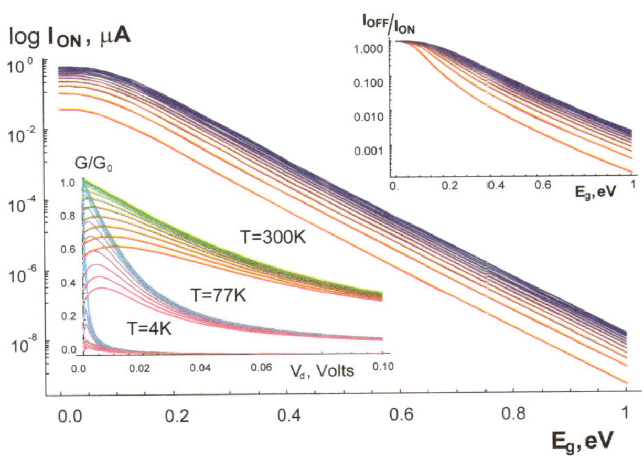

Fig. 1.11. Logarithm of the room temperature OFF/ON current ratio versus the opened gap. The width of a local gate is 50 nm. The upper right inset shows the logarithm of the OFF/ON current ratio at $T = 4\,\mathrm{K}$ and the gate width 15 nm. Each curve from bottom to top corresponds to increasing drain voltage from 1 to 20 meV. In the lower left inset the drain voltage dependence of the conductance at given temperature ($T = 4\,\mathrm{K}$, $77\,\mathrm{K}$ and $295\,\mathrm{K}$ from bottom to top) and gate voltage (from 0 to 0.2 eV from top to bottom) is presented

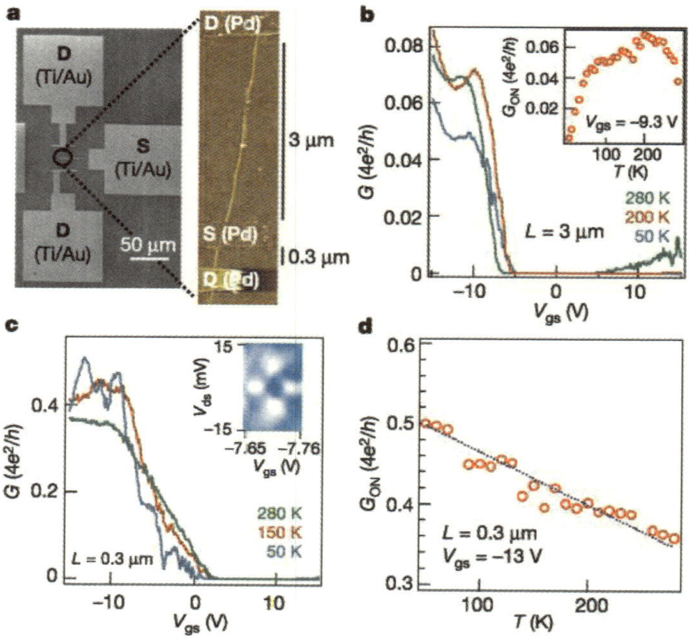

Fig. 4.1. (**a**) A scanning electron microscope (SEM) image (left) and atomic force microscope (AFM) image (right) of representative Pd-contacted long ($L <$ 3 μm) and short ($L <$ 300 nm) back-gated SWNT devices formed on the same nanotubes directly grown on SiO2/Si substrates using the CVD approach. Ti/Au metal bonding pads were used to connect to the Pd source (S) and drain (D) electrodes. The devices were annealed in Ar at 225°C for 10 min after fabrication. (**b**) G (at low S–D) bias V_{ds} versus gate voltage V_{gs} for a 3-mm-long SWNT ($d =$ 3.3 nm) device recorded at various T. Inset, G_{ON} versus T for the device. (**c**) G versus V_{gs} for a 300-nm-long tube section on the same tube as for (**b**) at various T. Differential conductance dI_{ds}/dV_{ds} versus V_{ds} and V_{gs} (inset, measured by a lock-in technique) at $T < 1.5$ K shows a Fabry-Perot-like interference pattern (bright peak $G < 4e^2/h$, dark region $G < 0.5 * 4e^2/h$). (**d**) G_{ON} versus T for the $L <$ 300 nm semiconducting tube down to 50 K. Data and caption reproduced from [10]

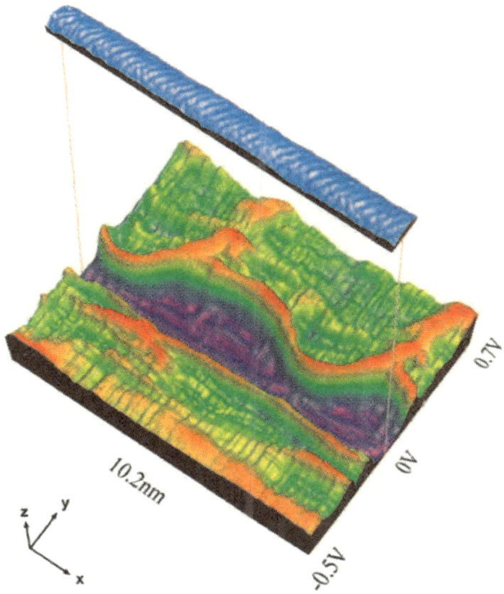

Fig. 5.7. 3D representation of a (11,9) SWNT topograph, 10.2 nm long in cyan. The corresponding 512 dI/dV spectra at the center of the tube along the tube. The x-axis indicates the position along the tube, the y-axis the energy and the z-axis dI/dV. The value of dI/dV (local density of state) is high with descending order of red-green-blue [10]

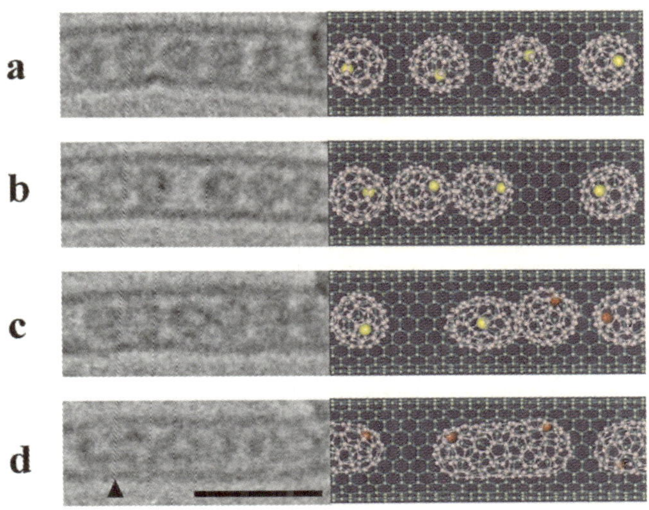

Fig. 5.11 Time evolution of the HRTEM images of (Sm@C_{82})@SWNTs (left) and the schematic illustrations (right). The HRTEM images were observed after (**a**) ~ 0 min, (**b**) ~ 4 min, (**c**) ~ 10 min and (**d**) ~ 20 min irradiations, respectively. The produced nanocapsules can be clearly seen inside the SWNT (*arrows*). The yellow and red balls in the illustrations denote the divalent Sm^{2+} and the trivalent Sm^{3+} atoms, respectively. Scale bar = 3 nm

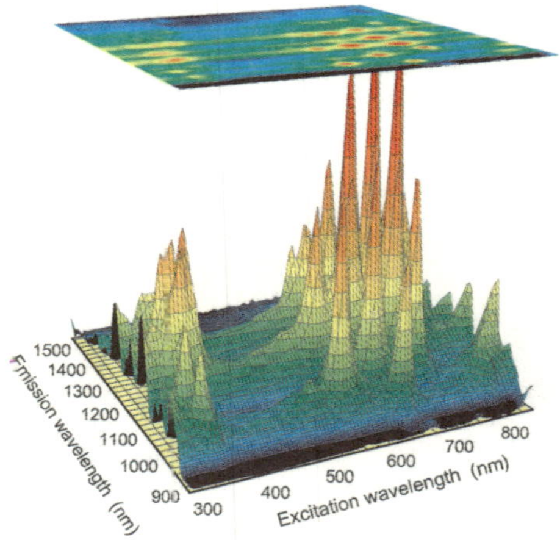

Fig. 7.2. Surface plot of photoluminescence (fluorescence) intensity measured as a function of emission and excitation wavelengths for a SWNT sample in aqueous SDS suspension. Each peak arises from a specific (n,m) semiconducting nanotube species

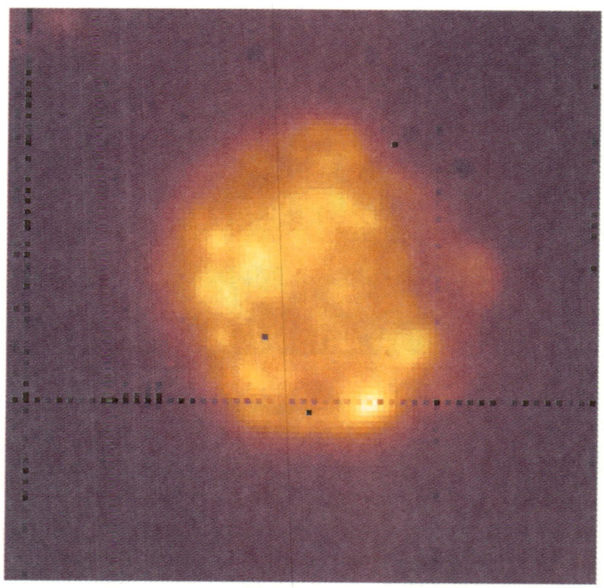

Fig. 7.10. Fluorescence micrograph of a single macrophage cell that had been incubated in a growth medium containing suspended SWNT. The cell was excited at 660 nm and nanotube emission was imaged only at wavelengths greater than 1125 nm. Bright regions (yellow in false color) show the highest nanotube concentrations; dark regions (blue in false color) show the lowest. The cell's diameter was approximately 20 μm. Black points and the horizontal dotted line arise from uncorrected defects in the InGaAs imaging camera

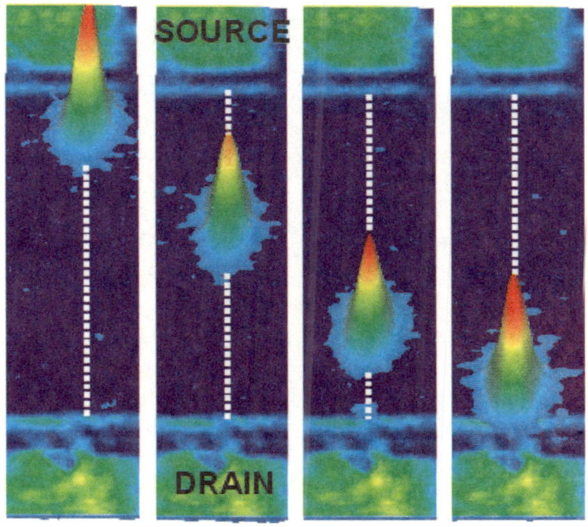

Fig. 9.15. Electroluminescence from a long ($\sim 50\,\mu m$) carbon nanotube field effect transistor. The emission spot can be translated along the axis of the nanotube by varying the gate bias under constant current ($18\,\mu A$) conditions. (From [75])

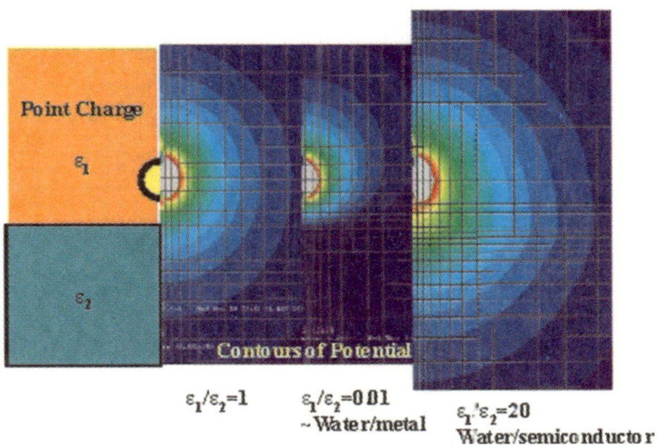

Fig. 10.6. Electrostatic potential (a.u.) surrounding a point charge near a bi-material interface. A comparison of the homogenous case ($\epsilon_1 = \epsilon_2$), water/metal, and water/semiconductor interfaces shows that the presence of the metal strongly attenuates the influence of the charge away from the nanotube. In contrast, the presence of the semiconductor enhances the influence of the point charge

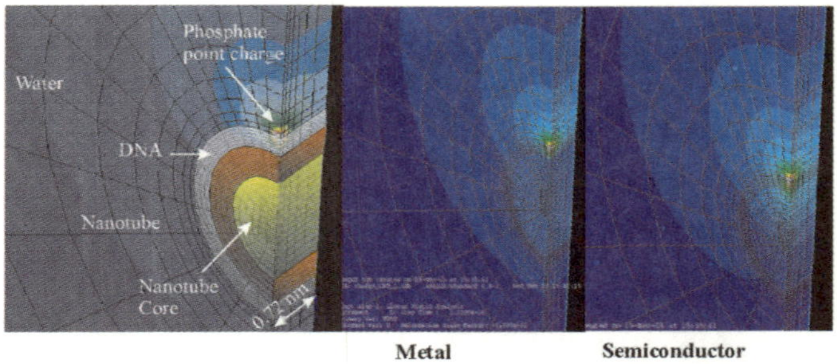

Fig. 10.7. Numerically computed electrostatic potential (a.u.) surrounding a point charge near a metallic or semiconducting nanotube. While the geometry and periodicity of charges reduces the difference due to metallic and semiconducting cores, it remains sufficient to explain differentiated interaction with external (e.g., fixed) charges

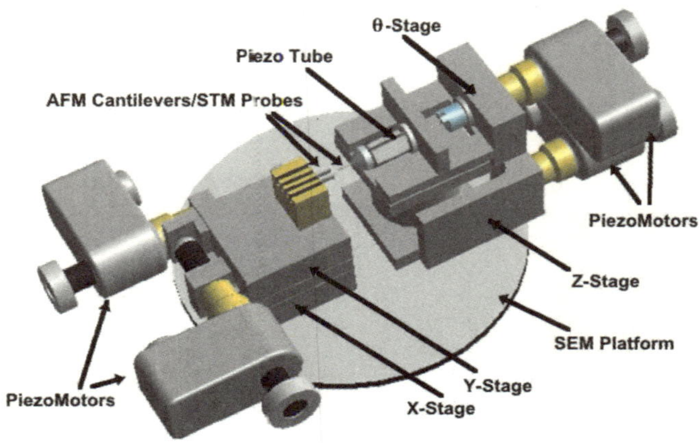

Fig. 12.1 Schematic of the first-generation nanomanipulator for use within an SEM. (Reproduced with permission from M.F. Yu, M.J. Dyer, G.D. Skidmore, H.W. Rohrs, X. Lu, K.D. Ausman, J.R. von Ehr, and R.S. Ruoff, *Nanotechnology*, 10, p. 244, 1999)

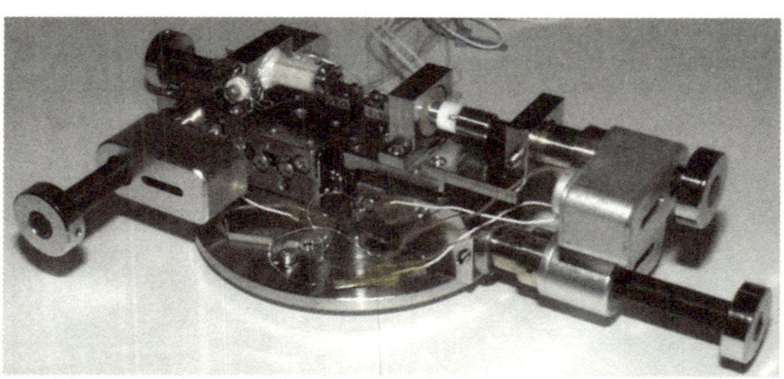

Fig. 12.2. Photograph of the first-generation nanomanipulator device

Index

Printing: Krips bv, Meppel
Binding: Stürtz, Würzburg